Principles of Heat Treatment of Steel

Principles of Heat Treatment of Steel

GEORGE KRAUSS
AMAX Foundation Professor
Department of Metallurgical Engineering
Colorado School of Mines

American Society for Metals
Metals Park, Ohio 44073

Library of Congress Cataloging in Publication Data

Krauss, George, 1933–
Principles of heat treatment of steel.
Edition of 1964 by M. A. Grossmann and E. C. Bain, published under title: Principles of heat treatment.
Includes bibliographical references.
1. Steel — Heat treatment. I. Grossmann, Marcus Aurelius, 1890-1952. Principles of heat treatment. II. Title.
TN751.G7 1980 672.3′6 80-16118
ISBN 0-87170-100-6

PRINTED IN THE UNITED STATES OF AMERICA

To the next generation

Matt, Jon, Ben and Tom

Preface

This book is a completely rewritten version of *Principles of Heat Treatment* by M. A. Grossmann and E. C. Bain. It is a pleasure to acknowledge the contributions of these authors. Much of their work, especially that concerning the development of hardenability concepts, is incorporated here. *Principles of Heat Treatment* covered developments between 1935 and 1964, and Grossmann, Bain and their contemporaries did their work so well that the heat treatment and metallurgy of carbon steels was almost taken for granted. Steels, however, are wonderfully complex, and continued effort in the last twenty years has brought deeper understanding of their response to thermal and mechanical treatments. New theoretical approaches to diffusion-controlled and martensitic transformations, the characterization of fine structure by transmission electron microscopy, fractography with the scanning electron microscope, new electron beam microanalysis techniques, fracture toughness testing, continued examination of hardenability, and the relationship of microstructure and fine structure to strength, toughness and ductility are all areas, highly developed only in the last twenty years, that I have attempted to build onto the solid foundations of steel heat treatment developed by earlier workers. My approach has been to develop the structure-property-processing relationships that underlie the many heat treatments applied to steels. The origin and characterization of microstructures are emphasized because they are so often forgotten as the source of the handbook graphs and tables of processing parameters and properties.

The give-and-take of many conferences and the contributions of many investigators to the literature have been the basis for our growing understanding of steels and their behavior. I have drawn widely from published sources and through the cited references hope to recognize at least some university and industrial scientists and their contributions. The reference lists are by no means complete, but every paper opens up an area by listing tens or even hundreds of additional references.

Especially rewarding has been my association with other investigators. Professor Morris Cohen initiated my interest in steel, and the enthusiasm of my colleagues and students has sustained that interest. I have learned from every thesis investigation, and examples of the work of my students are shown throughout this book.

In recent years I have benefited much from association with the heat treatment activities of the American Society for Metals, first through membership on the Heat Treatment Technical Division Council, and more recently as editor of the Journal of Heat Treating. These associations have made me aware of the scope and sophistication of new approaches to the heat treatment of steels, and I am especially grateful to Norman Kates, Dale Breen, Jon Dossett, and Joe Riopelle for their introduction to the demanding, practical world of heat treatment.

A university research effort is very much dependent on outside support. I gratefully acknowledge the Army Research Office, the National Science Foundation, the Bethlehem Steel Corporation, the AMAX Foundation, and the American Iron and Steel Institute for the support that has enabled me and my students to remain actively involved in research on the behavior of steels. I acknowledge also fruitful discussions with Professor Glenn Edwards, Tom Bell, and Norman Breyer concerning parts of the text and am grateful for micrographs supplied by Professors R. W. K. Honeycombe, Robert Hehemann, and Marvin Wayman. Mark Geib deserves special mention for his help with some of the figures. Finally, I am deeply grateful to my wife, Ruth, for her competent assistance, her unwavering support, and the typing of the manuscript — all given between a busy schedule of rehearsals and performances of the Central City Opera and the Evergreen Chorale.

George Krauss

Evergreen, Colorado
September 25, 1979

Contents

CHAPTER 1

Phases and Structures

Steel can be heat treated to produce a great variety of microstructures and properties. The desired results are accomplished by heating *in* temperature ranges where a phase or combination of phases is stable (thus producing changes in the microstructure or distribution of phases), and/or heating or cooling *between* temperature ranges in which different phases are stable (thus producing beneficial phase transformations). The iron-carbon equilibrium phase diagram is the foundation on which all heat treatment of steel is based. This diagram defines the temperature-composition regions where the various phases in steel are stable, as well as the equilibrium boundaries between phase fields. This chapter describes the iron-carbon diagram and the phases found in steels and iron-carbon alloys.

The Iron-Carbon Equilibrium Diagram

The iron-carbon (Fe-C) diagram is a map that can be used to chart the proper sequence of operations for a given heat treatment. The iron-carbon diagram should be considered only a guide, however, because most steels contain other elements that modify the positions of phase boundaries. The effects of alloying elements on the phase relations shown in the iron-carbon diagram are described later in this chapter. Use of the iron-carbon diagram is further limited because some heat treatments are specifically intended to produce nonequilibrium structures whereas others barely approach equilibrium. Nevertheless, knowledge of the changes that take place in a steel as equilibrium is approached in a given phase field, or of those that result from phase transformations, provides the scientific basis for the heat treatment of steels.

Figure 1.1 shows the Fe-C equilibrium diagram for carbon contents up to 7%. Steels are alloys of iron, carbon and other elements that contain less than 2% carbon — most frequently 1% or less. Therefore, the portion of the diagram below 2% carbon is of primary interest for steel heat treatment. Alloys containing more than 2% carbon are classified as cast irons. Actually, two diagrams are shown in Fig. 1.1: the solid lines show the equilibrium between Fe_3C and the several phases of iron, whereas the dashed lines show the equilibrium between graphite and the other phases. Graphite is a more stable form of carbon than Fe_3C and, given very long periods of time, Fe_3C will decompose to graphite. Graphitization, however, rarely occurs in steels, and thus the Fe-Fe_3C diagram is the more pertinent for understanding the heat treatment of steel. In cast irons, high carbon content and the usual high silicon additions promote graphite formation, and accordingly cast iron technology is based much more on the Fe-graphite diagram.

The diagram in Fig. 1.1 is strictly valid only at a pressure of one atmosphere. At very high pressures, the boundaries shift and new phases appear. For example, in pure iron a close-packed hexagonal crystal form of iron, epsilon iron, has been produced at high pressures (Ref 1.1). The triple point in pure iron between alpha iron, gamma iron and epsilon iron occurs at 770 K and 110 kbars (11 GPa).

Compositions of the Fe-C alloys and phases represented by the Fe-C diagram are conventionally given in weight percent. The percent symbol (%), unless otherwise identified, is understood to represent weight percent, a convention that is followed in this text. Sometimes it is useful to determine compositions in atomic percent. Conversion from weight percent to atomic percent carbon in an Fe-C alloy is accomplished by the following equations:

$$\text{at. \% C} = \frac{\text{C atoms}}{\text{C atoms} + \text{Fe atoms}} \times 100 \qquad \text{(Eq 1.1)}$$

or

$$\text{at. \% C} = \frac{\dfrac{\text{wt \% C}}{\text{at. wt C}}}{\dfrac{\text{wt \% C}}{\text{at. wt C}} + \dfrac{\text{wt \% Fe}}{\text{at. wt Fe}}} \times 100 \qquad \text{(Eq 1.2)}$$

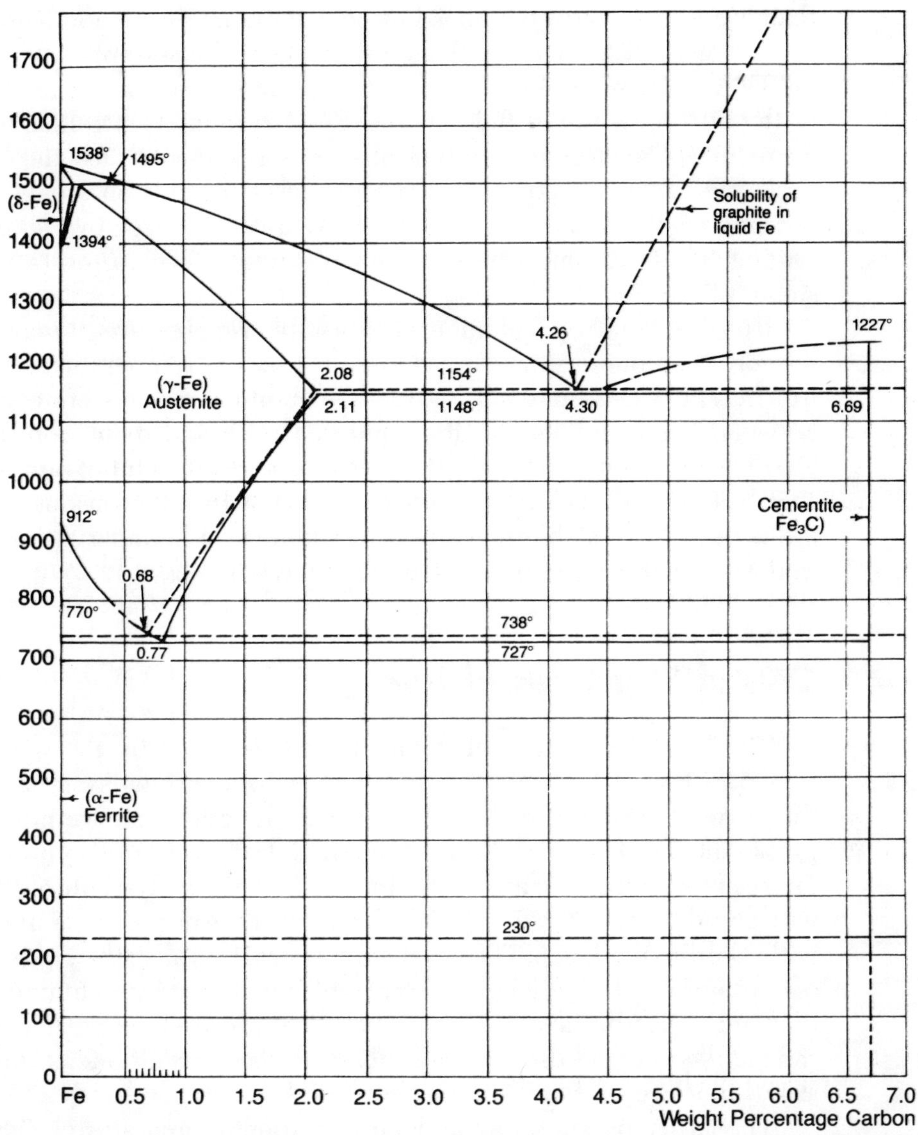

Fig. 1.1 The Fe-C equilibrium diagram up to 7% carbon. Solid lines indicate Fe-Fe_3C diagram; dashed lines indicate Fe-graphite diagram. (Ref 1.12)

Application of this calculation to an Fe-0.4C alloy shows that 0.4% carbon is equivalent to 1.8 atomic percent carbon, a reflection of the much lighter atomic weight of carbon (12) compared with that of iron

(56). Conversion to atomic percent for steels containing elements other than Fe and C requires an additional term in the denominator of Eq 1.1 or Eq 1.2 for each of the other elements present.

The art and science of steel heat treatment is based on the existence of the austenite phase field in the Fe-C system. Controlled transformation of austenite to other phases is responsible for the great variety of microstructures and properties attainable in steels. Hot working of heavy sections into useful shapes and sizes by rolling or forging is also accomplished at temperatures where austenite is the stable phase.

Iron is an allotropic element: at atmospheric pressure, it may exist in more than one crystal form depending on the temperature. Alpha iron (ferrite) exists up to 912 °C (1674 °F); gamma iron (austenite) exists between 912 and 1394 °C (1674 and 2541 °F); and delta iron (delta ferrite) exists from 1394 °C (2541 °F) to the melting point of pure iron, 1538 °C (2800 °F). The temperature ranges in which the various crystal forms of iron are stable make up the left vertical boundary (the pure iron end) of the Fe-C phase diagram shown in Fig. 1.1.

Crystal Structures of Iron

The crystal structure of ferrite is characterized by the unit cell shown in Fig. 1.2. Ferrite belongs to the cubic crystal system — all three axes of the unit cell are of the same length a and are mutually perpendicular. The space lattice of ferrite is body-centered cubic (bcc). There are a total of two atoms per unit cell — the body-centered atom with coordinates $a/2$, $a/2$, $a/2$, and the atom at the origin of the unit cell with coordinates 0, 0, 0. The latter atom represents all of the equivalent corner atoms of the unit cell, each of which is shared by eight unit cells that come together at a corner. The one-eighth atom per corner times the eight corners of the unit cell therefore accounts for one of the two atoms in a bcc unit cell.

The lattice parameter of alpha iron at room temperature is 2.86 Å, or 0.286 nm (Ref 1.2). The body diagonals of the unit cell, corresponding to ⟨111⟩ directions, are the directions in which the iron atoms are in contact in the bcc structure. Figure 1.2 shows that the body-centered atom has eight nearest neighbor atoms at a center-to-center distance of one-half a body diagonal, or $a\sqrt{3}/2$. Crystal structures in which the atoms are packed as closely together as possible have twelve nearest neighbor atoms, and therefore the bcc form of iron is a more open or less dense structure than the gamma iron structure described below. The difference in atomic packing between alpha and gamma is re-

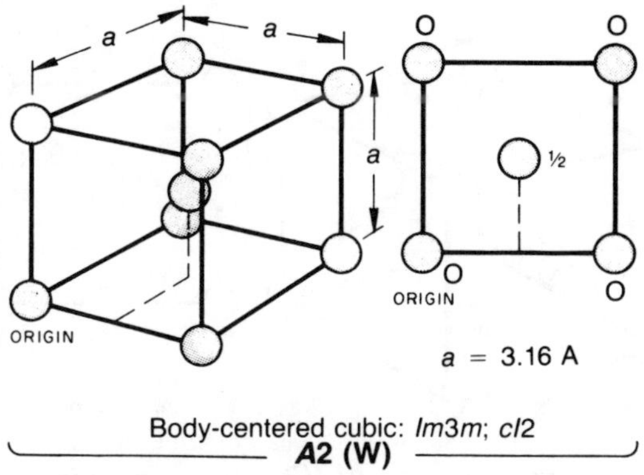

Fig. 1.2. Body-centered cubic (bcc) crystal structure. *A2* is structure (Strukturbericht) symbol, and W is prototype metal with bcc structure. Ferrite in steel is bcc. (Ref 1.12)

sponsible for the volume expansion that occurs when the higher density gamma iron transforms to alpha iron on cooling.

The unit cell of gamma iron or austenite is shown in Fig. 1.3. Austenite also belongs to the cubic crystal system, but has a face-centered cubic lattice. There are a total of four atoms per unit cell with coordinates 0, 0, 0; 0, $a/2$, $a/2$; $a/2$, $a/2$, 0; $a/2$, 0, $a/2$, corresponding to a corner atom and an atom in the center of each face of the unit cell. Each face atom is shared by two adjacent unit cells; the six faces of the cubic cell thus contribute three atoms. As described above for the bcc cell, the eight corners together contribute only one atom.

The lattice parameter of austenite, about 3.56 Å (0.356 nm), is larger than that of ferrite. However, the close-packed structure and the 4 atoms per unit cell make the density of austenite greater than that of ferrite. The face diagonals, corresponding to ⟨110⟩ directions, are the close-packed directions in the fcc structure, and establish the center-to-center atom spacing of the 12 nearest neighbor atoms as $a\sqrt{2}/2$.

Austenite also may be characterized as a structure made up of planes of closest atomic packing stacked in a sequence that repeats every three layers. The orientation of the close-packed {111} planes relative to the unit cell may be readily identified because each {111} plane is defined by three face diagonals of the unit cell. The close-packed planes in austenite are extremely important: the dislocation motion that makes mechanical deformation of austenite possible occurs on {111} planes, and microstructural features within grains known as twins have {111} planes as boundaries. Twins are characterized by

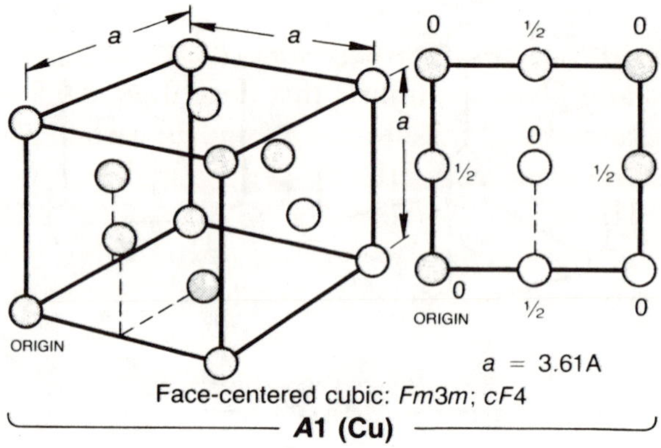

Fig. 1.3 Face-centered cubic (fcc) crystal structure. A1 is structure (Strukturbericht) symbol, and Cu is prototype metal with fcc structure. Austenite in steel is fcc. (Ref 1.12)

mirror symmetry of atoms across the planes separating the twins and the adjacent matrix (Ref 1.3). In austenite twins frequently form as a result of growth accidents in the stacking of {111} planes — accidents caused by recrystallization and grain growth during heating or annealing in the temperature range where austenite is stable.

Finally, the third phase that may form in pure iron is delta ferrite, a body-centered cubic structure that is crystallographically identical to that of alpha iron. Delta ferrite forms only at temperatures close to the melting point of iron. It is generally only of academic interest in the heat treatment of carbon steels because it is replaced at lower temperatures by austenite, the usual starting structure for commercial heat treatment. However, because delta ferrite is the first phase to form during solidification of iron and steel ingots and welds, it may be associated with interdendritic segregation patterns or concentration gradients of alloying and/or impurity elements (Ref 1.4). Hot working and homogenizing steels in the austenite range generally significantly reduce the segregation produced during solidification, and some degree of segregation may be tolerated in many applications. Occasionally, however, a problem that arises during heat treatment may be traced back to the segregation produced as delta ferrite first formed from the molten steel.

Effects of Carbon

The addition of carbon to iron produces several important changes in the phases and phase equilibria just described. Differences in the ability of ferrite and austenite to accommodate carbon result not only in important characteristics of the Fe-C diagram but also in the formation of Fe_3C. The crystal structures of the bcc ferrite and fcc austenite are modified by introducing carbon atoms into the interstices or interstitial sites between iron atoms. Austenite and ferrite in Fe-C alloys and steels are, therefore, interstitial solid solutions.

Carbon is an element that stabilizes austenite and thereby increases the range of austenite formation in steels. Figure 1.1 shows that, with the addition of carbon, the austenite field greatly expands from 912 to 1394 °C (1675 to 2540 °F) — the range in pure iron — to a wide range of temperatures and compositions. The maximum solubility of carbon in austenite reaches 2.11% at 1148 °C (2018 °F). Ferrite has a much lower ability to dissolve carbon than does austenite: the solubility decreases continuously from a maximum of only 0.02% at 727 °C (1340 °F). The limited solubility of carbon in ferrite is emphasized by the very small ferrite field (see Fig. 1.1). An expanded portion of the low-carbon end of the Fe-C diagram showing the temperature-composition range of ferrite and the decreasing solubility of carbon in ferrite with decreasing temperature is shown in Fig. 1.4. The room temperature solubility of carbon in ferrite is almost negligible.

When the solubility limit for carbon in austenite is exceeded, a new phase — iron carbide or cementite — forms in iron-carbon alloys and steels. Cementite assumes many shapes, arrangements and sizes that together with ferrite contribute to the great variety of microstructures found in steels. The various forms of cementite depend directly on thermal history or heat treatment. The crystal structures of cementite and of ferrite and austenite solid solutions will be discussed below, and the association and formation of those phases to produce characteristic microstructures will be discussed in later chapters.

Crystal Structures in Fe-C Alloys

The major difference between the structures of ferrite and austenite in steel and the corresponding phases in pure iron is the introduction of carbon atoms. There are two types of interstitial voids that may

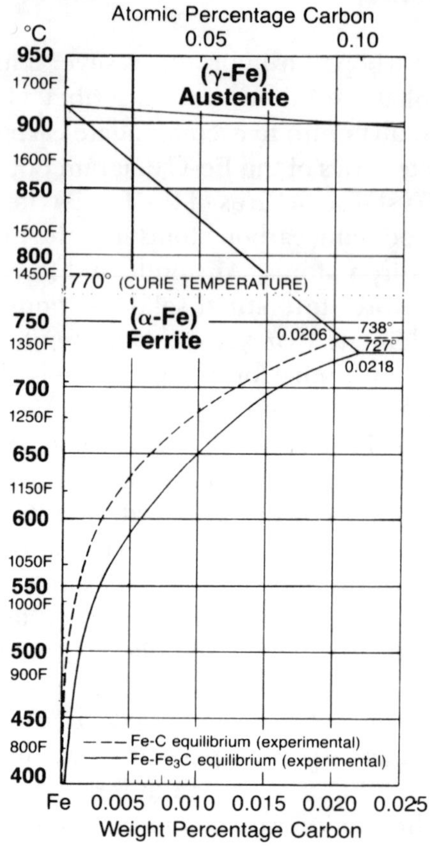

Fig. 1.4. Fe-rich side of Fe-C diagram, showing extent of ferrite phase field and decrease of carbon solubility with decreasing temperature. (Ref 1.12)

become sites for carbon atoms in bcc and fcc structures. Figures 1.5 and 1.6 show the octahedral and tetrahedral voids in the fcc and bcc structures, respectively. The two types of voids derive their names from the number of sides of the polyhedron formed by the iron atoms that surround a given site. A carbon atom has six nearest neighbor iron atoms if in an octahedral site and four if in a tetrahedral site.

The sizes of the different voids vary considerably. In austenite, assuming spherical iron atoms in contact, an octahedral site could accommodate an atom 0.052 nm in radius, but a tetrahedral site could only accommodate an atom 0.028 nm in radius (Ref 1.5). Carbon atoms

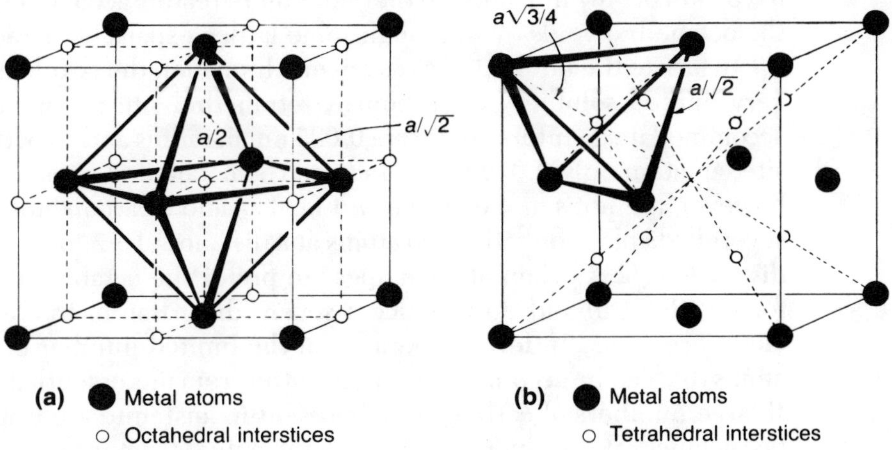

Fig. 1.5. Octahedral (a) and tetrahedral (b) interstitial voids in fcc structure. (Ref 1.5)

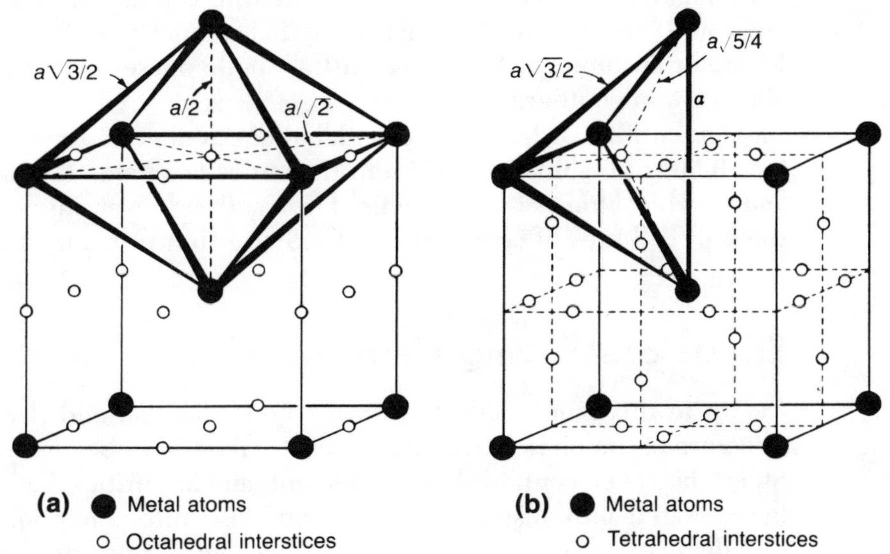

Fig. 1.6. Octahedral (a) and tetrahedral (b) interstitial voids in bcc structure. (Ref 1.5)

have radii of 0.07 nm, and are therefore more readily accommodated in the octahedral voids even though some lattice expansion is required.

In ferrite the interstitial sites are much smaller, thus explaining the very limited solubility of carbon. A tetrahedral site in ferrite could accommodate an interstitial atom 0.035 nm in radius and an octahedral site, an atom only 0.019 nm in radius. The octahedral sites in ferrite, however, are not symmetrical (see Fig. 1.6) and a carbon atom would severely displace only the two atoms at a distance of $a/2$, not those at a distance $a/\sqrt{2}$. Carbon atoms appear to prefer the octahedral sites in ferrite (Ref 1.5) and do produce a severe distortion of the lattice in ⟨100⟩ directions. In ferrite, because of the limited number of carbon atoms that can be accommodated, the lattice remains essentially cubic. If large numbers of carbon atoms present in austenite are trapped in bcc octahedral sites by rapid cooling, the cubic structure may actually become tetragonal. The latter structure typifies the phase "martensite", and its formation is the object of the very important hardening heat treatments to be described in later chapters.

Cementite, the phase that forms when the solubility of carbon in ferrite and austenite is exceeded, is a significantly different phase from the interstitial solid solutions described above. Cementite is a compound with a specific ratio of one carbon atom to three iron atoms, and is frequently referred to as Fe_3C. Cementite contains 6.67% carbon and could exist alone only in an alloy at that composition, in contrast to ferrite or austenite, which may exist as single phases over a range of alloy carbon content.

Cementite is orthorhombic, with lattice parameters a = 0.452 nm, b = 0.509 nm, and c = 0.674 nm. Its unit cell contains 12 iron atoms and 4 carbon atoms. The positions of iron and carbon atoms relative to the unit cell axes of cementite (Ref 1.5) are shown in Fig. 1.7.

Effects of Alloying Elements

Up to this point, only the binary Fe-C diagram and the crystal structure of the phases that form in Fe-C alloys have been described. Steels, however, contain alloying elements and impurities that must be incorporated into austenite, ferrite and cementite. Incorporation is usually by replacement of iron atoms if the alloy or impurity atoms are roughly the same size as iron atoms, but sometimes the atoms go into interstitial sites if they are significantly smaller than iron, as is nitrogen. In some cases, if sufficient quantities of alloying elements are present, solubility limits are exceeded and phases other than those already discussed may form. For example, small additions of chro-

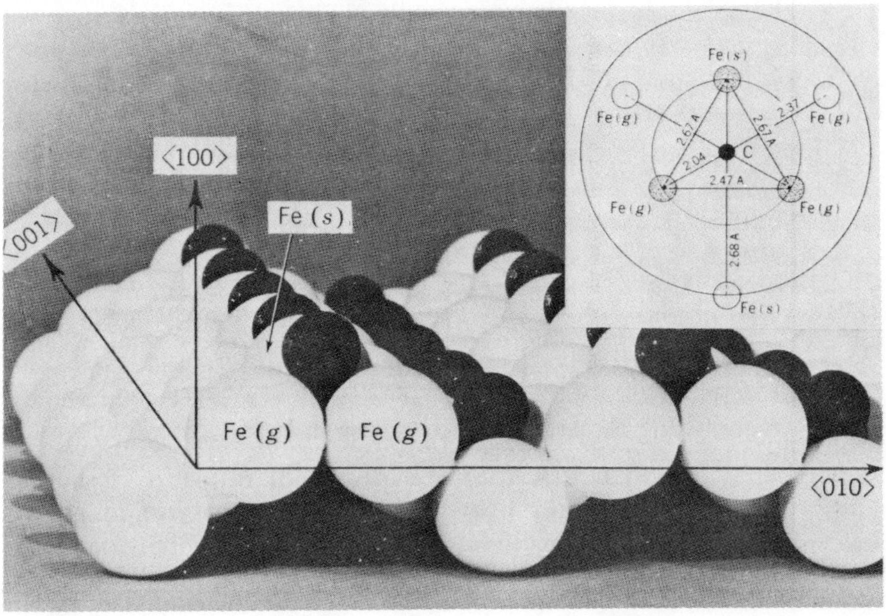

Fig. 1.7. Model of cementite structure that forms in steel. Insert is stereogram of iron nearest and next-nearest neighbor atoms around a carbon atom. (Ref 1.5)

mium to Fe-C alloys at 890 °C (1634 °F) maintain the cementite structure, M_3C, (M standing for a combination of chromium and iron atoms); larger additions cause the carbide M_7C_3 to form; and still larger additions produce the carbide $M_{23}C_6$ (Ref 1.6).

Some of the elements present in steels are austenite stabilizers (manganese and nickel, for instance), some are ferrite stabilizers (silicon, chromium and niobium), and some are strong carbide formers (titanium, niobium, molybdenum and chromium, if present in sufficient quantity). Ferrite and austenite stabilizers expand the respective phase fields. One measure of the effect of an alloying element on the Fe-C phase diagram is whether or not the eutectoid temperature (indicated by the horizontal line at 727 °C [1340 °F] in Fig. 1.1) is raised or lowered by an alloying addition. Austenite stabilizers lower the eutectoid temperature and thereby expand the temperature range over which austenite is stable. Figure 1.8 shows the change in eutectoid temperature with increasing amounts of several common alloying elements (Ref 1.7). Figure 1.9 shows a related effect of alloying elements on the Fe-C phase diagram: the decrease in carbon content of austenite of eutectoid composition. The type of evidence on which Fig.

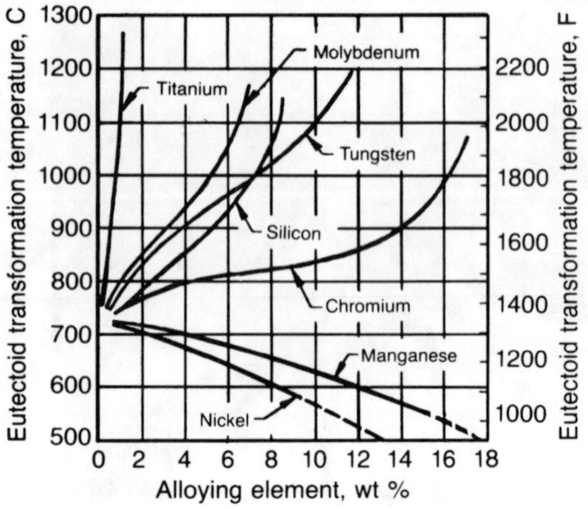

Fig. 1.8 Effect of substitutional alloying elements on eutectoid transformation temperature in steel. (Ref 1.7)

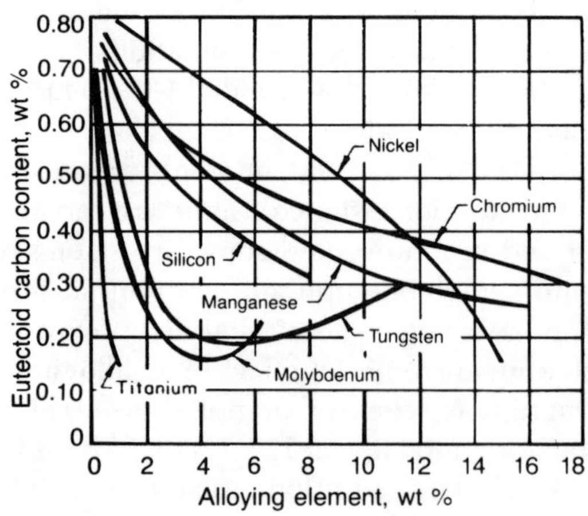

Fig. 1.9. Effect of substitutional alloying elements on the eutectoid carbon content in steel. (Ref 1.7)

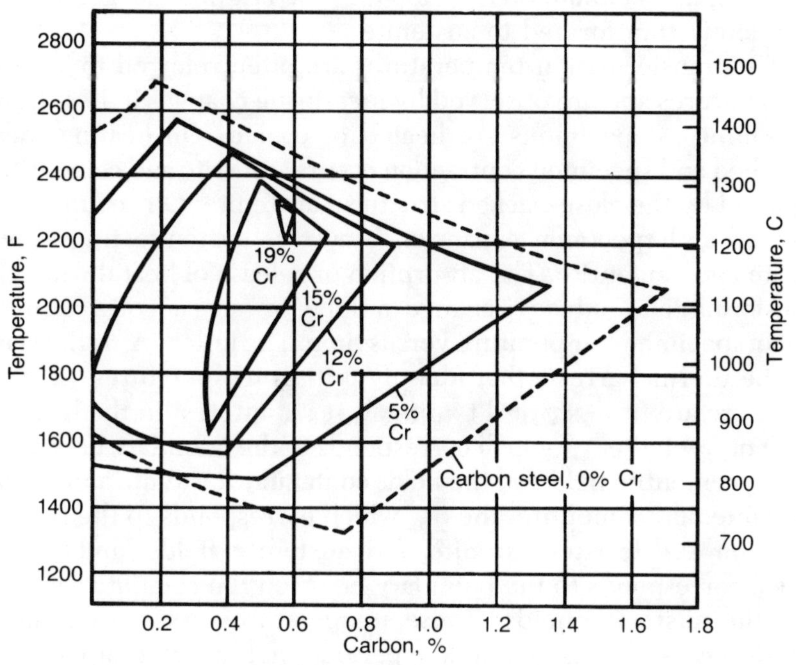

Fig. 1.10. Effect of chromium content on size of austenite phase field. (Ref 1.7)

1.8 and 1.9 were based is shown in Fig. 1.10 for the Fe-Cr-C system. Here, the austenite phase field shrinks with increasing chromium content. Associated changes in eutectoid composition and temperature are also graphically represented. The effect of alloying elements on the austenite phase field is usually determined by experimental techniques, but thermodynamic data have been used to calculate ranges of austenite stability in ternary alloys (Ref 1.8).

Critical Temperatures

The boundaries between phase fields of the Fe-C diagram shown in Fig. 1.1 identify temperatures for the various phase transformations that may occur in Fe-C alloys. For example, if an Fe-0.5C alloy were heated from room temperature at an extremely low rate, some of the ferrite and all of the cementite would transform to austenite at 727 °C

(1340 °F), and at about 860 °C (1580 °F) the last bit of ferrite would be completely transformed to austenite.

The transformation temperatures are often referred to as critical temperatures and are observed by measuring changes in heat transfer or volume as specimens are heated or cooled. On heating, heat is absorbed and specimen contraction occurs as ferrite and cementite are replaced by the close-packed structure, austenite. On cooling, heat is evolved and specimen expansion occurs as austenite transforms to ferrite and cementite. The absorption or release of heat during phase transformation produces a change in slope, or "arrest", on a continuous plot of specimen temperature versus time. The letter "A" is the symbol for the thermal arrests that identify critical temperatures.

There are three critical temperatures of interest in the heat treatment of steel: the A_1, which corresponds to the boundary between the ferrite-cementite field and the fields containing austenite and ferrite or austenite and cementite; the A_3, which corresponds to the boundary between the ferrite-austenite and austenite fields; and the A_{cm}, which corresponds to the boundary between the cementite-austenite and the austenite fields. These temperatures assume equilibrium conditions — that is, extended periods of time at temperature or extremely slow rates of heating or cooling. Sometimes A_1, A_3 and A_{cm} are designated as Ae_1, Ae_3 and Ae_{cm}, respectively, the letter "e" indicating assumed equilibrium conditions.

The transformations that occur at A_1, A_3, and A_{cm} are diffusion controlled. Therefore, the critical temperatures are sensitive to composition and to heating and cooling rates. Rapid heating allows less time for diffusion and tends to increase the critical temperatures above those associated with equilibrium. Likewise, rapid cooling tends to lower the critical temperatures. The effect of heating or cooling rate is defined practically by a new set of critical temperatures designated "Aĉ" or "Ar̂" (for the arrests on heating or cooling, respectively). The terminology was developed by the French metallurgist, Osmond (Ref 1.9) — Ac stands for *arrêt chauffant* and Ar for *arrêt refroidissant*. As a result of heating and cooling effects, therefore, there are two other sets of critical temperatures: Ac_1, Ac_3, and Ac_{cm}, and Ar_1, Ar_3, and Ar_{cm}. These sets of critical temperatures are shown schematically in Fig. 1.11.

Generally, the critical temperatures for a given steel are determined experimentally. However, some empirical formulas that show the effects of alloying elements on the critical temperatures have been developed by regression analysis of large amounts of experimental data. For example, Andrews (Ref 1.10) has developed the following formulas for Ac_3 and Ac_1 in degrees Celsius:

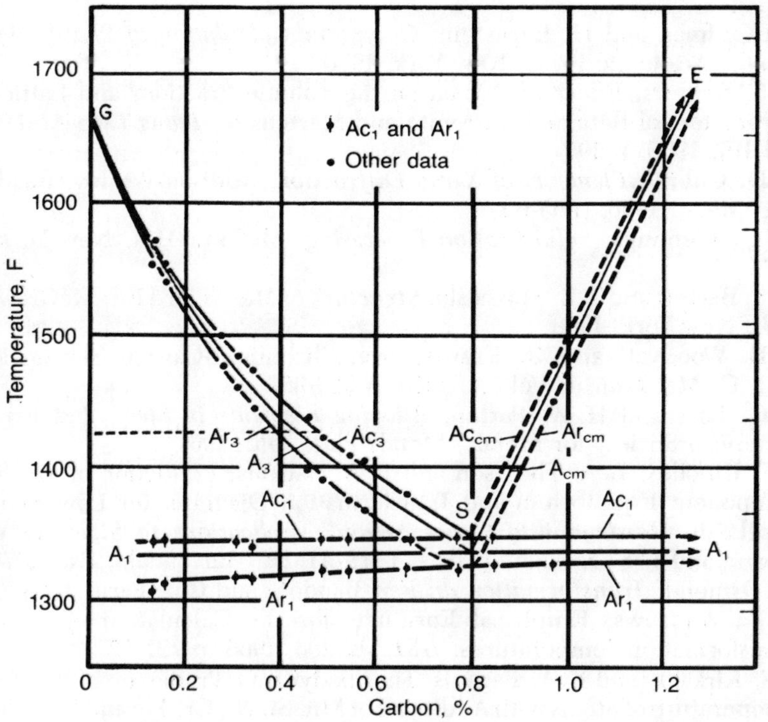

Fig. 1.11 Cooling (Ar), heating (Ac) and equilibrium (A) temperatures in Fe-C alloys. Heating and cooling at 0.125 °C/min. (Ref 1.7)

$$Ac_3 = 910 - 203\sqrt{C} - 15.2Ni + 44.7Si + 104V + 31.5\,Mo + 13.1W \qquad \text{(Eq 1.3)}$$

$$Ac_1 = 723 - 10.7Mn - 16.9Ni + 29.1Si + 16.9Cr + 290As + 6.38\,W \qquad \text{(Eq 1.4)}$$

These formulas present another way of describing the effect of alloying elements on both the Fe-C diagram and the transformation behavior of steels. Elements that stabilize austenite lower the Ac_3 and Ac_1 as evidenced by their negative contributions to the corresponding equation, whereas elements that stabilize ferrite or carbide raise the Ac_3 and Ac_1 and make a positive contribution. The effect of alloying elements on the Ac_3 has also been determined by thermodynamic calculations (Ref 1.11).

References

1.1 L. Kaufman and H. Bernstein, *Computer Calculation of Phase Diagrams*, Academic Press, New York, 1970

1.2 C. S. Roberts, Effect of Carbon on the Volume Fractions and Lattice Parameters of Retained Austenite and Martensite, *Trans TMS-AIME*, Vol 197, 1953, p 203

1.3 B. D. Cullity, *Elements of X-ray Diffraction*, Addison-Wesley, Reading, Mass., 1956, p 55-60

1.4 M. C. Flemings, *Solidification Processing*, McGraw-Hill, New York, 1974

1.5 C. S. Barrett and T. B. Massalski, *Structure of Metals*, 3rd Ed. McGraw-Hill, New York, 1966

1.6 L. R. Woodyatt and G. Krauss, Iron-Chromium-Carbon System at 870 °C, *Met Tran A*, Vol 7A, 1976, p 983-989

1.7 E. C. Bain and H. W. Paxton, *Alloying Elements in Steel*, 2nd Ed., American Society for Metals, Metals Park, Oh., 1961

1.8 J. S. Kirkaldy, B. A. Thomson and E. A. Baganis, Prediction of Multicomponent Equilibrium and Transformation Diagrams for Low Alloy Steels, in *Hardenability Concepts and Applications to Steel*, D. V. Doane and J. S. Kirkaldy (Eds.), TMS-AIME, Warrendale, Pa., 1978

1.9 F. Osmond, *Transformation du Fer*, Baudoin and Co., Paris, 1888

1.10 K. W. Andrews, Empirical Formulae for the Calculation of Some Transformation Temperatures, *JISI*, Vol 203, 1965, p 721-727

1.11 J. S. Kirkaldy and E. A. Baganis, Thermodynamic Prediction of the Ae_3 Temperature of Steels with Additions of Mn, Si, Ni, Cr, Mo and Cu, *Met Trans A*, Vol 9A, 1978, p 495-501

1.12 *Metals Handbook*, 8th Ed., Vol 8, American Society for Metals, Metals Park, Oh., 1973, p 236, 275, 276

CHAPTER 2

Pearlite, Ferrite and Cementite

Chapter 1 described the crystal structures of the phases that form in steels and the Fe-C phase diagram which defines the temperature-composition ranges over which these phases may exist. This chapter shows how various arrangements of phases or microstructures are produced by austenite transformation to ferrite and cementite. Alloy composition and the rate at which austenite is cooled profoundly affect which microstructure forms. The emphasis in this chapter is on the microstructures produced by the diffusion-controlled transformations that occur in carbon steels during relatively slow cooling from the austenite phase field.

Eutectoid Transformation

The Fe-C diagram introduced in Chapter 1 provides the basic framework for understanding the phase transformations and microstructures of concern in this chapter. Figure 2.1 is an enlarged section of the Fe-C diagram that includes the areas most pertinent to the transformation of austenite in slowly cooled steels. Consider first the Fe-0.77C alloy, which would be completely austenitic at all temperatures down to the A_1 temperature (727 °C or 1340 °F). If held for a very long period of time at this temperature, or cooled very slowly through A_1 (that is, under conditions approaching equilibrium), the phase diagram shows that the austenite must be replaced by a mixture of ferrite and cementite. A phase transformation in which one solid phase

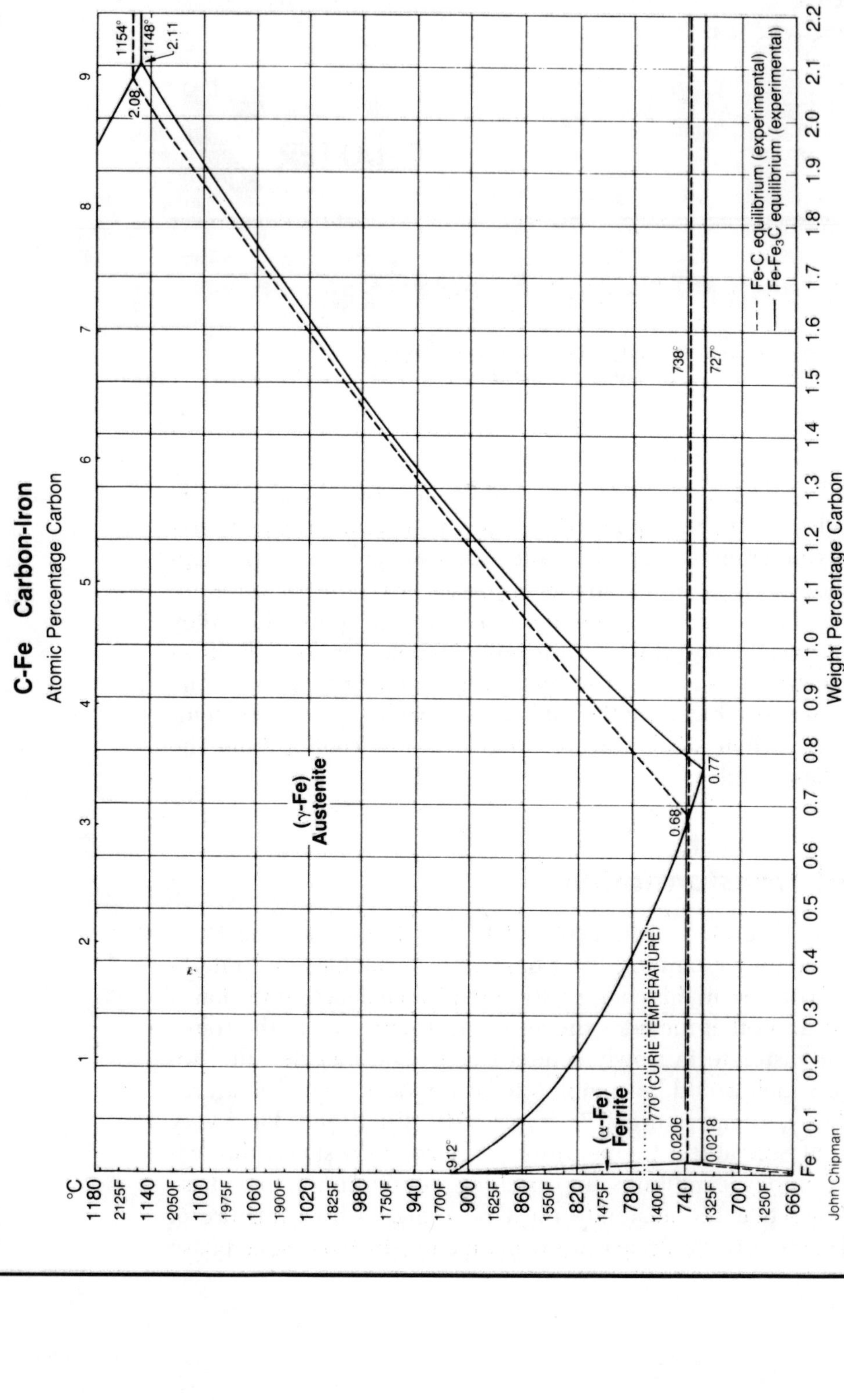

Fig. 2.1. Portion of the Fe-C diagram emphasizing regions of proeutectoid ferrite and cementite formation and the eutectoid transformation of austenite. (Ref 2.15)

is replaced by two different solid phases is classified as a eutectoid transformation, and in the Fe-C system may be written in the form of the following reaction:

$$\gamma\ (0.77\%\ \mathrm{C}) \underset{\text{heating}}{\overset{\text{cooling}}{\rightleftharpoons}} \alpha\ (0.02\%\ \mathrm{C}) + \mathrm{Fe_3C}(6.67\%\ \mathrm{C}) \quad \text{(Eq 2.1)}$$

This equation shows that the phases involved in the eutectoid reaction have fixed compositions and that the reaction is reversible depending upon whether heat is removed or added. Ideally the eutectoid reaction occurs isothermally at 727 °C (1340 °F). Equilibrium conditions, however, are rarely obtained in actual practice and the eutectoid reaction may in fact occur over a wide range of temperatures below A_1.

Structure of Pearlite

The eutectoid transformation in steels produces a unique microstructure termed "pearlite". Pearlite is made up of alternate lamellae of ferrite and cementite as shown in Fig. 2.2, a light micrograph of a furnace-cooled specimen of an Fe-0.75C alloy. Colonies of lamellae of various orientations and spacings characterize the microstructure. The spacing variations of the cementite lamellae in different areas may be partly due to differences in the angles that the lamellae make with the plane of polish, and partly due to the fact that the pearlite may have formed over a range of temperatures. Assuming all the pearlite was formed at about the same temperature, and that therefore all the lamellae had almost identical spacing, those colonies with lamellae perpendicular to the plane of polish would show the true spacing or closest spacing of the ferrite and cementite lamellae. Those lamellae at angles less than 90° would show a wider spacing. Determination of the true pearlite spacing from metallographically prepared specimens where the lamellae form a range of angles with the specimen surface requires special quantitative metallographic analyses (Ref 2.1, 2.2).

The origin of the term "pearlite" is related to the regular array of the lamellae in the colonies and the fact that etching attacks the ferrite phase more severely than the cementite. The raised and regularly spaced cementite lamellae of the colonies then act as diffraction gratings, and a pearl-like lustre is produced by diffraction of light of various wave lengths from the different colonies.

The amounts of cementite and ferrite in pearlite formed at 727 °C (1340 °F) can be determined by a calculation based on the lever rule.

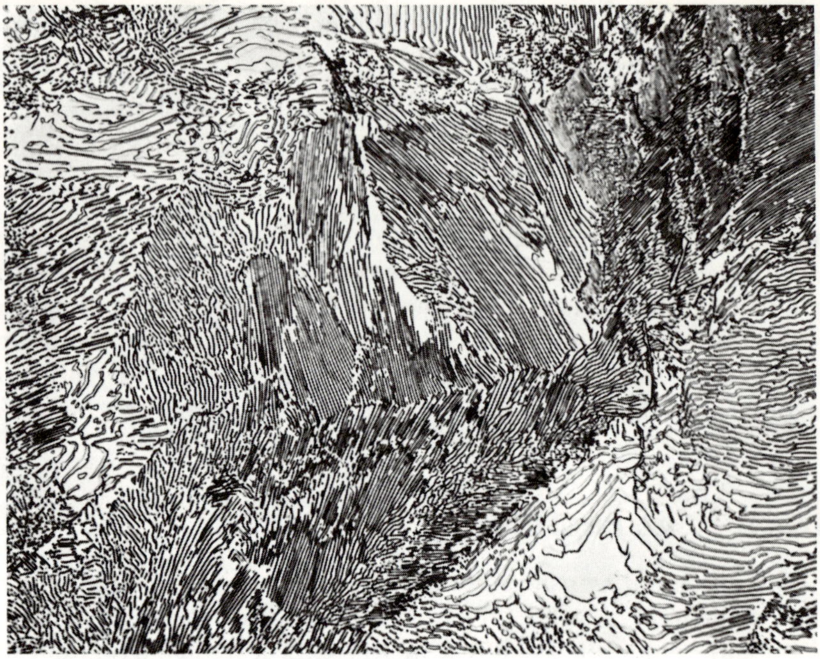

Fig. 2.2 Pearlite in a furnace-cooled Fe-0.75C alloy. Picral etch. Magnification, 500 ×. (Courtesy of A. R. Marder and A. Benscoter, Bethlehem Steel Corp., Bethlehem, Pa.)

The lever rule can be applied to any two-phase field of a binary phase diagram to determine the amounts of the different phases present at a given temperature in a given alloy. A horizontal line, referred to as a tie line, represents the lever, and the alloy composition its fulcrum. The intersection of the tie line with the boundaries of the two-phase field fixes the compositions of the coexisting phases, and the amounts of the phases are proportional to the segments of the tie line between the alloy and the phase compositions. For pearlite, assume a tie line immediately below 727 °C (1340 °F) that spans the ferrite-cementite phase field (see Fig. 1.1). Application of the lever rule calculation for the Fe-0.77C alloy, the alloy that transforms entirely to pearlite, shows that:

$$\begin{array}{l}\text{wt \% Fe}_3\text{C} \\ \text{in pearlite}\end{array} = \frac{0.77 - 0.02}{6.67 - 0.77} \times 100 = 13\% \qquad \text{(Eq 2.2)}$$

By difference, the weight percent ferrite in pearlite is 87%. Therefore, whenever austenite containing 0.77% carbon transforms to pearlite at or close to 727 °C (1340 °F), ferrite and cementite form in the fixed weight percentages as shown above. The densities of ferrite and cementite, 7.87 and 7.70 g/cm^3 respectively, are so close that the volume percentages of ferrite and cementite in pearlite are essentially the same as the weight percentages. Therefore, in Fe-C alloys, the amounts of phases calculated by the lever rule with compositions by weight should correlate well with the amounts of phases revealed in light micrographs. The amounts of phases visible in micrographs are related to area percentages, which in turn are directly related to their volume percentages if the phases are uniformly distributed.

The development of a pearlite colony has been shown to initiate from either ferrite or cementite crystals (Ref 2.3, 2.4). Originally, the lamellar structure was thought to develop only by sidewise nucleation of separate lamellae; however, branching of a single cementite crystal into parallel lamellae with spacing characteristic of a given transformation temperature has also been shown to produce the lamellar structure. According to the latter mechanism, all of the cementite in a given colony is interconnected, and a colony of pearlite may be regarded as two single crystals of ferrite and cementite. The latter structure was strikingly revealed in a serial sectioning experiment in which a pearlite colony was repeatedly photographed as successive layers were removed by polishing in 1-μm steps (Ref 2.3). All of the apparently separate cementite lamellae were shown to have a common origin. Once a pearlite colony is established by sidewise nucleation and/or branching of the ferrite and cementite, the lamellae are considered to grow by extension of their edges into the austenite, a process frequently referred to as edgewise growth (Ref 2.5).

Pearlite Transformation Kinetics

The preceding section described the lamellar structure of pearlite and its formation by a eutectoid reaction at or close to 727 °C (1340 °F). In actual practice, however, the formation of pearlite rarely occurs close to the A_1. Figure 2.3 shows an isothermal transformation diagram for eutectoid 1080 steel. Curves for the beginning and end of pearlite formation, obtained by cooling from the austenite phase field and holding at various temperatures below A_1, are shown. The beginning of transformation curve is asymtotic to the A_1, thus indicating that pearlite would not form at temperatures close to A_1 unless the steel were

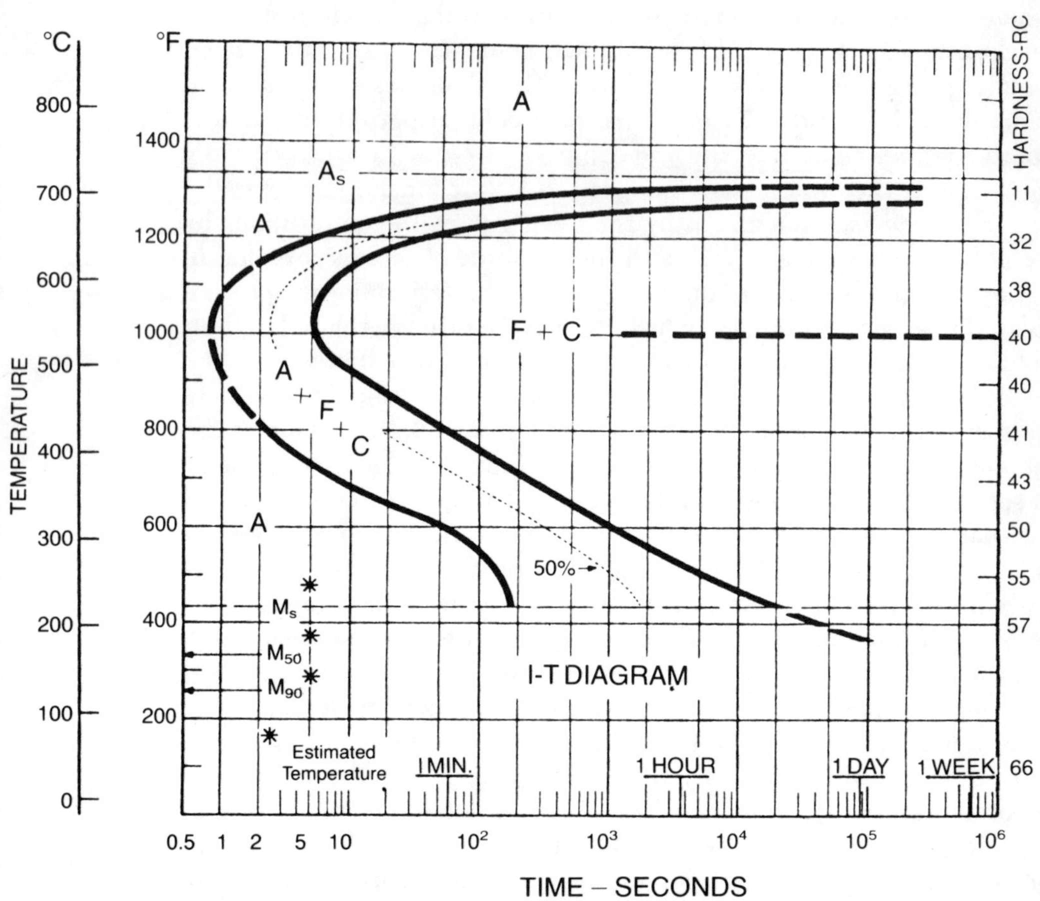

Fig. 2.3 Isothermal transformation diagram for 1080 steel containing 0.79% carbon and 0.76% manganese. Specimens were austenitized at 900 °C (1650 °F) and had an austenite grain size of ASTM No. 6. (Ref 2.16)

held at temperature for extended periods of time. In commercial heat treating practice, the slow rates of cooling that would permit pearlite formation close to the A_1 are approached only in very heavy sections or by furnace cooling. With increased undercooling below A_1, however, the time periods for the beginning and end of pearlite transformation are reduced substantially. At the nose of the transformation curve, 540 °C (1004 °F), the lowest temperature at which pearlite forms in this steel, only seconds are required for complete transformation.

Below 540 °C (1004 °F), bainite, a nonlamellar microstructure of ferrite and cementite, is formed.

A number of factors influence the rate of pearlite formation. Perhaps most important is the fact that substantial carbon atom rearrangement must take place to accomplish the transformation of austenite (containing nominally 0.77% carbon) to low-carbon ferrite and high-carbon cementite according to Eq 2.1. The diffusion of carbon, as characterized by its diffusion coefficient, is temperature dependent. One equation that has been developed (Ref 2.6) to show the temperature dependence of carbon diffusion in austenite is

$$D_C^\gamma = 0.12e^{-32,000/RT} \qquad \text{(Eq 2.3)}$$

where D_C^γ is the average diffusion coefficient (cm^2/s) of carbon in austenite, R is the gas constant (1.98 cal/g-mol/K), and T is the absolute temperature (°C + 273). Equation 2.3 shows that the diffusion coefficient decreases exponentially with decreasing temperature, a powerful effect that significantly lowers the diffusion coefficient for small decreases in temperature. At first glance, the temperature dependence of diffusion appears to contradict the experimentally established fact (see Fig. 2.3) that pearlite formation is faster at lower temperatures than at higher temperatures. This apparent anomaly is explained by the reduction of interlamellar spacing as the temperature of pearlite transformation decreases. Thus, the distance that carbon has to diffuse to distribute itself between the ferrite and cementite decreases, and despite the fact that diffusion becomes more sluggish at lower temperatures, the growth of pearlite colonies accelerates.

The interrelationships between diffusion and the lamellar structure of pearlite help to explain how the eutectoid transformation proceeds, but not why the transformation occurs. The stability of all phases and microstructures in metals and alloys is based on the principle of minimum free energy. If the free energy of a given microstructure or system is not a minimum, then either a phase transformation (for example, the austenite to pearlite transformation under consideration here) or microstructural rearrangement without a phase change (for example, grain growth or particle coarsening) would occur in order to lower free energy to the minimum possible value.

The free energy (G) per unit volume of a phase or combination of phases is defined in terms of other thermodynamic parameters, enthalpy or heat content (H), the absolute temperature (T) and entropy (S) as follows:

$$G = H - TS \qquad \text{(Eq 2.4)}$$

Enthalpy is the total energy of a phase (or microstructure composed of several phases) per unit volume of that structure. Entropy is a measure of the degree of order associated with a given structure at a given temperature. It may be influenced by the amplitude of atom vibration, the mixing of several component types of atoms and/or vacant lattice sites in a given phase, or the degree of order associated with a given solid or liquid structure. The TS term, therefore, is a measure of the energy associated with the order of a unit volume of a given structure at a given temperature, and is especially important in establishing phase stability at high temperatures. Equation 2.4 shows that the difference between the enthalpy and entropy terms defines free energy. A rigorous approach to the development of atomistic and classical thermodynamics is presented in the text by Swalin (Ref 2.7).

A helpful example of the applications of the principles of minimum free energy in establishing phase stability is the melting of a solid crystal structure. With increasing temperature, H increases, but TS increases much more if the liquid, with its high degree of atomic disorder, replaces the ordered crystal structure. Therefore, above the melting point, because of its higher entropy, the liquid has the lower free energy and is the stable phase relative to the solid. Similar considerations apply to transformations between solid phases such as the transformation of austenite to ferrite and cementite.

At the Ae_1 temperature in the Fe-C system, the free energy of austenite is exactly equal to the free energy of ferrite and cementite and there is no incentive for transformation to occur, especially if interfaces or boundaries between the austenite and pearlite must be created. Interfaces accommodate structural and chemical discontinuities between phases, and therefore make positive contributions to or raise the energy of a system. However, with decreasing temperature below Ae_1, the free energy of a unit volume of a mixture of ferrite and cementite becomes much less than that of austenite. This free energy difference is frequently referred to as the driving force for transformation and increases with decreasing temperature or undercooling below the Ae_1. A larger driving force makes possible not only the development of more colonies of pearlite but also a finer lamellar spacing within a pearlite colony, structural changes that result in increased interfacial area of two types. A higher density of pearlite colonies results in increased austenite/pearlite interfacial area, and a reduced interlamellar spacing results in increased ferrite-cementite interfacial energy within the colonies. The high driving force at low temperatures offsets the positive energy contributions due to the various interfaces produced during the austenite transformation to pearlite.

Many relationships for the change in interlamellar spacing with undercooling have been proposed, but the one most closely related to the above considerations was developed by Zener and Hillert (Ref 2.8) as presented in the following equation:

$$S = \frac{4\sigma^{\alpha/Fe_3C} T_E}{\Delta H_v \Delta T} \qquad \text{(Eq 2.5)}$$

where S is the interlamellar spacing defined by the combined width of the α and Fe_3C lamellae; σ^{α/Fe_3C} is the interfacial energy per unit area of α/Fe_3C boundary; T_E is the equilibrium temperature (Ae_1 in the case of Fe-C alloys and steels); ΔH_v is the change in enthalpy per unit volume between austenite and the mixture of ferrite and cementite; and ΔT is the undercooling below Ae_1. Figure 2.4 shows a set of measurements illustrating the decrease in pearlite spacing with increasing undercooling for a variety of steels.

The isothermal transformation kinetics of the eutectoid transformation, i.e., the progress of pearlite formation as a function of time at a constant temperature, are based on the nucleation and growth rates of pearlite colonies. Figure 2.5 shows circular cross sections of pearlite colonies in an Fe-C alloy of eutectoid composition that has been partially transformed to pearlite. A number of colonies of pearlite have been nucleated and are in the process of growing into the austenite at the reaction temperature. In contrast to the pearlite shown in Fig. 2.2, the individual lamellae are too closely spaced to be resolved at the magnification of the micrograph, and the pearlite colonies have a dark appearance. The balance of the microstructure is white-etching martensite, formed in any untransformed austenite when it was quenched from the reaction temperature. Martensite and its formation are described in Chapter 3.

Johnson and Mehl (Ref 2.9), assuming that the pearlite colonies are spherical and randomly nucleated as a function of time, developed the following equation for isothermal pearlite formation:

$$f(t) = 1 - \exp\left[-NG^3t^4/3\right] \qquad \text{(Eq 2.6)}$$

where f(t) is the volume fraction pearlite formed at any time t at a given temperature, N is the nucleation rate of the colonies and G is the rate at

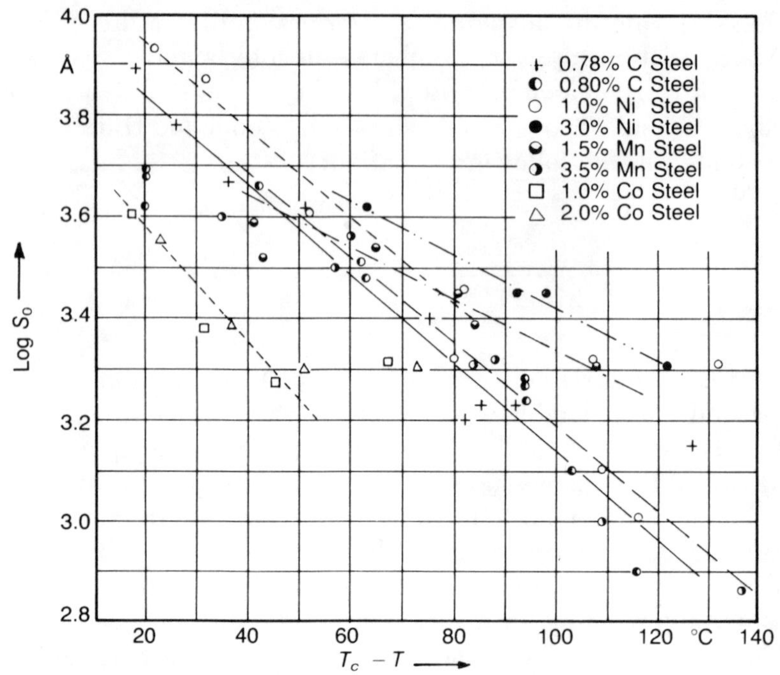

Fig. 2.4. Average true interlamellar spacings of pearlite, S_0, as a function of undercooling below Ae_1 for various steels as indicated. (Ref 2.1)

which the colonies grow into the austenite. The Johnson-Mehl equation describes mathematically the rate at which austenite is converted to a pearlitic microstructure by the nucleation and growth of pearlite colonies. At any given temperature, f(t) versus time fits an "S-shaped" or sigmoidal curve as shown in Fig. 2.6. The initial transformation rate is quite low and is associated with what is referred to as an incubation period, the time when the first stable nuclei develop. As more and more nuclei develop and are in various stages of growth, the rate of transformation increases. Finally the colonies impinge and the rate of transformation again slows as the microstructure gradually approaches complete transformation. The elapsed time periods needed to initiate and complete the pearlite transformation are directly related to the beginning and end of transformation curves in isothermal transformation diagrams shown schematically in Fig. 2.7. The exact beginning and end of pearlite formation at any given temperature is of course dependent on the sensitivity of the experimental techniques used to follow the transformation, but generally the accuracy is on the order of

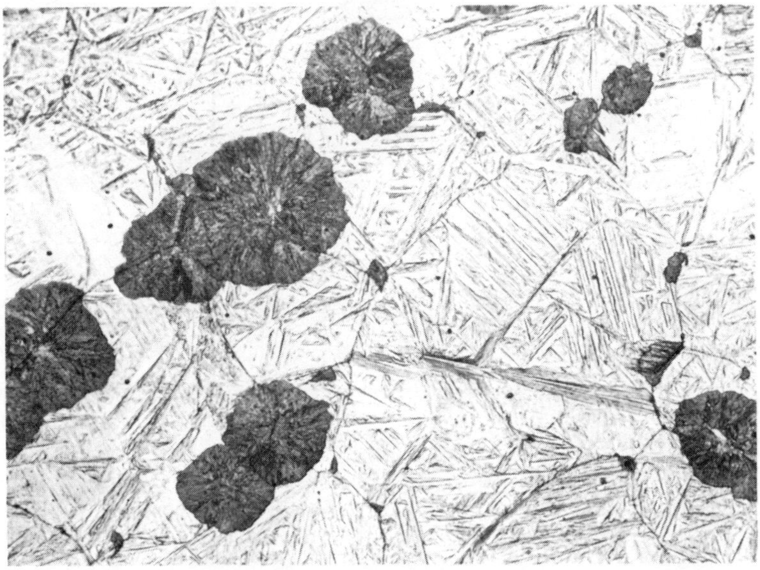

Fig. 2.5. Cross sections of spherical colonies of pearlite (dark) in eutectoid steel. Remainder of microstructure is martensite formed in austenite not transformed to pearlite at the reaction temperature. Magnification, 250 ×. (Courtesy of A. R. Marder and B. Bramfit, Bethlehem Steel Corp., Bethlehem, Pa.)

1%. Therefore, the beginning and end of transformation curves in Fig. 2.7 correspond to 1% and 99% transformation, respectively.

The shape of the isothermal transformation curve for eutectoid steel (see Fig. 2.3) is explained by the temperature dependence of the nucleation and growth rates of the pearlite colonies. Figure 2.8 shows that both N and G in an 0.78% carbon steel increase with decreasing transformation temperature, thus, according to the Johnson-Mehl equation, accelerating the eutectoid transformation at lower temperatures. As discussed earlier, the greater driving force associated with increased undercooling produces more nuclei and smaller interlamellar spacing. The latter in turn increases the growth rate of the pearlite colonies by effectively reducing the distance over which carbon must diffuse at the austenite/pearlite interface.

The above discussion shows that the Johnson-Mehl equation offers a highly effective approach to characterizing the kinetics of pearlite transformation. The assumption that the pearlite colonies nucleate

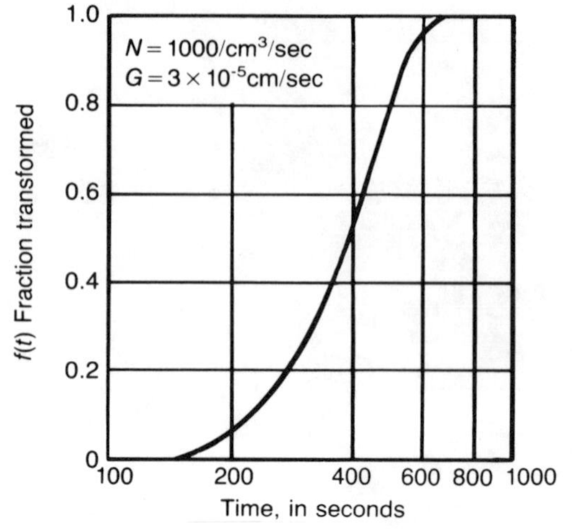

Fig. 2-6. Calculated fraction austenite transformed to pearlite as a function of time for the parameters shown. (Ref 2.1)

randomly in the austenite throughout the course of the transformation is not always valid, however. As shown in Fig. 2.5, the pearlite colonies invariably nucleate at austenite grain boundaries. Eventually, the grain boundaries become saturated with nuclei, nucleation terminates, and the balance of the transformation is accomplished solely by growth of the grain boundary nucleated colonies into the austenite (Ref 2.10).

The mechanism of pearlite formation continues to receive theoretical and experimental attention. Perhaps the most active considerations involve the way in which carbon and other alloying elements distribute themselves between the ferrite and cementite lamellae. Earlier in this chapter it was tacitly assumed that the growth of a pearlite colony is dependent on the diffusion of carbon atoms through the austenite ahead of the pearlite interface. Such diffusion through a crystal phase is referred to as bulk or volume diffusion. Another possibility, however, is that the carbon diffuses along the advancing interface between the pearlite and the austenite (Ref 2.5). Such interface or grain boundary diffusion occurs more rapidly than volume diffusion because of the more irregular or open packing of atoms at grain boundaries in comparison to the regular, close atom packing within a grain.

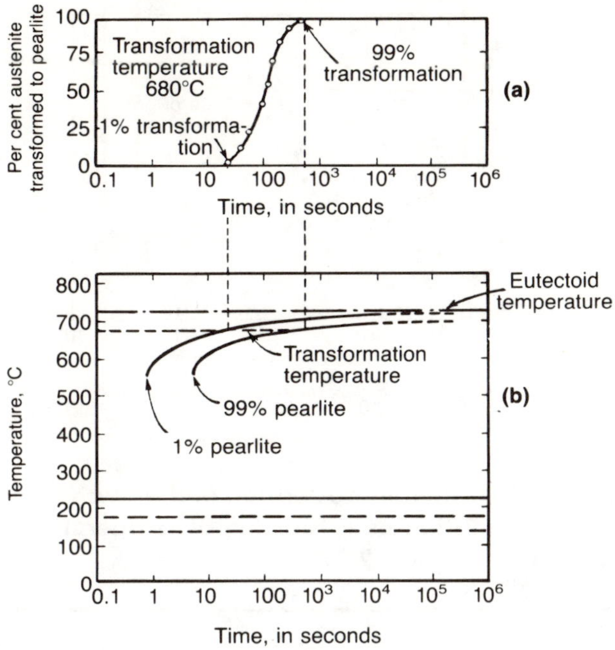

Fig. 2.7. Relationship of an isothermal reaction curve for pearlite formation (a) to the time-temperature-transformation diagram (b). (Ref 2.16)

In ternary systems and steels the effects of alloying elements must also be considered. Puls and Kirkaldy (Ref 2.8) suggest that manganese and nickel do not partition themselves between the ferrite and cementite and that therefore, pearlite formation in Fe-C-Mn and Fe-C-Ni alloys is dependent primarily on volume diffusion of carbon in austenite. Any reduction in the rate of pearlite growth in these systems is due to the effect of manganese and nickel on the diffusion of carbon in austenite. However, chromium and molybdenum, which are strong carbide-forming elements, are considered to partition to the carbide lamellae by interface diffusion. In the Fe-C-Cr and Fe-C-Mo systems, then, pearlite growth is retarded because chromium and molybdenum atoms must diffuse, a process that is much more sluggish than the diffusion of carbon because of the much larger size of the alloying element atoms compared to carbon atoms. Figures 2.9 and 2.10 show the effects of the various alloying elements on the growth rate of

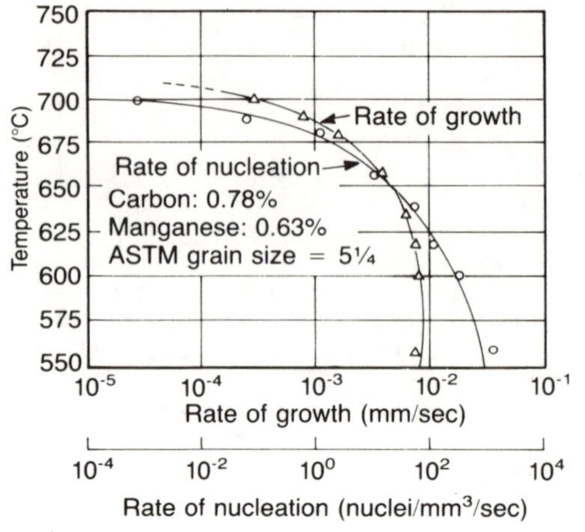

Fig. 2.8. Variation of nucleation and growth rates for pearlite formation as a function of temperature in a eutectoid steel. (Ref 2.17)

pearlite as a function of temperature. The alloying elements all slow the growth of pearlite, an effect that is extremely valuable when non-pearlitic microstructures are the desired objectives of heat treatment. The practical effects of alloying elements in retarding pearlite formation in steels are the basis of the topic of hardenability discussed in Chapter 6.

Interphase Precipitation

In alloy steels containing strong carbide-forming elements such as vanadium, niobium, titanium, molybdenum and tungsten, a special type of austenite decomposition occurs. Honeycombe and his colleagues (Ref 2.4) have shown that rows of very fine alloy carbide particles form along the interface between the decomposing austenite and newly formed ferrite. In view of the fact that the carbides nucleate and grow at the interface between austenite and ferrite, the reaction is termed interphase precipitation. However, one solid phase transforms into two other solid phases, and the reaction may also be classified as a eutectoid transformation that produces a special microstructure much different from lamellar pearlite.

The alloy carbides that make up the rows left behind by the moving austenite-ferrite interface are frequently on the order of 100 Å (10 nm)

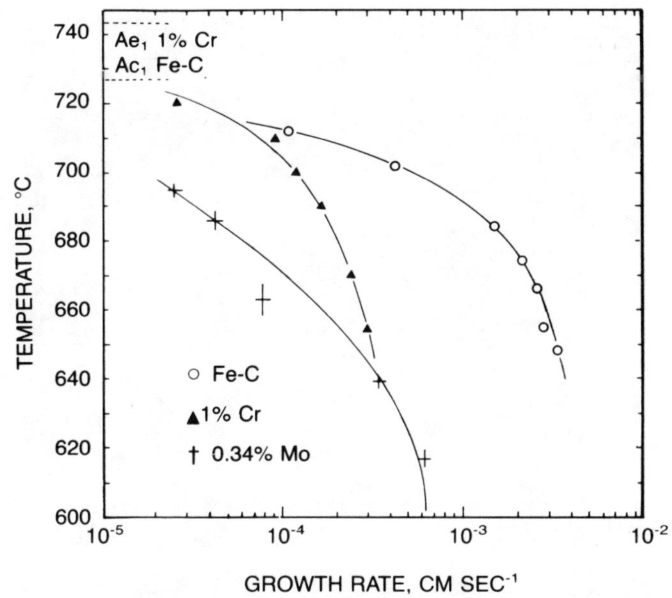

Fig. 2.9. Pearlite growth rates as a function of temperature for an Fe-0.81C alloy and eutectoid steels containing chromium and molybdenum. (Ref 2.8)

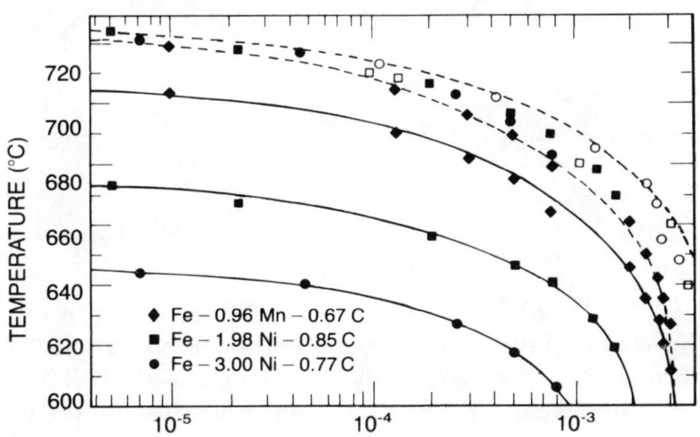

Fig. 2.10. Pearlite growth rates as a function of temperature for eutectoid steels with nickel and manganese. (Ref 2.8)

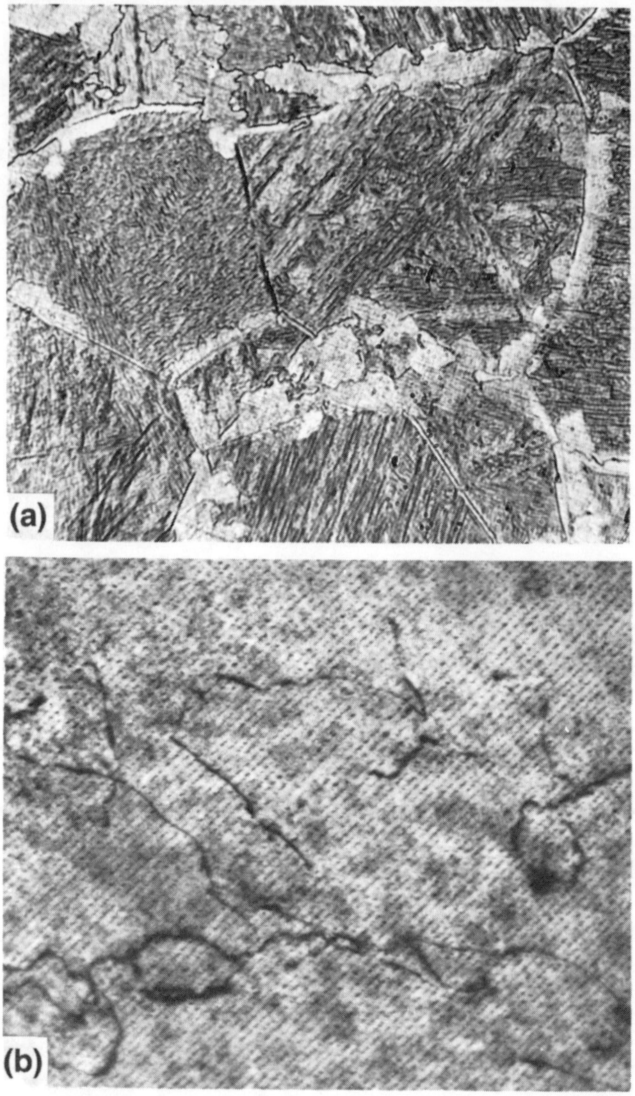

Fig. 2.11. (a) Colonies of interphase precipitation (light areas) nucleated at austenite grain boundaries of an Fe-0.75V-0.15C alloy held 10 s at 680 °C (1256 °F). Magnification, 125×. (b) Rows of fine alloy carbides within a colony of the same steel held 5 min at 725 °C (1340 °F). Magnification, 100 000×. (Courtesy of R. W. K. Honeycombe, University of Cambridge, U.K.)

or less and are too fine to be resolved in the light microscope. Figure 2.11(a) is a light micrograph that shows the initiation of interphase precipitation at austenite grain boundaries in an Fe-0.75V-0.15C alloy held for 10 s at 680 °C (1256 °F). The colonies of the interphase precipitate have curved interfaces and outline the austenite grain boundaries, but no structure is visible within the colonies. The balance of the microstructure is a low-carbon martensite formed on rapid cooling after the 10-s hold at 680 °C (1256 °F). The rows of fine precipitates present in the colonies of interphase precipitation are shown in Fig. 2.11(b), a transmission electron micrograph taken from the Fe-0.75V-0.15C steel held for 5 min at 725 °C (1335 °F).

An interesting aspect of the mechanism of interphase precipitation is the growth of the colonies by the extension of ledges in a direction parallel to the austenite-ferrite interface. As successively nucleated ledges complete their growth, a net extension of the colonies normal to the colony interface develops. Figure 2.12 shows examples of growth ledges in an Fe-12Cr-0.2C steel isothermally transformed for 36 min at 650 °C (1202 °F). Arrows point to the ledges. The planar interfaces left behind by the movement of the ledges are the sites for the carbide formation, while the ledges themselves are free of particles.

At this time, steels which transform solely by interphase precipitation are not used commercially, although alloying elements such as vanadium and niobium that form interphase carbides are being used increasingly in a new high-strength low-alloy (HSLA) or micro-alloyed steels. The very fine interphase alloy carbides and the high strength of microstructures with these particles are, however, quite attractive, and special applications may be developed in the future (Ref 2.4).

Proeutectoid Phases

In most steels — i.e., those not of eutectoid composition — austenite begins to transform well above the A_1 temperature. Figure 2.1 shows that ferrite forms below the A_3 temperature in steels that contain less than the eutectoid carbon content (hypoeutectoid steels), and cementite forms below the A_{cm} in steels containing more than the eutectoid carbon content (hypereutectoid steels). The ferrite and cementite that form prior to the eutectoid transformation are referred to as proeutectoid ferrite and cementite in order to indicate that they have formed by a mechanism other than the eutectoid transformation.

Proeutectoid ferrite and cementite are identical in crystal structure and composition to the ferrite and cementite of pearlite but are distrib-

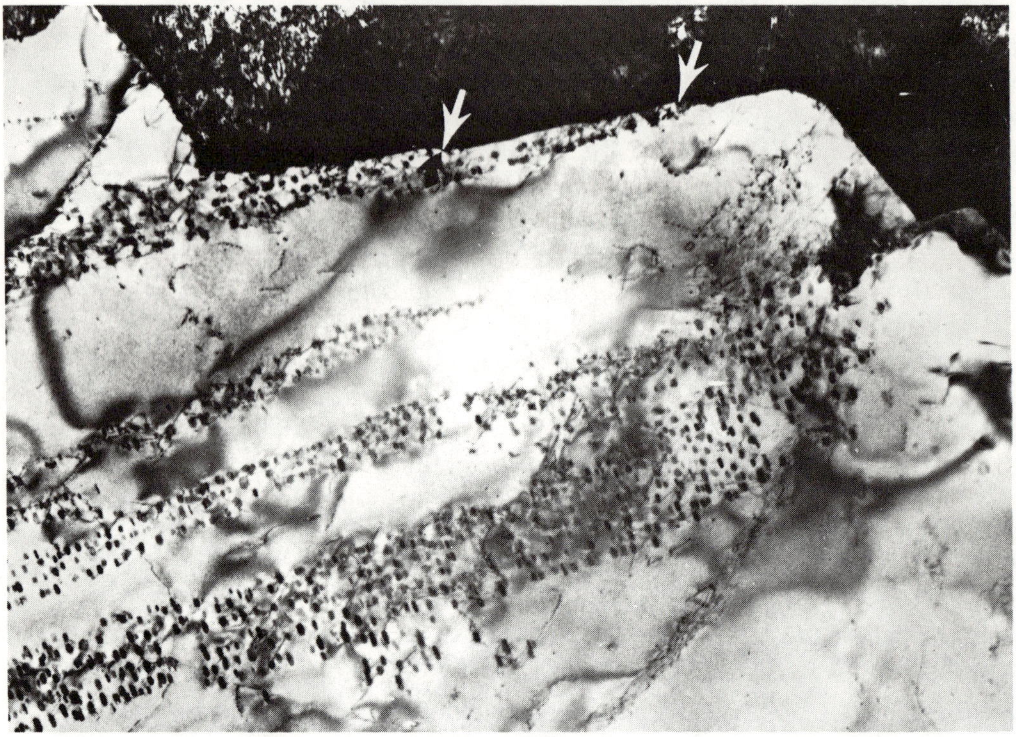

Fig. 2.12. Interphase precipitation and ledges in Fe-12Cr-0.2C steel isothermally transformed at 650 °C (1202 °F) for 36 min. Transmission electron micrograph. Magnification, about 70 000 ×. (Courtesy of K. Campbell and R. W. K. Honeycombe, University of Cambridge, U.K.)

uted quite differently in the microstructure than their lamellar arrangement in pearlite. Figure 2.13 shows a microstructure of proeutectoid ferrite and pearlite that formed in an Fe-0.4C alloy during slow cooling from the austenite phase field. The coarse network of white-etching proeutectoid ferrite is in marked contrast to the lamellar pearlite. Figure 2.1 shows that proeutectoid ferrite in a slowly cooled Fe-0.4C alloy begins to form just above 780 °C (1436 °F) and continues to grow until the A_1 temperature is reached. Tie lines through the ferrite-austenite field at successively lower temperatures and the applications of the lever rule to the Fe-0.4C alloy show that the amount of proeutectoid ferrite and the carbon content of the austenite increase continuously with decreasing temperature. The low solubility of carbon in ferrite requires that the carbon content build up in the

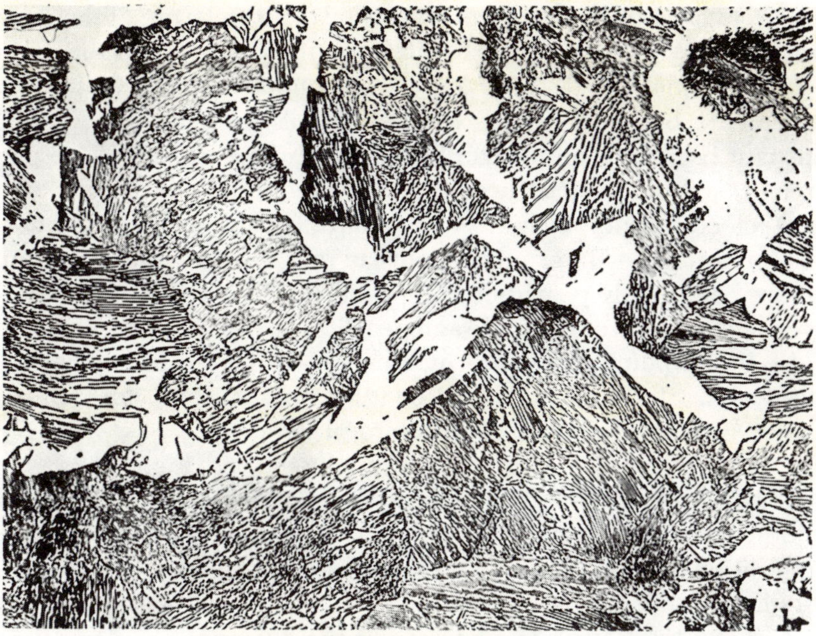

Fig. 2.13. Proeutectoid ferrite (white network) and pearlite in an Fe-0.4C alloy air cooled from the austenite field. Nital etch. Magnification, 500×. (Courtesy of A. R. Marder and A. Benscoter, Bethlehem Steel Corp., Bethlehem, Pa.)

austenite. At the A_1 temperature the carbon content of the austenite coexisting with the ferrite in the Fe-0.4C alloy, or any other hypoeutectoid steel for that matter, is 0.77%, which is just the right composition required for the eutectoid reaction, as written in Eq 2.1. Consequently any austenite coexisting with ferrite at the A_1 temperature transforms to pearlite on cooling, producing microstructures such as those shown in Fig. 2.13. The lever rule applied to the Fe-0.4C alloy in the ferrite-austenite phase field at 727 °C (1340 °F) shows that there should be about 50% by weight proeutectoid ferrite in the microstructure according to the following calculation:

$$\text{wt \% proeutectoid ferrite} = \frac{0.77 - 0.4}{0.77 - 0.02} \times 100 \qquad \text{(Eq 2.7)}$$

Alloys or steels with less carbon than 0.4% would contain more proeutectoid ferrite; those with more carbon would contain more pearlite. Depending on the carbon content of the steel, then, it is possible to have microstructures consisting of 100% ferrite (if the carbon content is less than or equal to 0.02%) or 100% pearlite (if the carbon content is equal to 0.77%) or any combination of proeutectoid ferrite and pearlite between these extremes. Steels designated for applications that require good formability — for example, automotive panel parts — have microstructures that are predominantly ferrite, while steels selected for applications where hardness and wear resistance are most important — for example, railroad rails — have microstructures that are completely pearlitic. The properties of steels heat treated to have microstructures of ferrite and pearlite are described in Chapter 5.

Up to this point only the formation of proeutectoid ferrite has been considered. Steels with carbon content greater than the eutectoid compositions form proeutectoid cementite if slowly cooled through or held in the cementite-austenite phase field (see Fig. 2.1). As the cementite (containing 6.67% carbon) forms, the carbon content of the austenite must decrease. With decreasing temperature, the austenite composition follows the A_{cm} until at the eutectoid temperature the austenite contains 0.77% carbon, again just the right composition for the eutectoid reaction. The balance of the austenite then transforms to pearlite.

Figure 2.14 shows a network of proeutectoid cementite that has formed by holding an Fe-1.22C alloy at 780 °C (1436 °F) for 30 min. Some colonies of pearlite are also present, the dark circular patches, and the balance of the microstructure is martensite formed during quenching from 780 °C (1436 °F). The carbon content of steels rarely exceeds 1.2%; therefore, little proeutectoid cementite ever forms. Application of the lever rule in the austenite-cementite field to the 1.2% carbon alloy at 727 °C (1340 °F) shows that only about 7% proeutectoid cementite could form. However, even though there can never be a large amount of proeutectoid cementite, the presence of a proeutectoid cementite network is considered to be very detrimental to the workability and toughness of high-carbon steels. Normalizing and spheroidizing heat treatments designed to modify or eliminate the cementite networks are discussed in Chapter 5.

Proeutectoid Phase Formation

Two major types of proeutectoid phase morphology develop in steels. One, shown in Fig. 2.13 and 2.14, consists of ferrite or cementite grains that nucleate and grow along the austenite grain boundaries. In a morphological classification system developed by Dubé and ex-

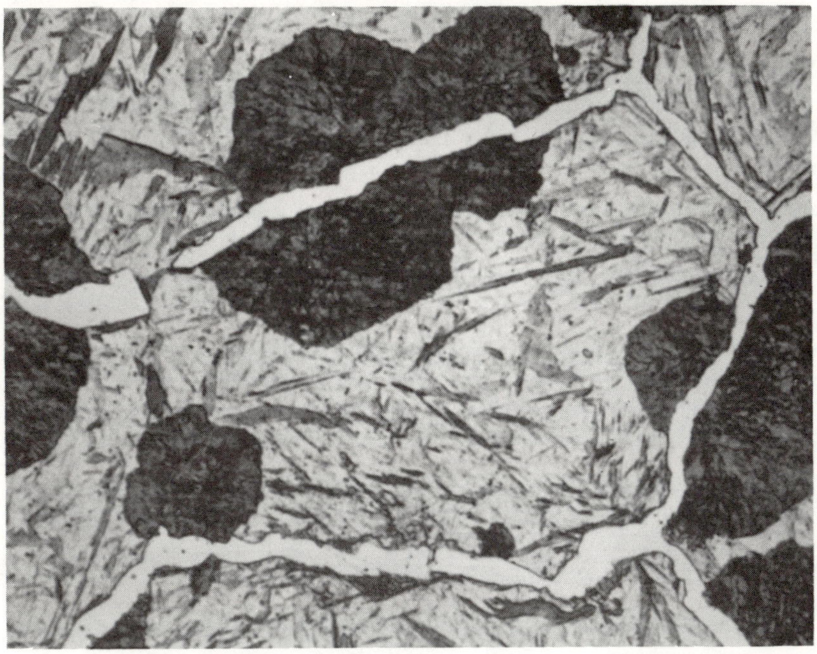

Fig. 2.14. Proeutectoid cementite (white network) formed at austenite grain boundaries in an Fe-1.22C alloy held at 780 °C (1436 °F) for 30 min. Dark patches are pearlite colonies and remainder of microstructure is martensite and retained austenite. Nital etch. Magnification, 600×. (Courtesy of T. Ando, Colorado School of Mines, Golden, Colo.)

tended by Aaronson (Ref 2.11), these crystals are referred to as grain boundary allotriomorphs. The second type of morphology consists of needlelike or plate-shaped grains that extend from the austenite grain boundaries. Examples of such grains of proeutectoid ferrite are shown in Fig. 2.15. In the Dubé classification system, the plate-shaped crystals are referred to as Widmanstätten side plates, the adjective "Widmanstätten" taken from the name of the French scientist, Alois de Widmanstätten. Widmanstätten is now widely used to describe any elongated or plate-shaped crystals that appear to form along specific crystallographic directions in a parent crystal.

Grain boundary allotriomorphs form under conditions approaching equilibrium, i.e., either by holding for extended periods of time in or by slow cooling through a two-phase field. Under these conditions,

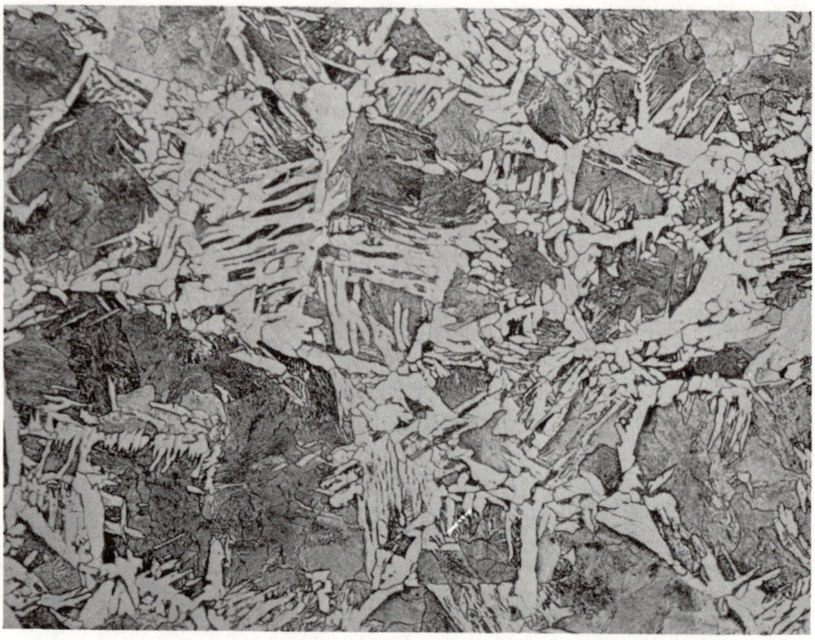

Fig. 2.15. Examples of proeutectoid ferrite with a Widmanstätten side plate morphology in an Fe-0.4C alloy. Nital etch. Magnification, 200×. (Courtesy of M. D. Geib, Colorado School of Mines, Golden, Colo.)

there is ample time for diffusion but the thermodynamic driving force is relatively low. Relatively large nuclei, therefore, form heterogeneously on the austenite grain boundaries, a process incorporating some of the existing austenite grain boundary and minimizing the interfacial energy increase. Nucleation of a proeutectoid phase within an austenite grain is prohibitive because all the new parent-product interface would have to be created. Widmanstätten side plates, on the other hand, are formed when the steel is cooled more rapidly and/or substantially undercooled below the A_3 or A_{cm}. Greater driving force is therefore available for the transformation of austenite to a proeutectoid phase, but diffusion is more limited. The latter disadvantage is partially offset by the fact that carbon atoms can diffuse in all directions around the tip of a growing side plate and that the diffusion distances in the austenite at the tips of plates are relatively short (Ref 2.12). The increased driving force apparently causes a shift to a mechanism that favors a closer crystallographic coupling between the austenite and the proeutectoid phase and some shear or cooperative movement of the iron atoms into the product crystal arrangement.

C. S. Smith (Ref 2.13) discussed the role of the crystallographic relationship between a proeutectoid crystal nucleated on an austenite grain boundary and the two adjacent austenite grains. He proposed that the crystal structure of one austenite grain might closely match the atom arrangement in the ferrite grain, i.e., a definite crystallographic orientation relationship might exist between the two crystals, and the resulting interface would have a high degree of coherency. The relatively good packing of atoms at the interface, however, would make transfer of the atoms across the interface difficult and result in a boundary with low diffusional mobility. The atom arrangement between the ferrite and the other austenite grain might not match nearly as well, thus an incoherent interface with a large degree of misfit would separate the two crystals. Atoms in such an interface would easily move from fcc packing to the bcc structure, thus producing a boundary with a high degree of diffusional mobility. At high transformation temperatures and low undercooling, the incoherent boundary could migrate and produce the typical grain boundary allotriomorphic morphology. With a high degree of undercooling, however, the migration of the incoherent boundary by diffusion would be restricted, and the high driving force would cause the ferrite with the coherent boundary in the other grain to propagate, resulting in a Widmanstätten side plate. The frequently observed growth of Widmanstätten side plates into only one grain is explained by the Smith hypothesis.

Figure 2.16 illustrates a microstructure that appears to have formed as a result of Smith's considerations. The microstructure was produced by water quenching an Fe-0.2C alloy with the intention of forming a completely martensitic structure. A number of proeutectoid ferrite crystals, however, nucleated at an austenite grain boundary prior to the martensite that makes up the bulk of the microstructure. The grains marked A and B have different orientations, as indicated by the grain boundary between them. The rapid cooling has prevented the motion of one interface of each grain, presumably because the interfaces were incoherent and the temperature was lowered too quickly for any significant amount of diffusional growth. Each ferrite grain, however, apparently had a coherent interface that could propagate into one of the austenitizing grains with sufficient undercooling. The fact that the Widmanstätten side plates are parallel to the fine martensite crystals and merge into the martensitic matrix supports the statement that some shear must be associated with Widmanstätten ferrite formation.

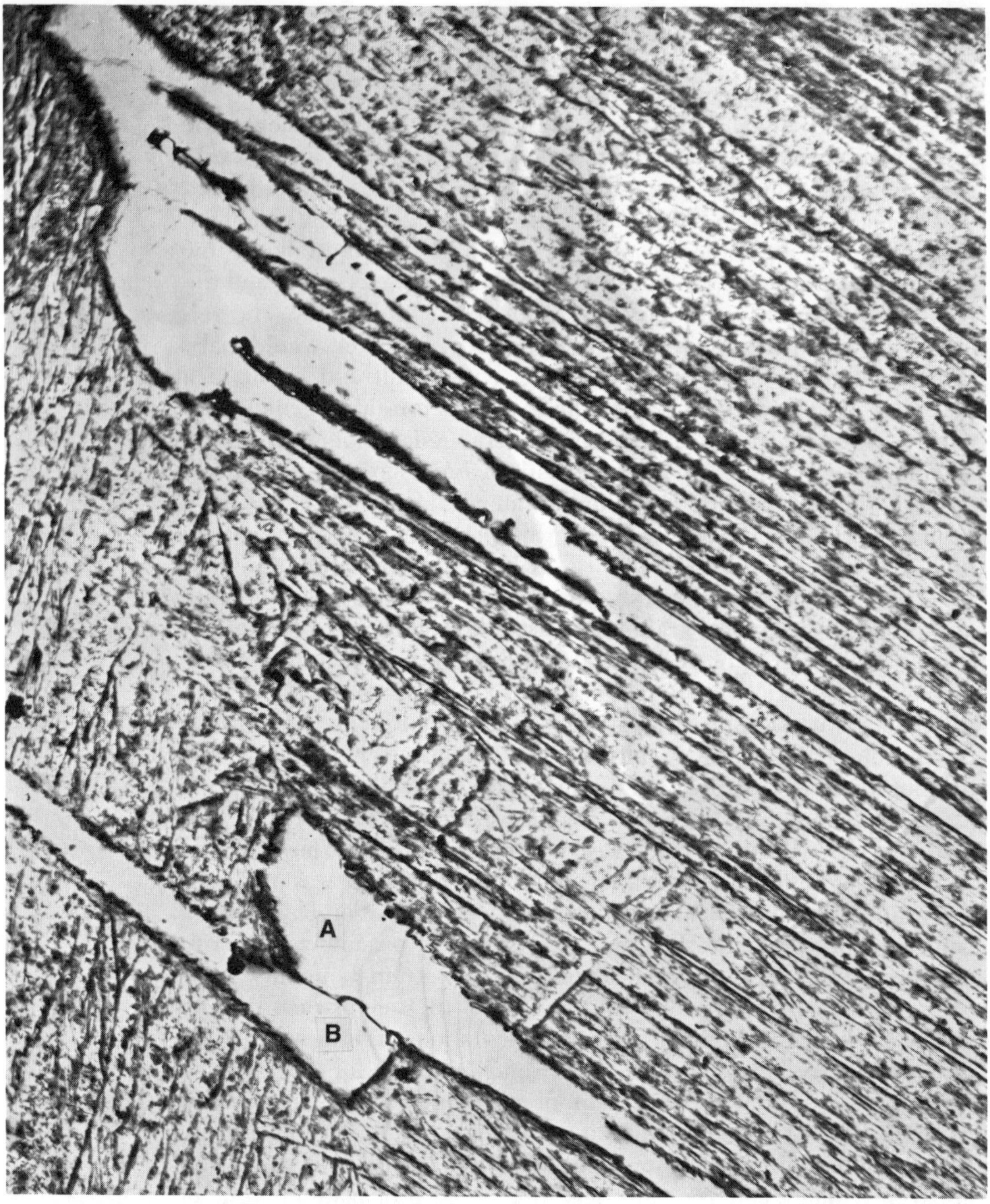

Fig. 2.16. Widmanstätten side plate formation in a quenched Fe-0.2C alloy. Grains A and B, separated by a grain boundary, have orientations that favor growth into different austenite grains. Electron micrograph from an extraction replica. Magnification, 7500 ×. (Courtesy of R.N. Caron, Olin Corp., New Haven, Conn.)

Ferrite formation requires the rejection of carbon, but the shear formation of martensite at temperatures below those at which ferrite forms traps the carbon in the martensitic structure.

Two other mechanisms of ferrite formation may operate under certain conditions or in certain alloy compositions. One involves the growth or thickening of both grain boundary allotriomorphs and Widmanstätten side plates by motion of ledges parallel to coherent interfaces (Ref 2.14). The ledges are considered to have incoherent interfaces with the austenite and therefore a high degree of mobility, as discussed relative to the Smith hypothesis. The other mechanism operates only in low-carbon steels containing less than 0.02% carbon. When these steels are cooled rapidly from the single-phase austenite field to the single-phase ferrite field, no change in composition is required. Carbon is, therefore, not rejected from the ferrite as is the case for higher carbon steels discussed earlier. The transformation is accomplished only by short range diffusion, on the order of one atomic distance, across the interface. This type of transformation has been classified as a massive or short range diffusion (SRD) transformation (Ref 2.15), the former term selected because massive, equiaxed grains are produced by this transformation mechanism.

References

2.1 R. F. Mehl and W. C. Hagel, The Austenite:Pearlite Reaction, in *Progress in Metal Physics*, B. Chalmers and R. King (Eds.), Vol 6, 1956, Pergamon Press, New York, p 74-134

2.2 D. Brown and N. Ridley, Rates of Nucleation and Growth and Interlamellar Spacing of Pearlite in a Low-Alloy Eutectoid Steel, *JISI*, Vol 204, 1966, p 811

2.3 M. Hillert, The Formation of Pearlite, in *Decomposition of Austenite by Diffusional Processes*, V. F. Zackay and H. I. Aaronson (Eds.), Interscience, New York, 1962, p 197-247

2.4 R. W. K. Honeycombe, Transformation from Austenite in Alloy Steels, *Met Trans A*, Vol 7A, 1976, p 915-936

2.5 B. E. Sundquist, The Edgewise Growth of Pearlite, *Acta Met*, Vol 16, 1968, p 1413-1427

2.6 C. Wells and R. F. Mehl, Rate of Diffusion of Carbon in Austenite in Plain Carbon, in Nickel and in Manganese Steels, *Trans AIME*, Vol 140, 1940, p 279

2.7 R. A. Swalin, *Thermodynamics of Solids*, John Wiley & Sons, New York, 1962

2.8 M. P. Puls and J. S. Kirkaldy, The Pearlite Reaction, *Met Trans*, Vol 3, 1972, p 2777-2796

2.9 W. A. Johnson and R. F. Mehl, Reaction Kinetics in Processes of Nucleation and Growth, *Trans AIME*, Vol 135, 1939, p 416-458

2.10 J. W. Cahn and W. C. Hagel, Theory of the Pearlite Reaction, in *Decomposition of Austenite by Diffusional Processes*, V. F. Zackay and H. I. Aaronson (Eds.), Interscience, New York, 1962, p 131-196
2.11 H. I. Aaronson, The Proeutectoid Ferrite and the Proeutectoid Cementite Reactions, in *Decomposition of Austenite by Diffusional Processes*, V. F. Zackay and H. I. Aaronson (Eds.), Interscience, New York, 1962, p 387-548
2.12 P. G. Shewmon, *Transformations in Metals*, McGraw-Hill, New York, 1969, p 220
2.13 C.S. Smith, *Trans ASM*, 1953, Vol 45, p 533
2.14 H. I. Aaronson, C. Laird and K. R. Kinsman, Mechanisms of Diffusional Growth of Precipitate Crystals, in *Phase Transformations*, American Society for Metals, Metals Park, Oh., 1970, p 313
2.15 T. B. Massalski, Massive Transformation Structures, in *Metals Handbook*, Vol. 8, 8th Ed., American Society for Metals, Metals Park, Oh., 1973, p 186-187
2.16 *Atlas of Isothermal Transformation and Cooling Transformation Diagrams*, American Society for Metals, Metals Park, Oh., 1977, p 28
2.17 R. F. Mehl and A. Dubé, *Phase Transformations in Solids*, John Wiley & Sons, New York, 1951, p 545

CHAPTER 3

Martensite and Bainite

This chapter describes the diffusionless, shear-type transformation of austenite to martensite. Athermal transformation kinetics, crystallographic features, and the development of fine structure are all special characteristics of the martensitic transformation. These features are described and related to the major morphologies and microstructural arrangements of martensite, lath and plate, that form in steel. Rapid cooling or quenching is required to form martensite, primarily to avoid the diffusion-dependent transformations described in Chapter 2, but the exact cooling conditions that will result in martensite in a given steel are strongly dependent on carbon content, alloying, and austenitic grain size, factors which determine hardenability, a subject discussed in Chapter 6. At the end of this chapter, bainite, a product of transformation that involves both shear and diffusion, is described.

General Considerations

Martensite, named after the pioneering German metallurgist, Adolf Martens, has long been used to designate the hard microstructure found in quenched carbon steel (Ref 3.1). More recently, however, emphasis has been placed on the nature of the transformation itself rather than the product. Martensitic transformation also occurs in many nonferrous systems (Ref 3.2), the Cu-Al and Au-Cd systems to name just two metal systems, and in oxides such as SiO_2 (Ref 3.3) and ZrO_2 (Ref 3.4). In fact, ceramists and geologists have independently identified the characteristics of martensitic transformation in nonmetal systems and used the term "displacive" to describe transformations which would be called martensitic by metallurgists (Ref 3.3). Martensite then becomes any phase produced by a martensitic or displacive transformation, even though that phase may have significantly different composition, crystal structure and properties than does martensite in steels.

In Fe-C alloys and steels, austenite is the parent phase that transforms to martensite on cooling. The martensitic transformation is diffusionless, and, therefore, the martensite has exactly the same composition as does its parent austenite, up to 2% carbon (see Fig. 1.1 and 2.1), depending on the alloy composition. Since diffusion is suppressed, usually by rapid cooling, the carbon atoms do not partition themselves between cementite and ferrite (see Chapter 2) but instead are trapped in the octahedral sites of a body-centered cubic structure, thus producing a new phase, martensite. The solubility of carbon in a bcc structure is greatly exceeded when martensite forms; hence martensite assumes a body-centered tetragonal (bct) unit cell (see Fig. 3.1) in which the c parameter of the unit cell is greater than the other two a parameters. With higher carbon concentration of the martensite, more interstitial sites are filled, and the tetragonality increases, as shown in Fig. 3.2.

Martensite is a unique phase that forms in steels. It has its own crystal structure and composition and is separated by well defined interfaces from other phases, but it is a metastable phase present only because diffusion has been suppressed. If the martensite is heated to a temperature where the carbon atoms have mobility, the carbon atoms diffuse from the octahedral sites to form carbides. As a result, the tetragonality is relieved, and martensite is replaced by a mixture of ferrite and cementite as required by the Fe-C phase diagram. The decomposition of martensite to other structures on heating is referred to as tempering and is the subject of Chapter 8.

Martensite forms by a shear mechanism. Many atoms move cooperatively and almost simultaneously to effect the transformation, a mechanism very much in contrast to atom-by-atom movement across interfaces during diffusion-dependent transformations. Figure 3.3 shows schematically a number of features of the shear or displacive transformation of austenite to martensite. The arrows point in the directions of shear on opposite sides of the plane on which the transformation was initiated. The martensite crystal formed is displaced partly above and partly below the surface of the austenite by the shear. Thus (as shown in Fig. 3.3) the originally horizontal surface of the parent phase is rotated or tilted into a new orientation by the shear

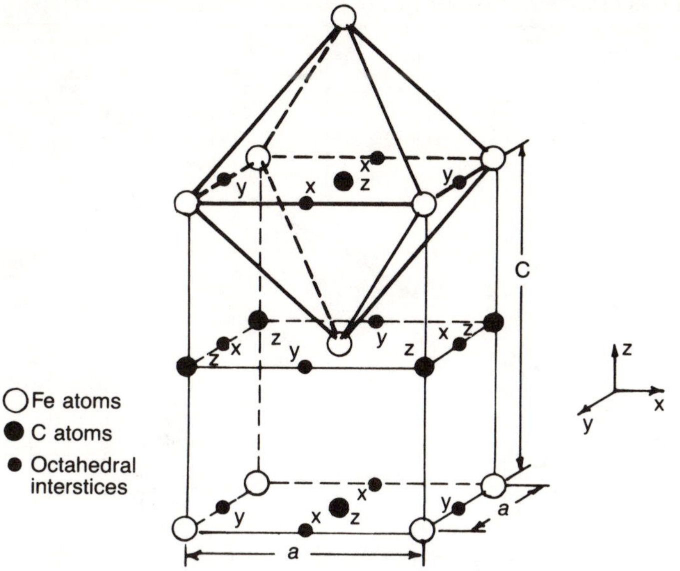

Fig. 3.1. Body-centered tetragonal crystal structure of martensite in Fe-C alloys. Carbon atoms are trapped in one set (z) of interstitial octahedral sites. The x and y sites are unoccupied. (Ref 3.5)

transformation. Surface tilting is an important characteristic of shear-type or martensitic transformation. The atom-by-atom transfer across interfaces by which diffusion-controlled transformations proceed does not produce tilting but tends to produce surfaces of the product phase parallel to the surface of the parent phase.

Figure 3.3 also shows that considerable flow or plastic deformation of the parent austenite must accompany the formation of a martensite crystal. Eventually the constraints of the parent phase limit the width of a martensite lath, and further transformation can proceed only by the nucleation of new plates. If the parent austenite could not accommodate the shape change produced by the martensitic shears, separation or cracking at the martensite/parent phase interface would occur. Fortunately austenite in steels has sufficient ductility to accommodate martensite formation. However in many ceramic systems, the parent phase cannot accommodate the shape change, and displacive transformations must be avoided.

Martensite crystals ideally have planar interfaces with the parent austenite (see Fig. 3.3). The preferred crystal planes of the austenite on which the martensite crystals form are designated as habit planes. The

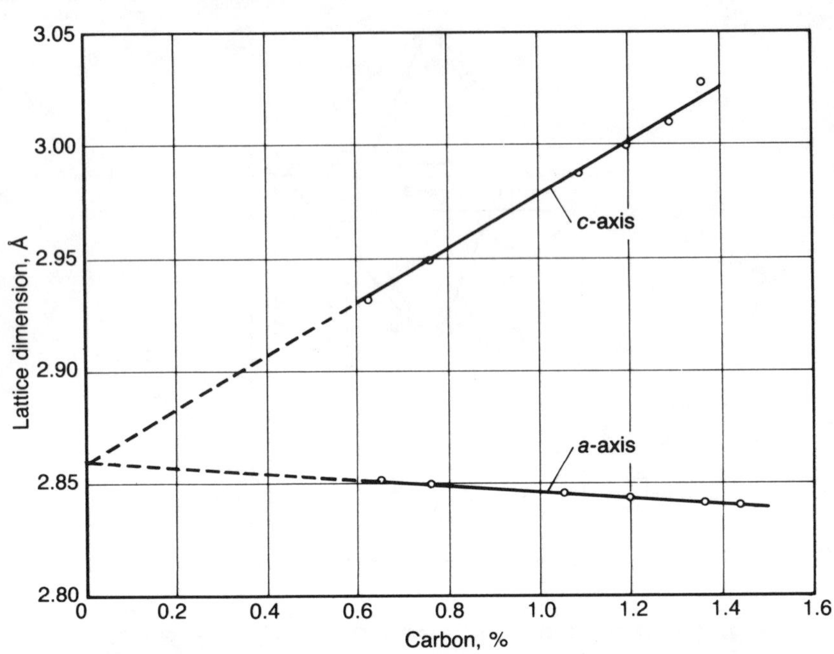

Fig. 3.2 Change in *c* and *a* parameters of martensite with carbon content in iron-carbon alloys. (Ref 3.47)

habit planes vary according to alloy composition, and some examples are presented in the section on morphology in this chapter. The midrib shown in Fig. 3.3 is generally considered to be the starting plane for the formation of a plate of martensite and may in fact have a different fine structure than other parts of the plate.

An example of surface relief and its relationship to martensitic microstructure is shown in Fig. 3.4. This series of light micrographs was obtained after a prepolished Fe-0.2C alloy specimen was austenitized and quenched in a hot stage microscope with an argon gas atmosphere. Figure 3.4(a) shows the surface relief associated with the formation of hundreds of martensite crystals. The surface tilting is emphasized in some areas by the dark shadows present on surfaces tilted away from the light source. In Fig. 3.4(b) the surface relief has been almost polished away, and in Fig. 3.4(c) the surface shown in (b) has been etched. Finally, Fig. 3.4(d) shows the microstructure after the surface has been polished to remove all relief and etched once again. Comparison of Fig. 3.4(c) and (d) with Fig. 3.4(a) shows the direct correspondence of the surface relief with the martensitic units in the polished and etched microstructure. In polished and etched sections,

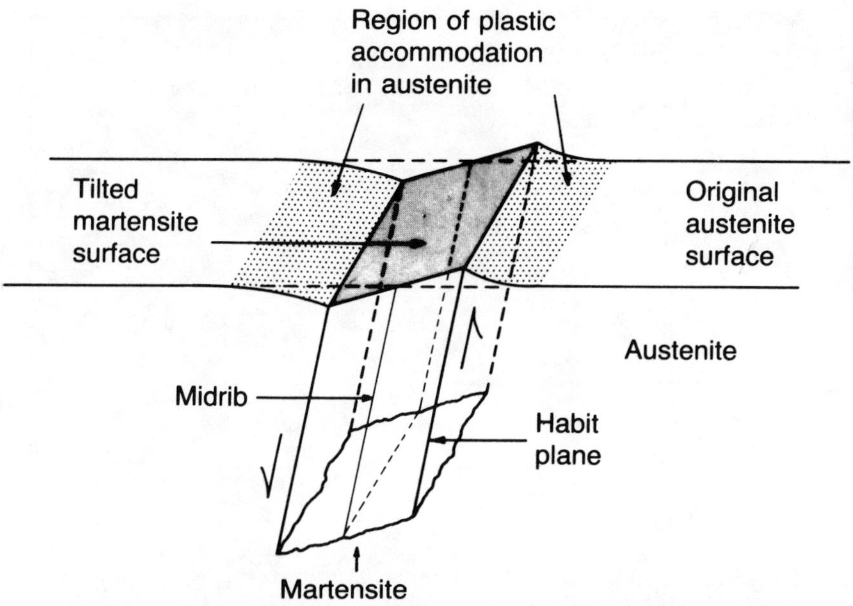

Fig. 3.3 Schematic of shear and surface tilt associated with formation of a martensite plate. (Ref 3.6; adapted courtesy of M.D. Geib, Colorado School of Mines, Golden, Colo.)

the individual crystals of martensite appear to be long and thin and are very often characterized as acicular or needlelike. In three dimensions, however, the crystals have a lath or plate shape with flat interfaces, as shown schematically in Fig. 3.3. The needlelike shapes visible on polished and etched surfaces, therefore, are cross sections through laths or plates.

Martensitic Transformation Kinetics

The conversion of an austenitic microstructure to a martensitie microstructure in many commercial steels takes place continuously with decreasing temperature during uninterrupted cooling. This mode of transformation kinetics is referred to as athermal (without thermal activation) in order to differentiate it from the isothermal kinetics that characterize thermally activated diffusion-controlled transformations. Pearlite formation, for example, occurs continuously as a function of time if austenite is held at a constant temperature below A_1. Martens-

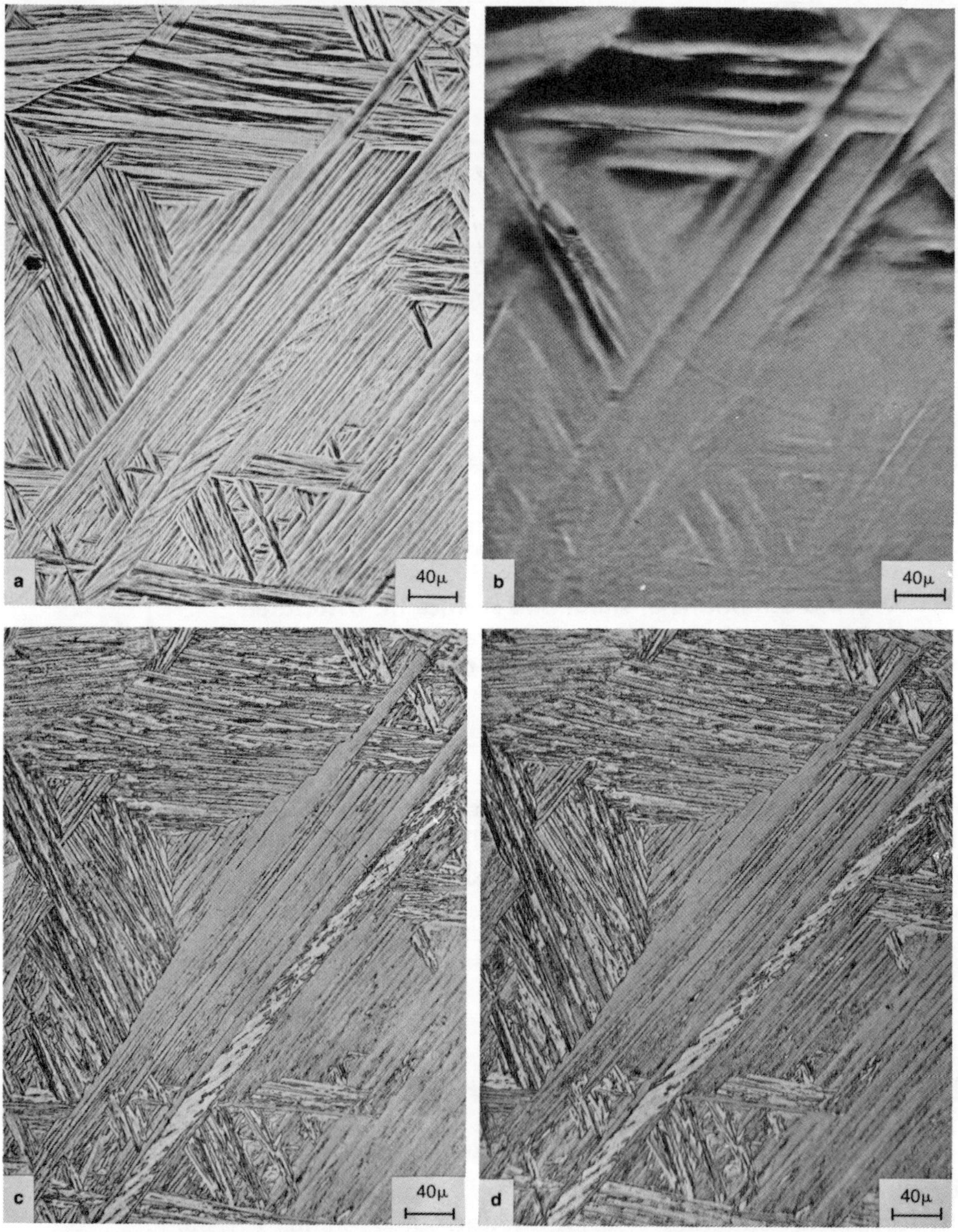

Fig. 3.4. Surface tilting and its relationship to martensitic structure in an Fe-0.2C alloy. (a) Surface tilting after quenching; (b) partially polished surface; (c) area in (b) after etching; (d) same area after polishing to remove all relief and re-etching. Nital etch. (Ref 3.7)

ite formation, however, is accomplished virtually as soon as a given temperature is reached; should cooling be stopped at that temperature, no further transformation to martensite will occur. Additional transformation, usually by means of the nucleation and rapid growth of new plates of martensite, is accomplished only by cooling to lower temperatures.

Figure 3.5 shows the progress of athermal transformation in an Fe-1.86C alloy. The austenite in this high-carbon alloy is quite stable, and martensitic transformation was initiated just above room temperature. Figure 3.5(a) shows a few very large plates of martensite that formed on cooling to room temperature. The balance of the microstructure is austenite. Figures 3.5(b) and (c) show how some of the austenite retained at room temperature is transformed to new plates on successive subzero cooling to −60 °C (−76 °F) and −100 °C (−148 °F), respectively. The new plates have nucleated within the framework of the initially formed plates (see Fig. 3.5a), and the parent austenite has been subdivided into smaller and smaller units with increasing amounts of martensitic formation. Clearly, the martensitic transformation effectively ceases on reaching a given temperature, and only additional undercooling drives the transformation further.

In contrast to the Fe-1.86C alloy, most hardenable steels transform to martensite at temperatures well above room temperature. Figure 3.6 shows the transformation of austenite to martensite in an Fe-1.94Mo alloy, an alloy in which austenite transforms to martensite in the same manner as in low- and medium-carbon steels. Hot stage cinephotomicrography was required to follow the high-temperature formation of the martensite in the continuously cooled Fe-1.94Mo alloy (Ref 3.8). Figure 3.6 shows a sequence of frames taken from a film of the transformation sequence. Frame 1 shows several austenite grains that are largely untransformed and the succeeding frames show the step-by-step formation of the martensitic microstructure. The martensite plates in Fig. 3.6 are visible only because of the surface tilting associated with transformation; it was obviously impossible to polish and etch (as in the case of the Fe-1.86C alloy) between frames during the cooling of the Fe-1.94Mo alloy. Figure 3.6 also shows that an important characteristic of the athermal transformation of the Fe-Mo alloy, and low- and medium-carbon steels that behave similarly, is the development of parallel groups of plates or laths by nucleation and growth of new plates parallel and adjacent to existing plates.

The temperature at which martensite starts to form in a given alloy is designated as the martensite start temperature (M_s). The M_s reflects the amount of thermodynamic driving force required to initiate the shear transformation of austenite to martensite. Figure 3.7 shows that

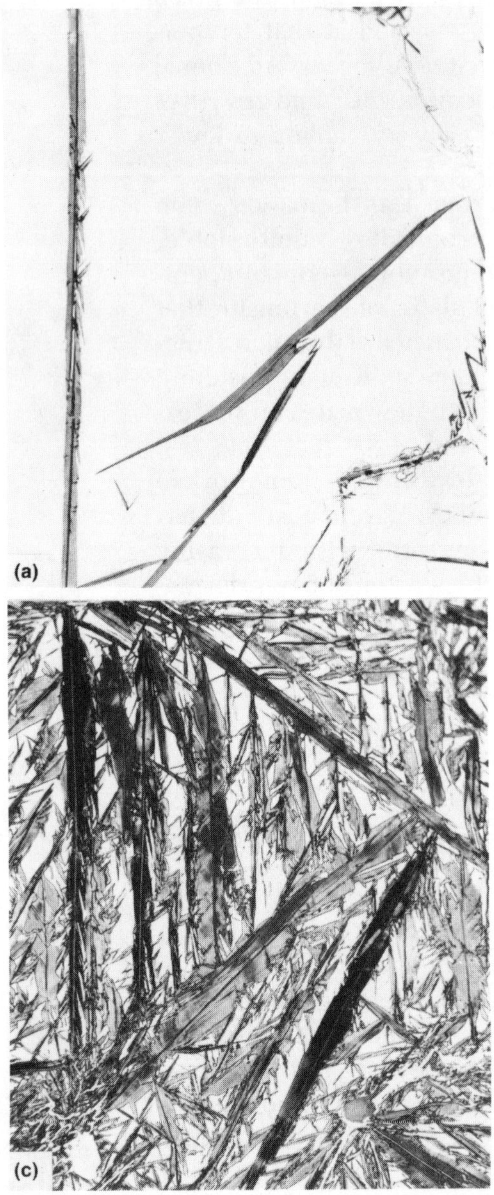

Fig. 3.5. Progress of athermal martensitic transformation in an Fe-1.8C alloy after cooling to: (a) 24 °C (75 °F); (b) −60 °C (−76 °F); and (c) −100 °C (−148 °F). Nital etch. Magnification, 500 ×; shown here at 74%. (Ref 3.8)

the M_s decreases significantly with increasing carbon content in Fe-C alloys and carbon steels. Carbon in solid solution increases the strength or shear resistance of the austenite and, therefore, greater undercooling or driving force is required to initiate the shear for martensite formation in higher carbon alloys. The martensite finish

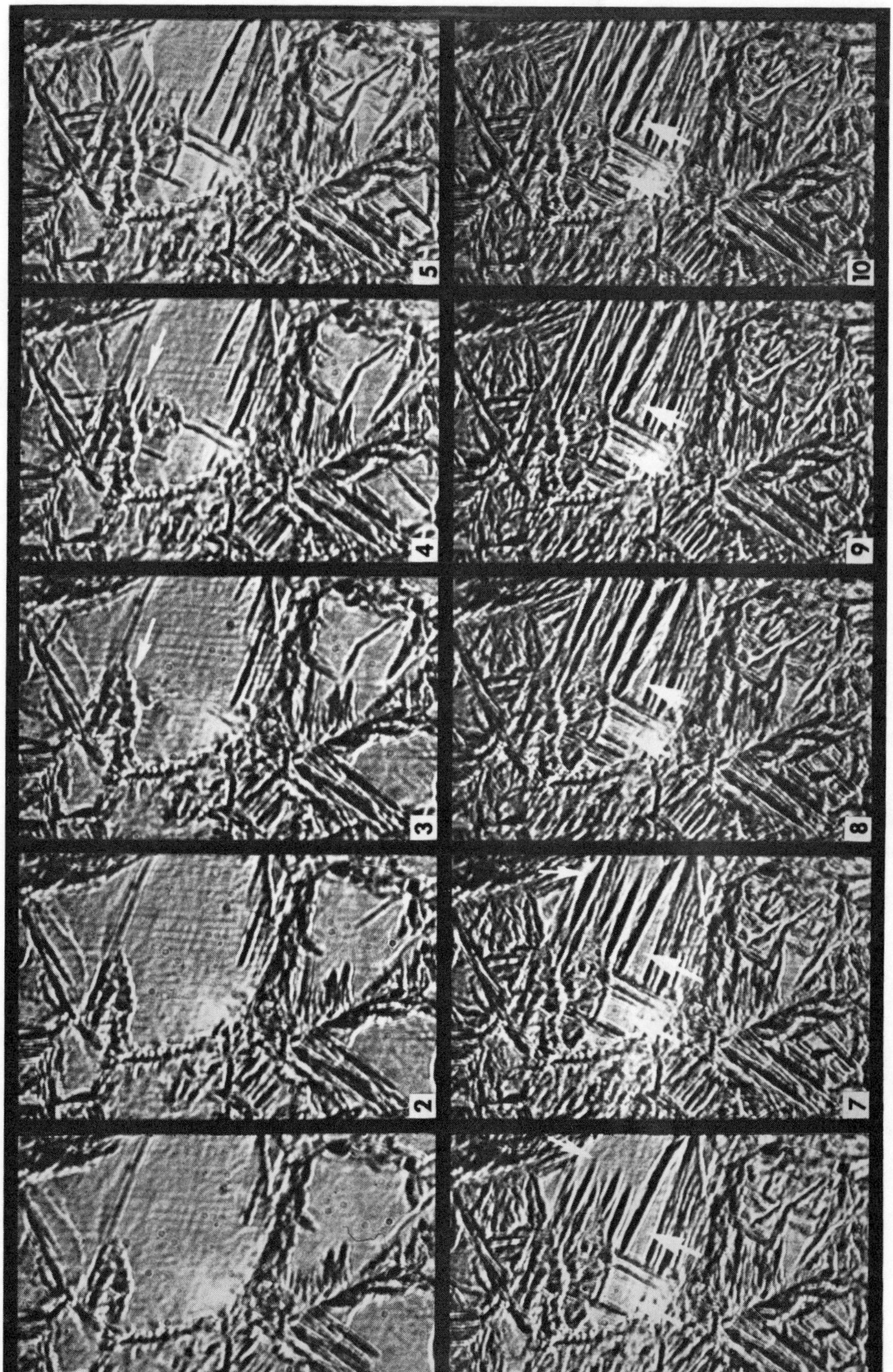

Fig. 3.6. Progress of athermal martensitic transformation in an Fe-1.94Mo alloy. Successive exposures taken of surface relief on a hot stage microscope. Magnification, 105 ×; shown here at 85%. (Ref 3.8)

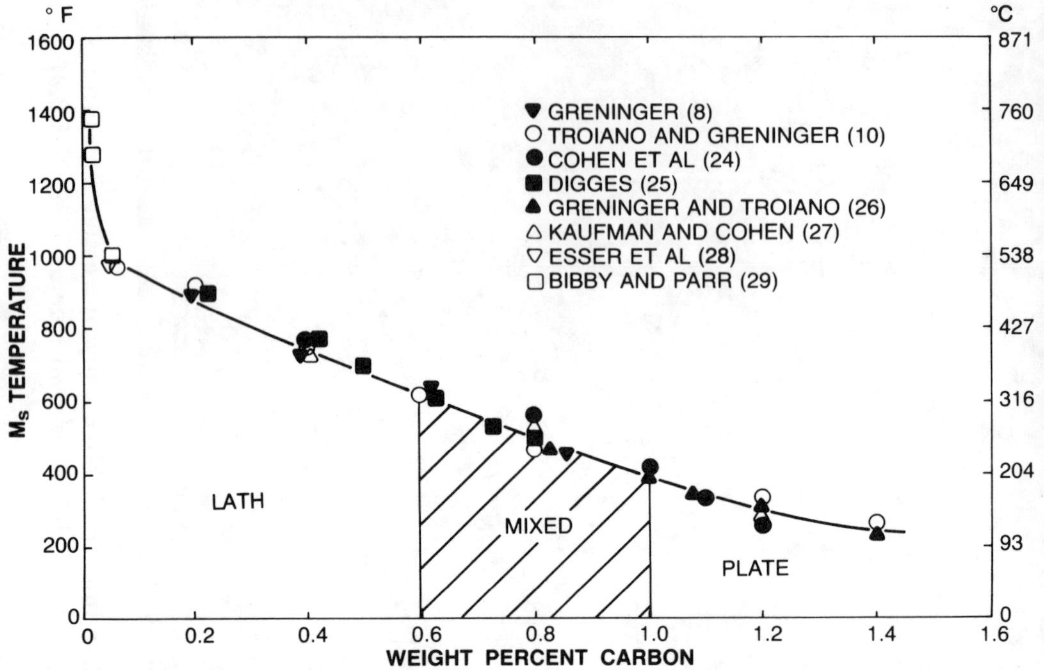

Fig. 3.7. M_s temperatures as a function of carbon content in steels. Composition ranges of lath and plate martensite in Fe-C alloys are also shown. (Ref 3.11; investigations indicated are identified by their numbers in this reference)

temperature (M_f), or the temperature at which the martensite transformation is complete in a given alloy, is also a function of carbon content. The detection of the last small amounts of untransformed austenite is experimentally difficult (Ref 3.9); therefore, the M_f curves based on results of early investigations are only approximate. The M_f drops below room temperature in alloys containing more than about 0.3% carbon. Therefore, significant amounts of untransformed austenite, especially in high-carbon steels, may be present with martensite at room temperature. Figure 3.8 shows that this is actually the case. Retained austenite content, measured by x-ray diffraction techniques (Ref 3.10, 3.11) at room temperature, is as high as 30 to 40% in Fe-C alloys containing 1.2 to 1.4% carbon. Even in alloys containing only 0.3 to 0.4% carbon, some small amount of austenite is retained. Alloying elements that stabilize austenite increase the amount of retained austenite at any given carbon level and temperature.

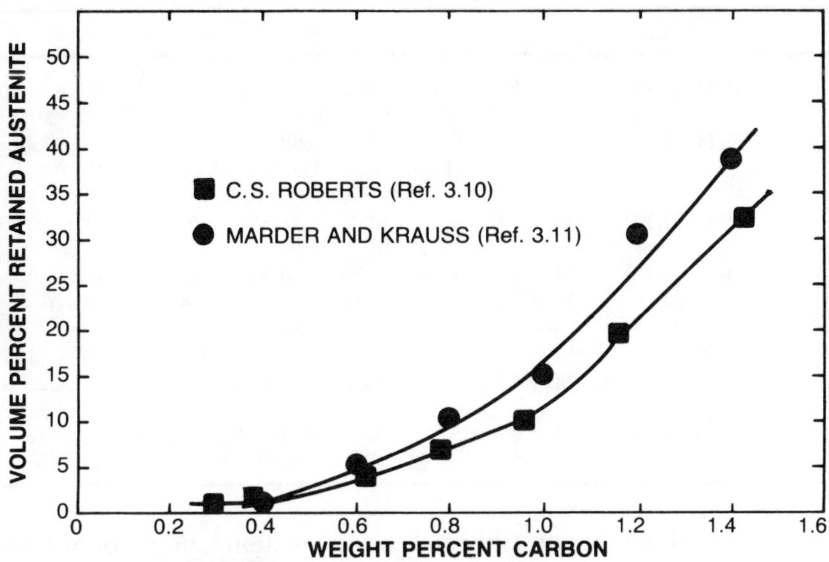

Fig. 3.8. Retained austenite as a function of carbon content in Fe-C alloys. (Ref 3.11)

Alloying elements also influence the M_s temperatures of steels, and a number of equations have been developed to relate M_s to steel composition. Table 3.1 lists various equations that have been developed over the years. All alloying elements, except cobalt, lower M_s temperatures. The equations developed by Andrews (Ref 3.18) are based on measurements of M_s temperatures and compositions of a large number of steels of British, German, French and American manufacture with maximum carbon content of 0.6%, manganese up to 4.9%, chromium up to 5%, nickel up to 5%, and molybdenum up to 5.4%. Andrews showed that 92 and 95% of measured M_s temperatures for the steels were within ± 25 °C of the M_s temperatures calculated from their compositions according to the linear and product equations, respectively. A test of Andrews' equations with M_s measurements and steel compositions published in the 1970s shows that Andrews' equations continue to give good agreement between measured and calculated M_s values with the ± 25 °C limits (Ref 3.19).

Once the M_s of a steel is known, the extent of the athermal transformation of austenite to martensite is dependent only on the amount of undercooling below the M_s temperature. Two equations have been

Table 3.1. List of Formulas for M_s Calculation from Alloy Composition

Investigators	Date (Ref)	Equation
Payson and Savage	1944 (3.12)	M_s (°F) = 930 − 570C − 60Mn − 50Cr − 30Ni − 20Si − 20Mo − 20W
Carapella	1944 (3.13)	M_s (°F) = 925 × (1 − 0.620C)(1 − 0.092Mn)(1 − 0.033Si)(1 − 0.045Ni)(1 − 0.070Cr)(1 − 0.029Mo)(1 − 0.018W)(1 + 0.120Co)
Rowland and Lyle	1946 (3.14)	M_s (°F) = 930 − 600C − 60Mn − 50Cr − 30Ni − 20Si − 20Mo − 20W
Grange and Stewart	1946 (3.15)	M_s (°F) = 1000 − 650C − 70Mn − 70Cr − 35Ni − 50Mo
Nehrenberg	1946 (3.16)	M_s (°F) = 930 − 540C − 60Mn − 40Cr − 30Ni − 20Si − 20Mo
Steven and Haynes	1956 (3.17)	M_s (°C) = 561 − 474C − 33Mn − 17Cr − 17Ni − 21Mo
Andrews (linear)	1965 (3.18)	M_s (°C) = 539 − 423C − 30.4Mn − 12.1Cr − 17.7Ni − 7.5Mo
Andrews (product)	1965 (3.18)	M_s (°C) = 512 − 453C − 16.9Ni + 15Cr − 9.5Mo + 217(C)^2 − 71.5(C)(Mn) − 67.6(C)(Cr)

developed to describe the athermal transformation kinetics of martensite formation:

$$f = 1 - 6.96 \times 10^{-15}(455 - \Delta T)^{5.32} \quad \text{(Eq 3.1)}$$

$$f = 1 - \exp - (1.10 \times 10^{-2}\Delta T) \quad \text{(Eq 3.2)}$$

where f is the volume function of martensite and ΔT is the undercooling below M_s in degrees Centigrade. Equation 3.1 was developed by Harris and Cohen (Ref 3.20) for steels containing 1.1% carbon, and Eq 3.2 was developed by Koistinen and Marburger (Ref 3.21) from Fe-C alloys containing between 0.37 to 1.1% carbon. The data used to develop Eq 3.1 and Eq 3.2 together with measurements of Steven and Haynes (Ref 3.17) from hardenable steels containing 0.32 to 0.44% carbon are shown in Fig. 3.9. All sets of data agree closely for small amounts of undercooling, but they diverge significantly as undercooling increases. For example, at 100 °C (212 °F) below M_s, the Koistinen and Marburger data show only about 65% transformation while the Steven and Haynes data show 90% transformation. The discrepancy is probably due to experimental difficulties in determining the amount of austenite remaining at any given temperature. Koistinen and Marburger used x-ray analysis while the other investigators used light microscopy to determine the amount of retained austenite. The detection of small amounts of retained austenite by the latter technique is difficult in high-carbon steels and virtually impossible in medium-carbon steels. Therefore, the Koistinen and Marburger equation, based on the most accurate technique for determining small amounts of retained austen-

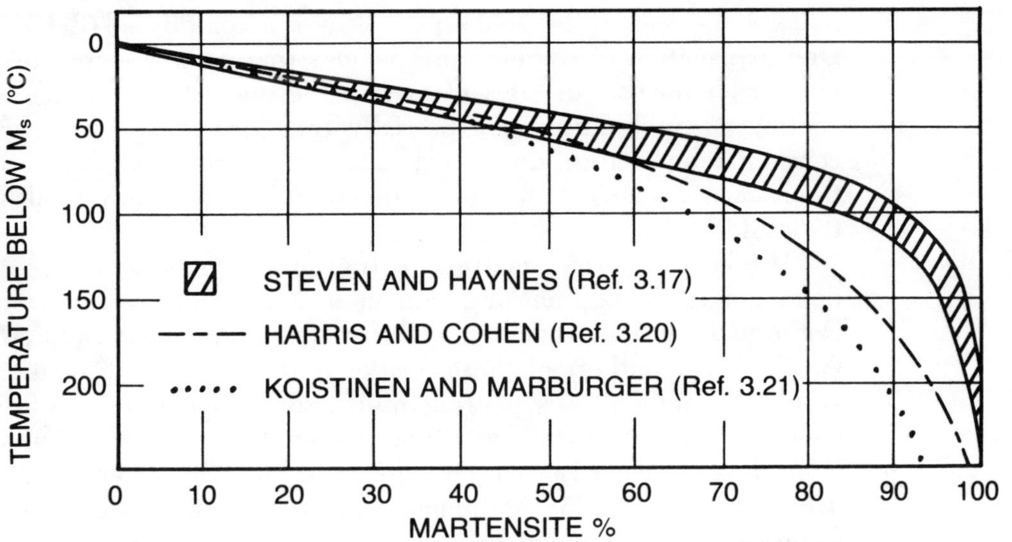

Fig. 3.9. Extent of martensite formation as a function of undercooling below M_s according to three different investigations as shown. (Ref 3.19)

ite, is considered to give the best representation of martensite transformation over the entire range of undercooling.

During the course of the athermal martensite formation discussed up to this point, two types of anomalies may develop: bursting and stabilization. The burst phenomenon occurs in Fe-Ni and Fe-Ni-C alloys with subzero M_s temperatures. Large numbers of martensite plates, sometimes enough to transform 70% of the austenite, form in a "burst" at a temperature designated the M_B (Ref 3.22). This transformation behavior is related to the ability of plates of martensite to nucleate other plates of martensite, a process called autocatalysis. The stimulus to nucleation is the stress, generated at the tips of plates, that helps to initiate the shear transformation process on other favorably oriented variants of the habit plane (Ref 3.23). The habit plane variants that are activated are generally not parallel to that of the initiating plate, and frequently zig-zag arrays of martensite plates are observed in alloys susceptible to autocatalytic nucleation or bursting.

Stabilization, a phenomenon that reduces the ability of austenite to transform into martensite, occurs during slow cooling or interruption of cooling before complete transformation. For example, an oil-quenched steel may contain more retained austenite than the same steel water quenched, and if transformation of a steel is interrupted by holding at some temperature between M_s and M_f, no martensite transformation may occur when cooling is resumed until substantial undercooling below the hold temperature is accomplished (Ref 3.24). One explanation of stabilization assumes that carbon segregates to potential embryos, or sites of martensitic nucleation, during slow cooling or on holding of a partially transformed specimen at a constant temperature. Once segregated, the carbon atoms increase the shear resistance of the austenite, thereby effectively stabilizing the austenite.

Although athermal martensite transformation kinetics are the dominant mode of transformation in heat treatable carbon steels, isothermal transformation has been observed in Fe-Ni-Mn and Fe-Ni-Cr alloys. The isothermal transformation is time-dependent and occurs at subzero temperatures; plotting this transformation frequently forms C-curves. Figure 3.10 shows a time-temperature-transformation diagram developed for isothermal martensite formation in an Fe-23Ni-3.7Mn alloy (Ref 3.25). Mathematical modeling of isothermal transformation kinetics has made possible a separation of the effects of preexisting nucleation sites or embryos and those produced by autocatalysis (Ref 3.22, 3.26). Also, the studies of isothermal transformation have shown a relationship between embryo size and kinetic mode of transformation. Alloy systems with large embryos or lattice sites predisposed to transformation require little or no thermal activation and therefore transform to martensite athermally. Systems with smaller embryos require thermal activation to produce martensite nuclei of size sufficient to initiate transformation, a process which leads to the time-dependent isothermal kinetics. Also, it has been shown that the activation energy for isothermal martensitic nucleation in Fe-Ni-Mn alloys is inversely proportional to the chemical driving force for the transformation, i.e., the greater the driving force the lower the activation energy (Ref 3.26, 3.27).

Crystallography of Martensitic Transformation

The diffusionless, shear mechanism of martensitic transformation requires good crystallographic coupling between the parent and product phases. Two important crystallographic parameters or characteristics emphasize this interrelationship between austenite and martensite in ferrous alloys. One is the orientation relationship between the crystal structure of the parent and the product martensite. The orientation relationship specifies planes and directions of the parent phase

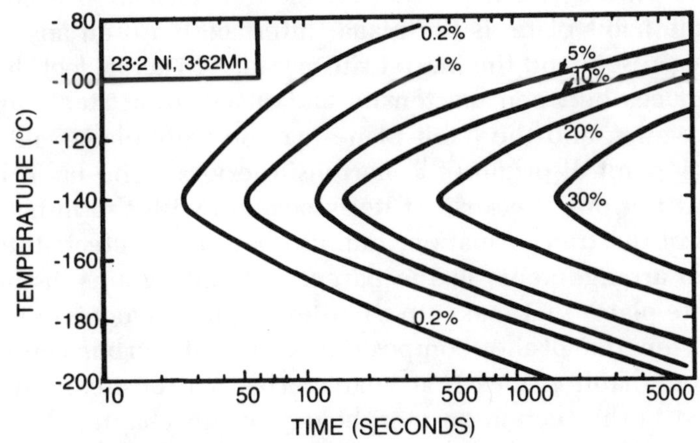

Fig. 3.10. Isothermal transformation curves for martensite formation in an Fe-23Ni-3.6Mn alloy. Curves are identified by the percentage of martensite formed. (Ref 3.25)

and the planes and directions in the product martensite to which they are parallel. Two well-known orientation relationships have been determined in ferrous alloy systems by means of x-ray diffraction techniques (Ref 3.28). The Kurdjumov-Sachs orientation relationship:

$$\{111\}_A \parallel \{101\}_M$$

$$\langle 110\rangle_A \parallel \langle 111\rangle_M$$

is valid for high-carbon steels with $\{225\}_A$ habit planes. The other orientation relationship, which was determined by Greninger and Troiano and is also attributed to Nishiyama, is:

$$\{111\}_A \parallel \{011\}_M$$

$$\langle 112\rangle_A \parallel \langle 011\rangle_M$$

This relationship is observed in alloys where the martensite plates have $\{259\}_A$ habit planes.

The other crystallographic parameter that emphasizes the interrelationship of the parent and product phases is the habit plane, already mentioned in the discussion of Fig. 3.3. In steels, the habit plane is the plane in the parent austenite on which the martensite forms and grows. When the martensitic transformation is complete, ideally the habit plane is the planar interface between any retained austenite present and the martensite crystals. In actual fact, however, the interfaces between martensite and austenite in steels might be quite irregular, and the habit plane may be truly planar only at the midrib or point of origin of a martensite crystal. The habit plane is important not only because of its association with the initiation and progress of the transformation, but also because it affects the microstructural arrangements in the parent austenite grains of the many martensite plates that make up a hardened microstructure. The habit plane is a function of alloy composition, especially carbon content, and the various habit planes that characterize martensite in steels are presented in the section on morphology in this chapter.

The orientation relationship and habit plane in a given steel are parameters that relate the crystallography of austenite to martensite after transformation. Crystallography is also important in describing the martensitic transformation itself. A crystallographic theory of martensitic transformation was developed in the 1950s by Wechsler, Lieberman and Read (Ref 3.29) and by Bowles and MacKenzie (Ref 3.30). The significance of the crystallographic theory is the understanding it provides for the origin of the internal fine structure found within any martensite crystal. The fine structure may consist of dislocations, twins or a mixture of the two, depending on alloy composition. The presence of fine structure is a unique result of martensitic transformation but occurs to some extent in any transformation where shear and diffusion are required to form the new phase. Widmanstätten ferrite and bainite formation are examples of the latter type of transformation.

The crystallographic theory of martensite formation is based on two important microstructural (macroscopic relative to atomic dimensions) observations: that the habit plane is unrotated and undistorted, and that the shape change that produces the surface tilting in Fig. 3.3 is homogeneous and a result of plane strain. The plane strain may be visualized as shear or displacement on planes parallel to the habit plane. Greninger and Troiano (Ref 3.31) first noted that the shape change could not be produced merely by the lattice deformation (i.e., in steels, the change in lattice from fcc austenite to bct martensite). Another deformation was necessary to satisfy the requirement of plane strain and the undistorted habit plane. This additional deformation, lattice invariant deformation, involves deformation of the bct martensite crystal by twinning or slip but not a change of the lattice or crystal structure itself.

The major elements of the crystallographic theory for martensite formation in steels are shown schematically in Fig. 3.11 and 3.12.

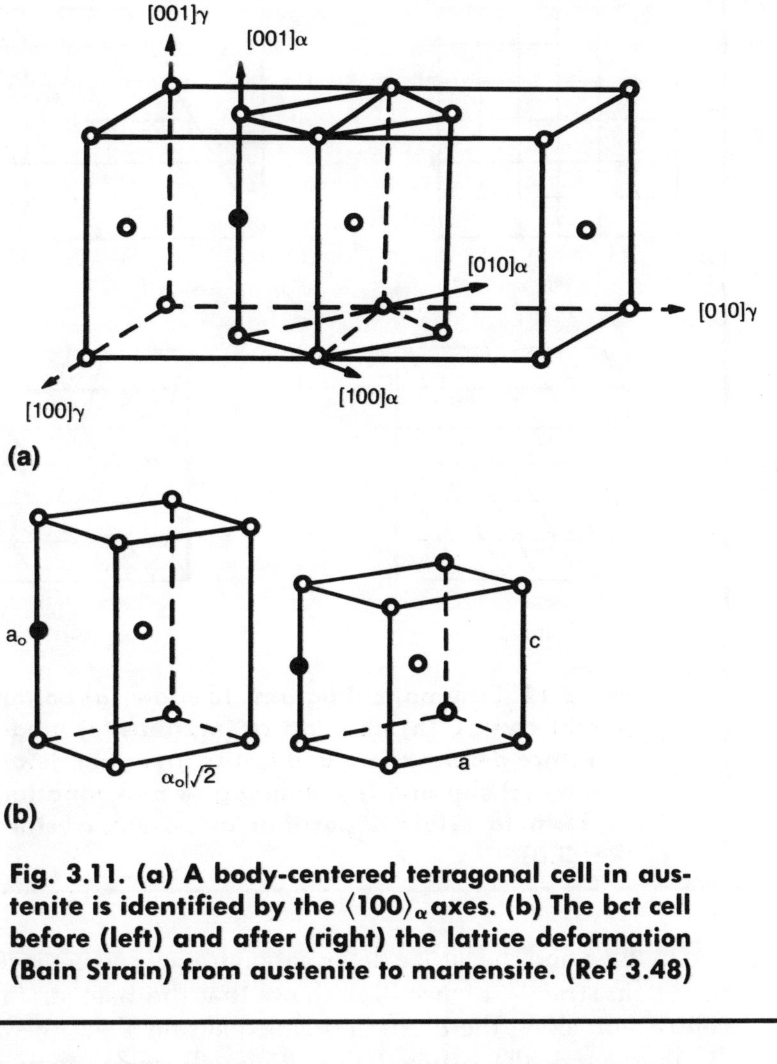

Fig. 3.11. (a) A body-centered tetragonal cell in austenite is identified by the $\langle 100 \rangle_\alpha$ axes. (b) The bct cell before (left) and after (right) the lattice deformation (Bain Strain) from austenite to martensite. (Ref 3.48)

Figure 3.11(a) shows two adjacent fcc unit cells of austenite in which a bct unit cell has been identified. This identification of a set of atoms in the parent phase that will transform to a set of atoms in the product phase is referred to as the lattice correspondence. The atoms identified in Fig. 3.11(a) have been isolated in the unit cell schematic on the left of Fig. 3.11(b). At this stage, the dimensions of the bct cell are still those derived from the austenitic lattice parameter. The unit cell on the right of Fig. 3.11(b) is that of martensite with lattice parameters *a* and *c*, corresponding to given carbon content (see Fig. 3.2). A lattice deformation was required to produce the martensite from austenite. In steels, the lattice correspondence shown in Fig. 3.11(a) was first identi-

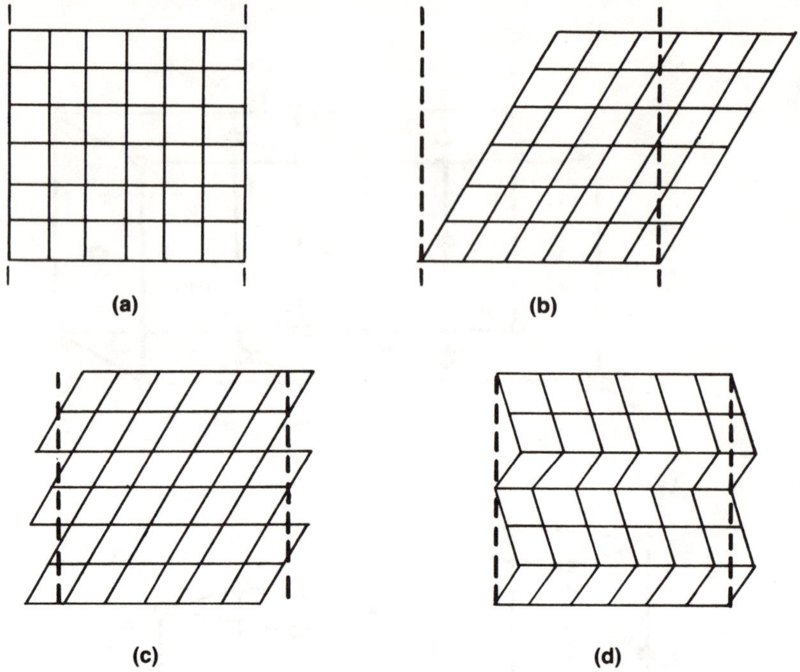

Fig. 3.12. Schematic diagrams to show (a) portion of parent crystal; (b) new lattice (martensite) produced by lattice deformation; and lattice invariant deformation by (c) slip and (d) twinning to make martensite conform to original position of parent crystal (a). (Ref 3.6)

fied by Bain and the lattice deformation from fcc to bct is referred to as the "Bain strain". Figure 3.11 shows that the Bain strain produces a contraction along the *c* axis and an expansion along the *a* axes.

In general, the lattice deformation will cause rotation away from the habit plane, as shown in Fig. 3.12(a) and (b), where it is assumed a number of cells of a parent crystal (a) are transformed to a new lattice (b). The vertical dashed lines represent the unrotated, undistorted habit plane. The constraints of the surrounding parent phase, however, cause the martensite unit to accommodate or deform by a lattice invariant deformation to the original boundaries (a) as required by the crystallographic theory. Figures 3.12(c) and (d) show, respectively, the martensite deformed by slip (dislocation movement) or twinning to satisfy on a macroscopic scale the requirement of an unrotated, undistorted habit plane. Figure 3.13 is another schematic

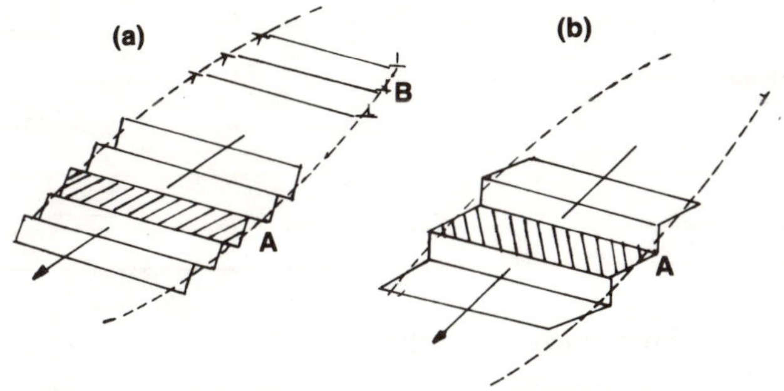

Fig. 3.13. Schematic representations of lattice invariant deformation by (a) slip and (b) twinning in martensite plates. (Ref 3.6)

representation of the slip (a) and twinning (b) modes of lattice invariant deformation within martensite plates. These sketches, of course, are idealized to demonstrate the concept of the lattice invariant shear. In actual crystals, when slip is the mechanism of accommodation, not only are dislocations introduced at the austenite-martensite interface, but also a high dislocation density remains in the fine structure within the plates. Examples of the latter are presented in the next section of this chapter.

The crystallographic theory of the martensitic transformation is well developed mathematically and has successfully predicted crystallographic parameters in a number of alloys. For example, if the lattice and lattice invariant deformations are specified, the habit plane may be predicted. For development of the theory and its application see Bilby and Christian (Ref 3.6) and Wayman (Ref 3.2). Successes and limitations of the theory are reviewed by Dunne and Wayman (Ref 3.32).

Morphology of Ferrous Martensites

Two major morphologies of martensite, lath and plate, develop in heat treatable carbon steels. Figure 3.14 shows the carbon ranges of formation and M_s temperature of the two morphologies. The boundaries of the various regions are based on characterization of high-purity Fe-C alloys and may shift in alloy steels. The designations of the two morphologies originate from the shape of the individual units of martensite. The lath designation is used to describe the board-shaped units of martensite that form in low- and medium-carbon steels, while the plate designation accurately describes the shape of the martensite

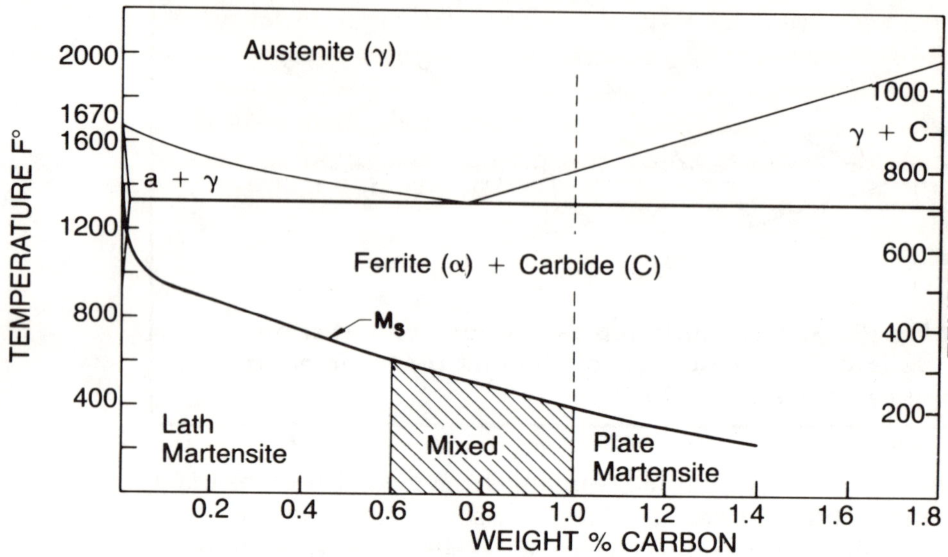

Fig. 3.14. Ranges of lath and plate martensite formation in iron-carbon alloys. (Ref 3.11)

units that form in high-carbon steels. The terms lath and plate, therefore, refer to the three-dimensional shapes of individual martensite crystals. In metallographic specimens, sections through the martensite laths or plates are revealed by polishing and etching. Generally, these cross sections will appear to be needlelike or acicular, and the latter adjectives are often used to describe martensite microstructures. Other terms based on one or another feature of the different forms of martensite have been used to describe the two morphologies of martensite, but the terms lath and plate are preferred (Ref 3.8).

Until the advent of the electron microscope, the plate martensites, which could be readily resolved by light microscopy, received the most emphasis in the literature. The units of plate martensite are well within the size range resolvable in the light microscope, and frequently the retained austenite that coexists with the martensite in high-carbon alloys helps to sharply define the plates in the light microscope. On the other hand, as will be demonstrated, many of the individual units of lath martensite are below the resolution of the light microscope, and any retained austenite present is also too fine to be resolved. Although plate martensites are important in some heat treated applications (such as the case microstructure of carburized steels), most hardenable steels have low- or medium-carbon content and, therefore, microstructures composed of lath martensite. As a result, lath martensites have overwhelming industrial significance. However, in order to follow the historical development of the understanding of martensitic microstructures, plate martensite will be described first in the following sections. The characteristics of individual units and the arrangement of the units to produce the microstructures that are put into service as a result of good heat treatment practice will be described for both lath and plate martensite.

Plate Martensite

Many other ferrous systems show the same transition from lath to plate martensite (see Fig. 3.14) with increasing alloying as does the Fe-C system (Ref 3.8). Figure 3.15 shows plate martensite that was produced by cooling a single crystal of Fe-33.5Ni austenite in liquid nitrogen (−196 °C or −321 °F). Subzero cooling was required because the high nickel content had lowered the M_s to −30 °C (−22 °F). The specimen was not polished or etched after the liquid nitrogen treatment, and therefore all features shown in Fig. 3.15 are due to the surface relief generated by the martensitic transformation. On the scale shown, the surface tilting is indeed quite homogeneous except for small dark bands visible in some of the martensite plates. These bands are deformation twins formed in response to the constraints of the austenite matrix. The deformation twins, however, are micron sized and irregularly distributed in contrast to the much finer and more regularly distributed fine structure that results from the lattice invariant deformation.

Figure 3.16 is a transmission electron micrograph of the fine structure that formed in a single plate of martensite in the Fe-33.5Ni alloy. Fine transformation twins (small dark bands), dislocations (the fine linear features) and a large deformation twin band are present. Figure 3.17 shows the dislocation fine structure at a higher magnification. Relatively straight dislocation lines in two directions are visible. By selected area diffraction techniques (Ref 3.33), the preferred directions were shown to correspond to ⟨111⟩ directions and, therefore, the dislocations are mostly screw dislocations. This type of dislocation array is characteristic of those formed by deformation of bcc iron at low temperatures and/or high strain rates, and illustrates one type of fine structure formed by the lattice invariant deformation of bcc martensite as a result of the austenitic constraints. Figure 3.18 shows another type of fine structure, very fine transformation twins, about 100Å in thick-

Fig. 3.15. Plate martensite formed in an austenitic single crystal of an Fe-33.5Ni alloy by cooling to −196 °C (−321 °F). Plates are visible only because of surface relief generated by martensitic transformation. Magnification, 200×. (Ref 3.33)

ness, in Fe-33.5Ni plate martensite. The twins lie on $\{112\}_m$ planes and represent another plastic deformation mode that forms in bcc crystals deformed at low temperatures and high strain rates. Also shown in Fig. 3.18 is a larger deformation twin across which the fine transformation twins have changed their orientation to a $\{11\bar{2}\}_m$ orientation in the twin (Ref 3.34).

Examples of plate martensite in Fe-C alloys are shown in Figures 3.5 and 3.19. Many different orientations of the martensite plates are apparent in the microstructures shown. This characteristic appearance of a plate martensitic microstructure is directly related to the habit planes of the plate martensite and the tendency of adjacent plates to assume different variants of the habit plane. Plate martensites have irrational habit planes, i.e., the planes are not defined by low number indices such as (100) or (111). Early work by Greninger and Troiano (Ref 3.35) (see Fig. 3.20) showed that Fe-1.78C alloys had habit planes best characterized as $\{259\}_A$, and lower carbon alloys, containing 0.92% and 1.4% carbon, had $\{225\}_A$ habit planes. The Fe-Ni plate martensites have been the subject of extensive crystallographic studies. Figure 3.21 shows that there is considerable scatter of the habit plane in Fe-Ni alloys containing 29.0 to 35.0% nickel. The scatter may be due to compositional variations, the habit plane shifting toward $\{225\}_A$ with

Fig. 3.16. Fine structure within martensite plates shown in Fig. 3.15. A deformation twin, fine transformation twins, and dislocations are shown. Transmission electron micrograph; magnification, 20 000×, shown here at 75%. (Ref 3.33)

decreasing nickel content, or to mixtures of lattice invariant deformations such as combinations of twinning and dislocations in a given alloy (Ref 3.33).

The many orientations of martensite plates in the microstructures shown in Fig. 3.5 and 3.19 are due to the many variants of the irrational habit planes. A variant is merely a different orientation of a given {hkl} plane as defined by a different arrangement of the same hkl indices. For example, $(925)_A$, $(592)_A$ and $(952)_A$ are all variants of the $\{259\}_A$ plane. Any plane where h, k, l are all different, as is the case for $\{259\}_A$, has 24 different variants, and a plane with two indices equal, such as the $\{225\}_A$ plane, has 12 variants. Thus, the plate martensite microstructures, because of the large number of variants possible, and the fact that adjacent plates assume different variants, appear quite haphazardly arranged, despite the fact that there is only a single habit plane for all the plates in a given alloy.

An important consequence of the nonparallel plate formation in Fe-C alloys is the development of microcracks in the martensite plates

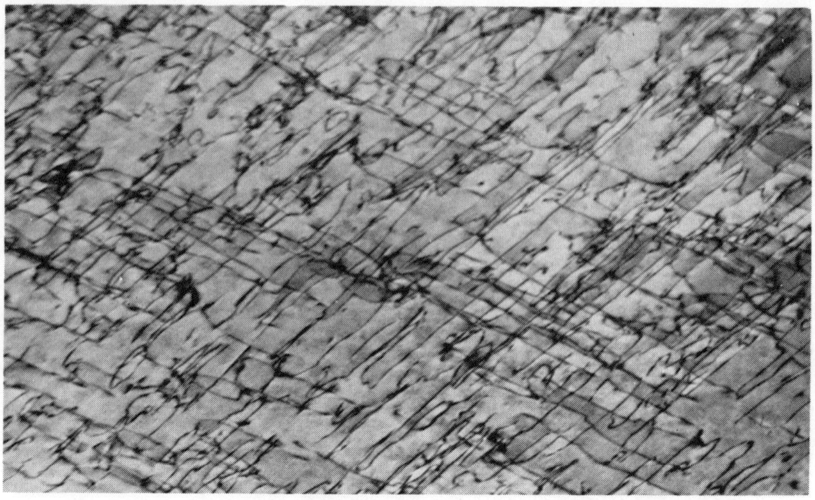

Fig. 3.17. Dislocation fine structure in martensite plates shown in Fig. 3.15. Transmission electron micrograph; magnification, 20 000×, shown here at 75%. (Ref 3.33)

Fig. 3.18. Fine transformation twins in plate martensite of an Fe-33.5Ni alloy. Note change in orientation of fine twins in large deformation twin. Transmission electron micrograph; magnification, 15 000×, shown here at 75%. (Ref 3.34)

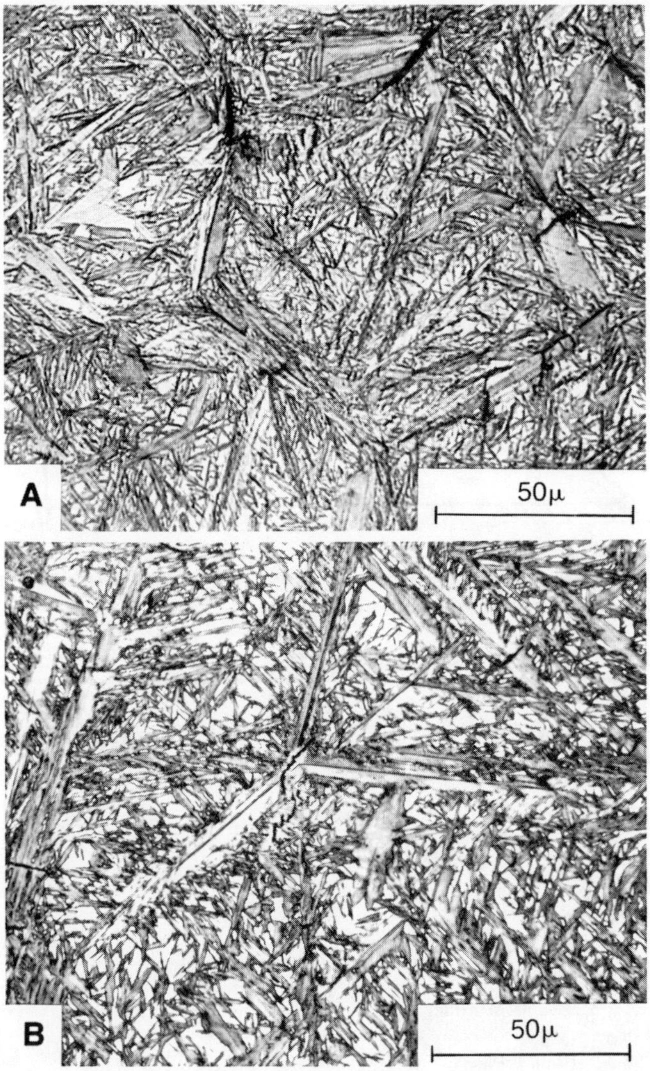

Fig. 3.19. Plate martensite and retained austenite (white patches) in (a) Fe-1.22C and (b) Fe-1.4C alloys. Light micrographs. (Ref 3.11)

as a result of the impingement of plates of different habit plane variants (Ref 3.37). Figure 3.22 shows an example of the microcracks in the plate martensite of an Fe-C alloy. The microcracks tend to form in the largest martensite plates and therefore are not present to any great extent in steels where austenite grain size, and accordingly martensite

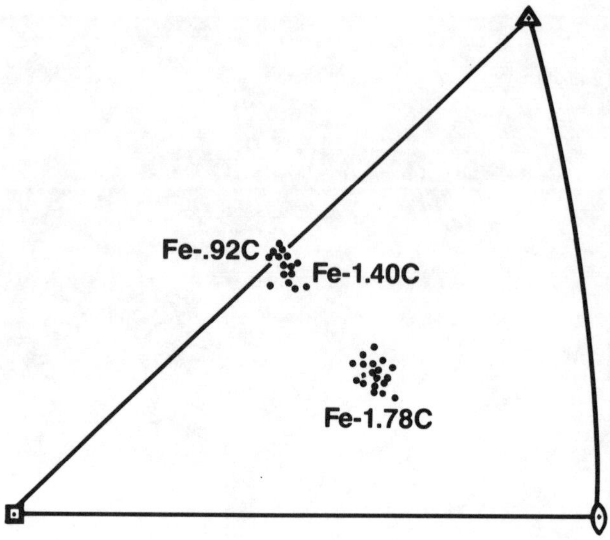

Fig. 3.20. Habit planes of Fe-C plate martensites in unit stereographic triangle. (Ref 3.35)

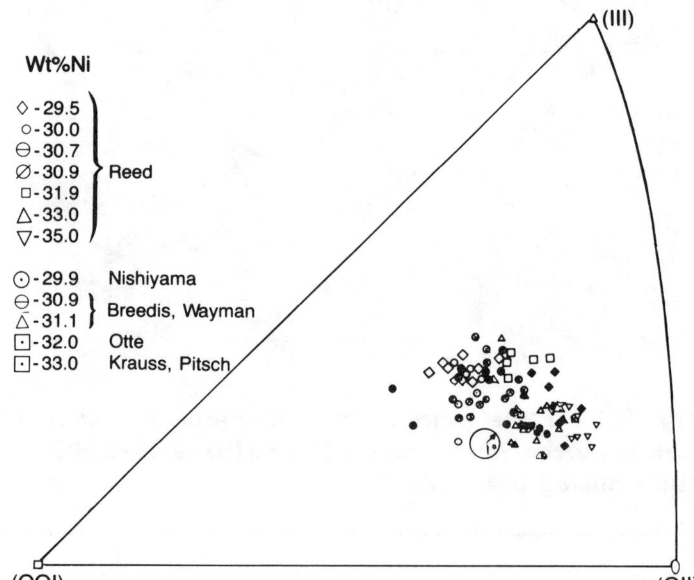

Fig. 3.21. Habit planes of plate martensite in Fe-Ni alloys containing 29 to 35% nickel. (Ref 3.36)

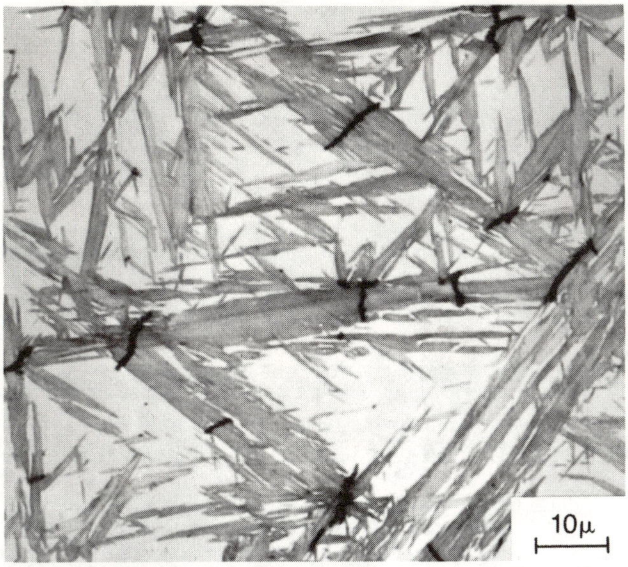

Fig. 3.22. Microcracks in plate martensite of an Fe-1.4C alloy. (Ref 3.38)

plate size, are fine (Ref 3.39). Also in lower carbon steels, the morphology shift to lath martensite eliminates the impingements and the development of microcracks (Ref 3.40). The high carbon plate martensites are quite brittle and sensitive to microcracking. However, in Fe-Ni alloys where the martensite is much more ductile, the impingement of martensite plates is accommodated by deformation twinning rather than cracking.

Lath Martensite

Light micrographs of lath martensite in Fe-C alloys are shown in Fig. 3.23. The lath martensite units tend to be quite fine but the characteristic acicularity of a martensitic microstructure is apparent. An important microstructural characteristic of lath martensites is the tendency of many laths to align themselves parallel to one another in large areas of the parent austenite grain. These regions of parallel lath alignment are referred to as packets and tend to develop most prominently in lower carbon alloys, as shown in Fig. 3.23(a). The packets are delineated because of the different etching characteristics of the different variants or orientations of the laths in the various packets.

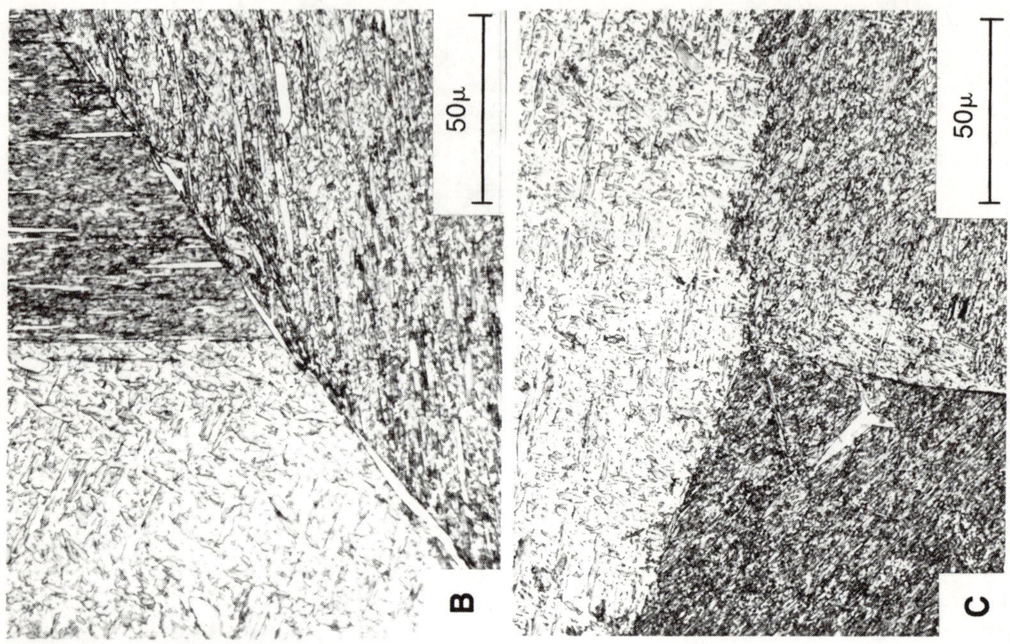

Fig. 3.23. Lath martensite microstructures in (a) Fe-0.2C, (b) Fe-0.4C, and (c) Fe-0.6C alloys. Light micrographs. (Ref 3.11)

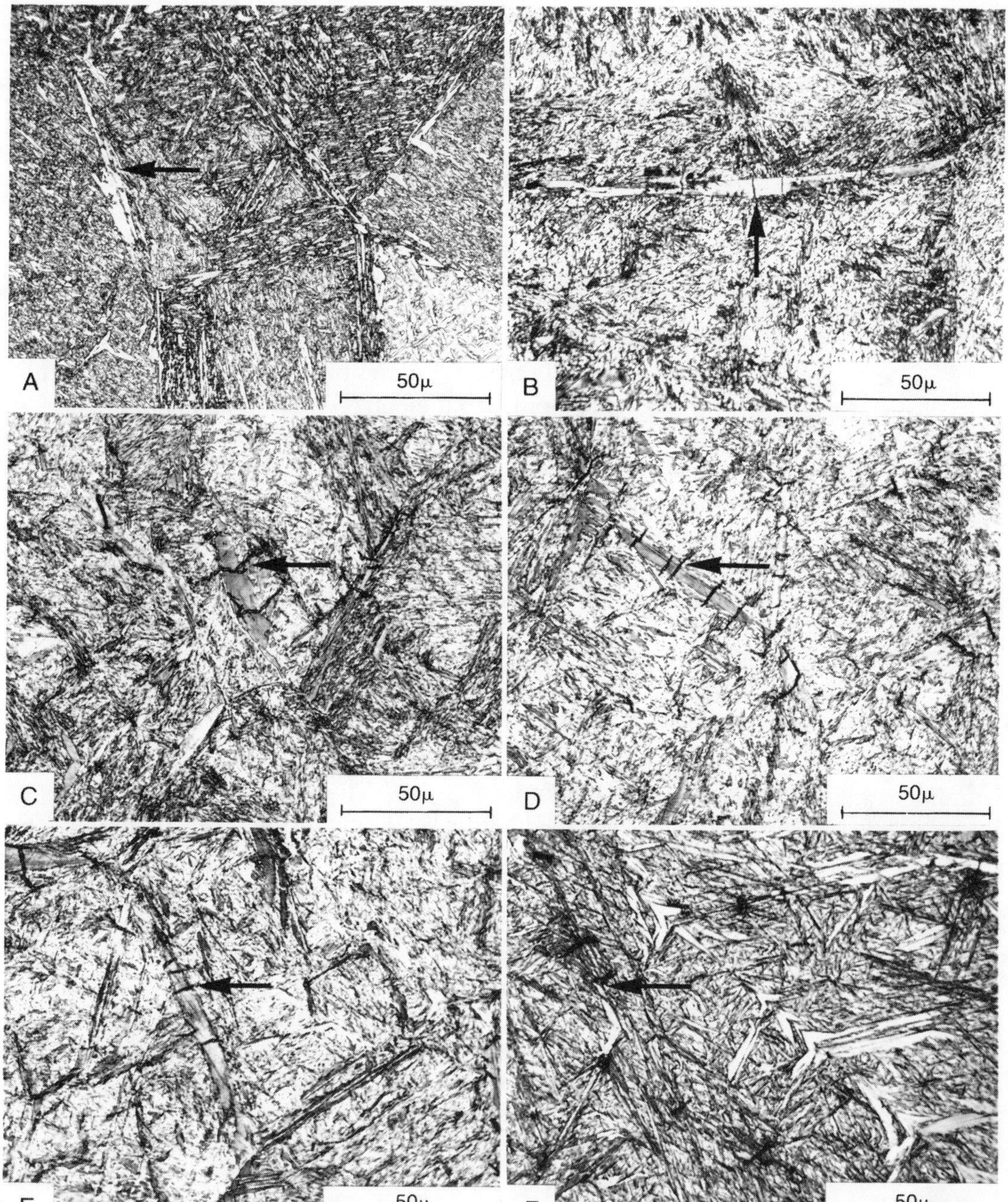

Fig. 3.24. Transition from lath to plate martensite microstructures in Fe-C alloys between 0.67 and 1.00% carbon. A: 0.67% carbon; B: 0.75%; C: 0.82%; D: 0.85%; E: 0.93%; F: 1.00%. Light micrographs. Sodium bisulfite etch. (Ref 3.11)

Figure 3.24 shows the transition in martensite morphology that develops as the carbon content of the Fe-C alloys increases from 0.67 to 1.00% carbon. With increasing carbon content, more plates of martensite, differentiated from lath martensite by their larger size and their tendency to microcrack, are discernible in the microstructure.

A transmission electron micrograph of the lath martensite in an Fe-0.2C alloy is shown in Fig. 3.25. All of the laths, even the very thin ones, are resolved, in contrast to the light micrograph of the same structure in Fig. 3.23(a) where many units are not clearly defined. Parts of two packets are shown. In each packet there appear to be two major orientations or variants of the martensite laths, and there are many very fine laths. Figure 3.26 shows the distribution of lath widths obtained by measurements from electron micrographs obtained from thin foils and replicas of polished and etched metallographic specimens. The important result shown in Fig. 3.26 is that most of the laths have widths smaller than 0.5 μm, the resolution limit of the light microscope, and therefore cannot possibly be revealed by light metallography. There are some laths with widths up to almost 2 μm, and these larger laths would, of course, be visible in the light microscope as some are in Fig. 3.23. It is the very fine size of most of the laths in a packet of low- or medium-carbon martensite that have over the years made the light metallographic characterization of lath martensite difficult.

The habit plane of lath martensite, $\{557\}_A$, is irrational, a plane close to $\{111\}_A$, as shown in Fig. 3.27. There are three $\{557\}_A$ variants clustered about each of the four $\{111\}_A$ planes, and the angle between these variants is only 16°. Laths of different orientation within packets (see Fig. 3.25) frequently are observed to make angles of about 16° with each other, leading to the conclusion that the variants in a given packet all have variants close to the same $(111)_A$ plane. This coupling of variants, the small angles between variants, and the fine size of the laths give the microstructural impression that lath martensite has a $\{111\}_A$ habit with only four variants. Lath martensitic microstructures, therefore, appear much more orderly (see Fig. 3.23) than do the plate martensitic microstructures (see Fig. 3.19) with as many as 24 variants.

Although there may be several variants of laths in a packet of lath martensite, one variant tends to be dominant (Ref 3.41). This characteristic of a packet means that most of the laths, separated by low-angle boundaries or perhaps retained austenite, have the same crystal orientation and that a packet may be considered as a single grain or crystal, albeit a grain divided by many low-angle boundaries and containing a fine structure of many dislocations. The packet structure of lath martensite has no counterpart in plate martensite and is impor-

Fig. 3.25. Lath martensite in an Fe-0.2C alloy. Two packets, each with two variants of laths, are shown. Transmission electron micrograph. (Ref 3.41)

tant in determining mechanical properties and fracture behavior of the martensite that forms in low- and medium-carbon steels.

The fine structure of lath martensite consists predominantly of a very high density of dislocations, too high to be resolved even by

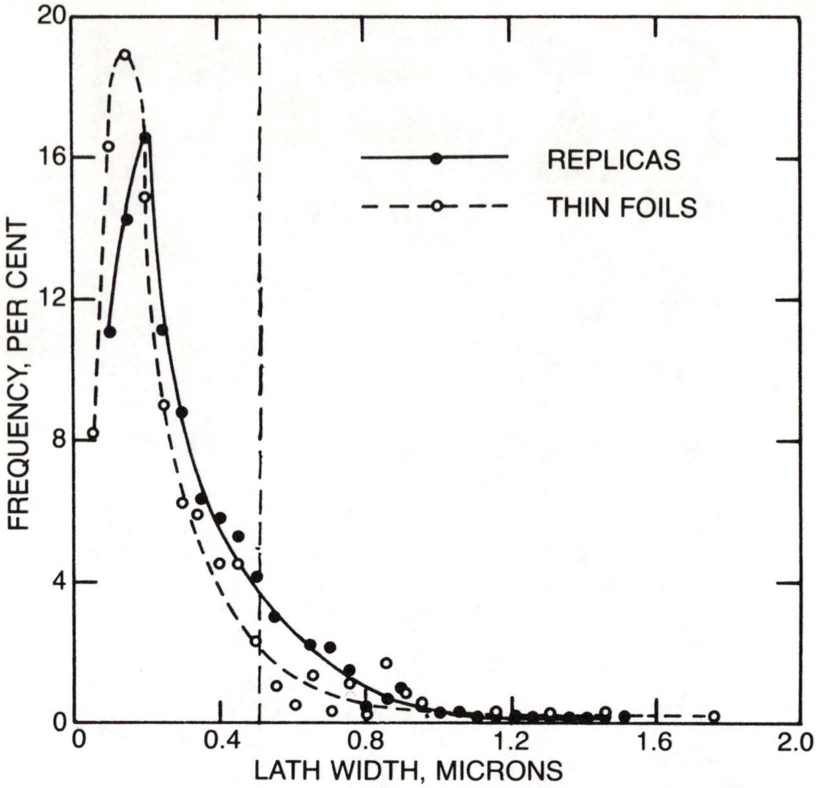

Fig. 3.26. Distribution of lath widths in Fe-0.2C martensite. Vertical line shows limit of resolution of the light microscope. (Ref 3.41)

electron microscopy of thin foils. However, Speich (Ref 3.42) was able to determine, indirectly by electrical resistivity measurements, a dislocation density of almost 10^{12} dislocations per square centimetre in low-carbon lath martensite.

An example of the fine structure of lath martensite in an Fe-0.2C alloy is shown in Fig. 3.28. The dislocations are tangled and arranged in incipient dislocation cells, a structure much different from the essentially straight, uniformly distributed dislocations of the Fe-Ni plate martensite shown in Fig. 3.17. The dislocation tangles are a result of a plastic deformation mode consistent with the high M_s and high-temperature range of formation of the low-carbon lath martensite (see Fig. 3.14) whereas, as already noted, the straight dislocations of the Fe-Ni martensite are consistent with low-temperature deformation of bcc iron alloys.

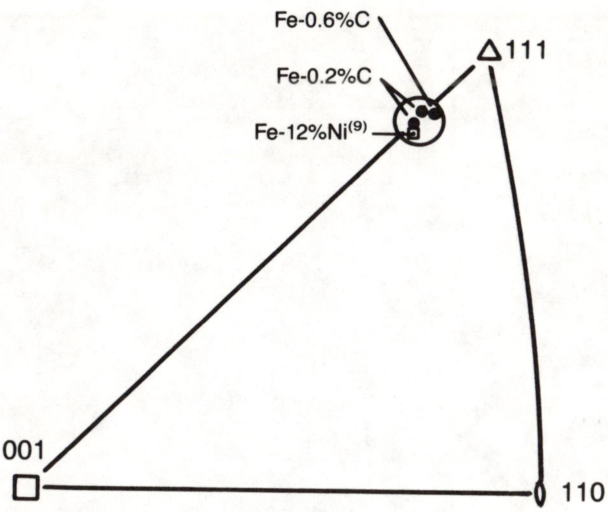

Fig. 3.27. Habit planes of lath martensite. (Ref 3.7)

Although dislocations are the major fine structural component in lath martensite, fine transformation twins, a low temperature mode of plastic accommodation, are also found to some extent in Fe-C lath martensite. The amount of fine twinning increases in accord with the decreasing M_s temperatures and lower athermal transformation ranges of lath martensite formation as carbon content increases.

Bainite

The formation and structure of the various microstructural constituents that form during the heat treatment of steels (in particular, proeutectoid ferrite and cementite, pearlite, and martensite) have already been discussed. The description of bainite, the last microstructural constituent to be discussed, has been deferred to this point because it forms under continuous cooling or isothermal transformation conditions intermediate to those of pearlite and martensite formation and because there are transformation and structural similarities to both pearlite and martensite.

Similar to pearlite, bainite is a mixture of the phases ferrite and cementite, and is therefore dependent on the diffusion-controlled partitioning of carbon between ferrite and cementite. However, unlike pearlite, the ferrite and cementite are present in nonlamellar arrays

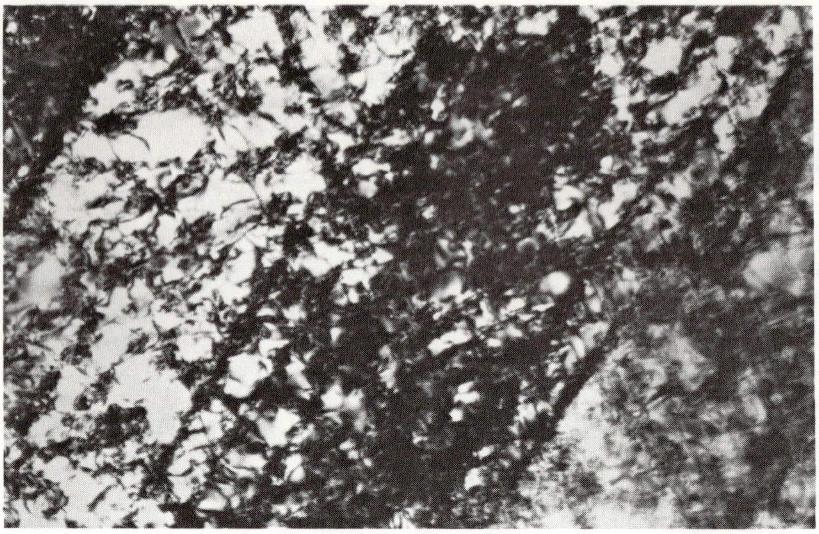

Fig. 3.28. Dislocation fine structure in lath martensite of an Fe-0.2C alloy. Transmission electron micrograph; magnification, 82 500×. (Ref 3.43)

whose characteristics are dependent on alloy composition and transformation temperature. Similar to martensite, the ferrite of bainite may be in the form of laths or plates containing a dislocation structure, and to some extent therefore, the mechanism of bainite formation involves shear as well as diffusion.

There are two major morphologies or forms of bainite: upper bainite that forms in the temperature range just below that of pearlite formation, and lower bainite which forms at temperatures closer to M_s. Figure 3.29 shows upper bainite in a 4360 steel partially transformed at 410 °C (770 °F) and at 495 °C (923 °F). In Fig. 3.29 the bainite is dark and the white-etching matrix is martensite formed from untransformed austenite on quenching. The bainite appears dark (i.e., has low reflectivity) because of roughness produced by etching around the carbide particles, but the carbides themselves in this example are too fine to be resolved in the light microscope. The feathery appearance of the clusters of ferrite laths is clearly shown in the light micrographs and is an important identifying feature of upper bainite. The carbide particles of upper bainite tend to be elongated and form between ferrite laths,

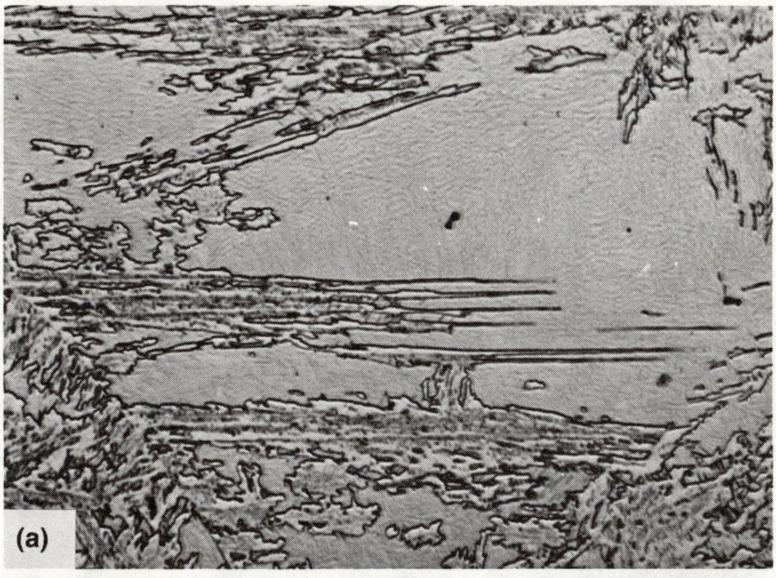

Fig. 3.29. Upper bainite in 4360 steel isothermally transformed at (a) 495 °C (923 °F) and (b) 410 °C (770 °F). Light micrographs. Picral etch. Magnification, 750×. (Ref 3.44)

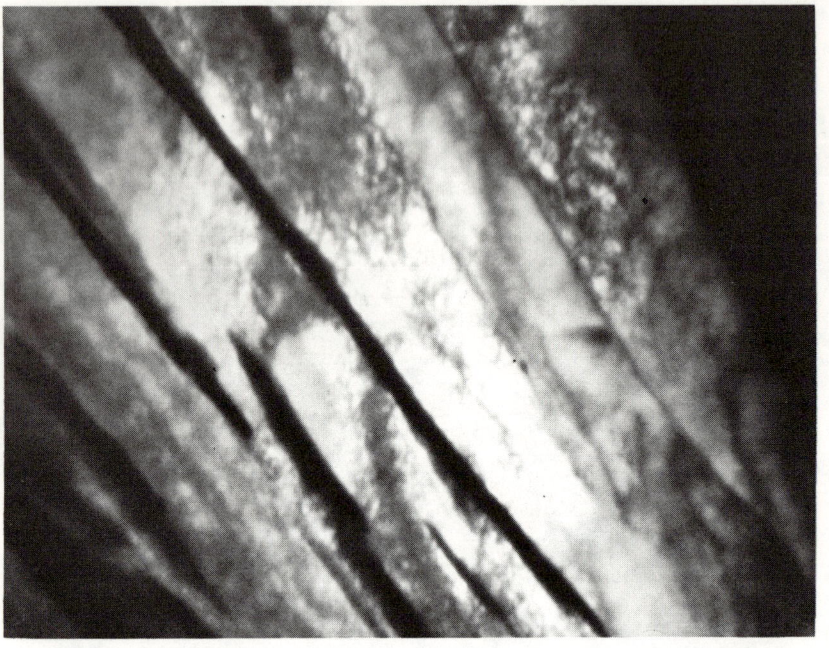

Fig. 3.30. Carbides (dark) formed between laths of upper bainite in 4360 steel transformed at 495 °C (923 °F). Transmission electron micrograph; magnification, 25 000×. (Ref 3.44)

areas to which carbon has been rejected during the growth of the ferrite. Figure 3.30 is a thin foil electron micrograph which shows the interlath cementite in a 4360 steel transformed to bainite at 495 °C (923 °F). In certain steels, in particular those with high silicon content, the carbon-enriched austenite between the bainite laths is quite stable and is retained between the ferrite laths at room temperature. Figure 3.31 shows such retained austenite, marked A, in bainite formed at 400 °C (752 °F) in a steel containing 0.6% carbon and 2.0% silicon.

Lower bainite in 4360 steel partially transformed at 300 °C (572 °F) and quenched is shown in Fig. 3.32. Again the bainite etches dark and the white etching matrix is martensite formed from austenite untransformed at 300 °C (572 °F). The lower bainite is composed of large plates and, analogously to plate martensitic microstructures, is often characterized as acicular. The carbides in lower bainite are responsible for its dark etching appearance but are much too fine to be resolved in the light microscope. Figure 3.33 shows the carbides in lower bainite

Fig. 3.31. Retained austenite between laths of upper bainite in 0.6% carbon steel containing 2.0% silicon transformed at 400 °C (752 °F). Retained austenite is identified by letter A. Transmission electron micrograph; magnification, 40 000×. (Ref 3.44)

formed in the 4360 steel transformed at 300 °C (572 °F). In contrast to upper bainite, fine carbides have formed within the plates, rather than between plates, and are significantly finer than the interlath carbides of upper bainite. In some lower bainites the fine carbides have been identified as a transition carbide, epsilon carbide, that forms prior to cementite at lower temperatures. The transition carbide is especially stable in steels containing aluminum and silicon.

Several studies have characterized crystallographic parameters such as habit planes and orientation relationships in the bainite of Fe-Cr-C alloys (Ref 3.45), and the mechanism and characterization of bainite has been the subject of a lively written debate between R. Hehemann, K. R. Kinsman and H. Aaronson (Ref 3.46). The interested reader is referred to the latter reference for more information regarding the somewhat controversial characterization of bainite and its formation.

Fig. 3.32. Lower bainite in 4360 steel transformed at 300 °C (572 °F). Light micrograph. Picral etch. Magnification, 750×. (Ref 3.44)

Fig. 3.33. Lower bainite with fine carbides within plates in 4360 steel transformed at 300 °C (572 °F). Transmission electron micrograph; magnification, 24 000×. (Ref 3.44)

References

3.1 F. Osmond, Méthode générale pour l'analyse micrographique des aciers au carbone, *Bulletin de la societe d'Encouragement pour l'Industrie National*, Vol 10, 1895, p 480

3.2 C. M. Wayman, *Introduction to the Crystallography of Martensite Transformations*, MacMillan, New York, 1964

3.3 W. D. Kingery, *Introduction to Ceramics*, John Wiley & Sons, New York, 1960

3.4 G. K. Bansal and A. H. Heuer, On a Martensitic Phase Transformation in Zirconia (ZrO_2) — I. Metallographic Evidence, *Acta Met*, Vol 20, 1972, p 1281-1289

3.5 M. Cohen, The Strengthening of Steel, *Trans TSM-AIME*, Vol 224, 1962, p 638-656

3.6 B. A. Bilby and J. W. Christian, The Crystallography of Martensite Transformations, *JISI*, Vol 197, 1961, p 122-131

3.7 A. R. Marder and G. Krauss, The Formation of Low-Carbon Martensite in Fe-C Alloys, *Trans ASM*, Vol 62, 1969, p 957-964

3.8 G. Krauss and A. R. Marder, The Morphology of Martensite in Iron Alloys, *Met Trans*, Vol 2, 1971, p 2343-2357

3.9 G. Thomas, Retained Austenite and Tempered Martensite Embrittlement, *Met Trans A*, Vol 9A, 1978, p 439-450

3.10 C. S. Roberts, Effect of Carbon on the Volume Fractions and Lattice Parameters of Retained Austenite and Martensite, *Trans AIME*, Vol 197, 1953, p 203-204

3.11 A. R. Marder and G. Krauss, The Morphology of Martensite in Iron-Carbon Alloys, *Trans ASM*, Vol 60, 1967, p 651-660

3.12 P. Payson and C. H. Savage, Martensite Reactions in Alloy Steels, *Trans ASM*, Vol 33, 1944, p 261-275

3.13 L. A. Carapella, Computing A^{11} or M_s (Transformation Temperature on Quenching) from Analysis, *Metal Progress*, 1944, Vol 46, p 108

3.14 E. S. Rowland and S. R. Lyle, The Application of M_s Points to Case Depth Measurement, *Trans ASM*, 1946, Vol 37, p 27-47

3.15 R. A. Grange and H. M. Stewart, The Temperature Range of Martensite Formation, *Trans AIME*, 1946, Vol 167, p 467-490

3.16 A. E. Nehrenberg, *Trans AIME*, 1946, Vol 167, p 494-498

3.17 W. Steven and A. G. Haynes, The Temperature of Formation of Martensite and Bainite in Low-alloy Steel, *JISI*, Vol 183, 1956, p 349-359

3.18 K. W. Andrews, Empirical Formulae for the Calculation of Some Transformation Temperatures, *JISI*, Vol 203, 1965, p 721-727

3.19 G. Krauss, Martensitic Transformation, Structure and Properties in Hardenable Steels, in *Hardenability Concepts with Applications to Steel*, AIME, Warrendale, Pa., 1978, p 229-248

3.20 W. H. Harris and M. Cohen, Stabilization of the Austenite-Martensite Transformation, *Trans AIME*, Vol 180, 1949, p 447-470

3.21 D. P. Koistinen and R. E. Marburger, A General Equation Prescribing the Extent of the Austenite-Martensite Transformation in Pure Iron-Carbon Alloys and Plain Carbon Steels, *Acta Met*, Vol 7, 1959, p 59-60

3.22 A. R. Entwisle, The Kinetics of Martensite Formation in Steel, *Met Trans*, Vol 2, 1971, p 2395-2407

3.23 J. C. Bokros and E. R. Porter, The Mechanism of the Martensite Burst Transformation in Fe-Ni Single Crystals, *Acta Met*, Vol 11, 1963, p 1291-1301
3.24 K. R. Kinsman and J. S. Shyne, Thermal Stabilization of Austenite in Iron-Nickel-Carbon Alloys, *Acta Met*, Vol 15, 1967, p 1527-1543
3.25 C. H. Shih, B. L. Averbach and M. Cohen, Some Characteristics of the Isothermal Martensitic Transformation, *Trans AIME*, Vol 203, 1955, p 183-187
3.26 V. Raghavan and M. Cohen, Measurement and Interpretation of Isothermal Martensitic Kinetics, *Met Trans*, Vol 2, 1971, p 2409-2418
3.27 S. R. Pati and M. Cohen, Nucleation of the Isothermal Martensitic Transformation, *Acta Met*, Vol 17, 1969, p 189-199
3.28 E. R. Petty, in *Martensite, Fundamentals and Technology,* Longman, London, 1979, p 6
3.29 M. S. Wechsler, D. S. Lieberman and T. A. Read, On the Theory of the Formation of Martensite, *Trans AIME*, Vol 197, 1953, p 1503-1515
3.30 J. S. Bowles and J. K. MacKenzie, The Crystallography of Martensite Transformations, *Acta Met*, Vol 2, 1954, p 129-137, 138-147, 224-234
3.31 A. B. Greninger and A. R. Troiano, The Mechanism of Martensite Formation, *Trans AIME*, Vol 185, 1949, p 590-598
3.32 D. P. Dunne and C. M. Wayman, The Crystallography of Ferrous Martensites, *Met Trans*, Vol 2, 1971, p 2327-2341
3.33 G. Krauss and W. Pitsch, The Fine Structure and Habit Planes of Martensite in an Fe-33 wt pct Ni Single Crystal, *Trans TMS-AIME*, Vol 233, 1965, p 919-926
3.34 G. Krauss and W. Pitsch, Deformation Twins in Martensite, *Acta Met*, Vol 12, 1964, p 278-279
3.35 A. B. Greninger and A. R. Troiano, Crystallography of Austenite Decomposition, *Trans AIME*, Vol 140, 1940, p 307-336
3.36 H. M. Ledbetter and R. P. Reed, On the Martensite Crystallography of Fe-Ni Alloys, *Mater Sci Eng*, Vol 5, 1969-70, p 341-349
3.37 A. R. Marder and A. O. Benscoter, Microcracking in Fe-C Acicular Martensite, *Trans ASM*, Vol 61, 1968, p 293-299
3.38 A. R. Marder, A. O. Benscoter and G. Krauss, Microcracking Sensitivity in Fe-C Plate Martensite, *Met Trans*, Vol 1, 1970, p 1545-1549
3.39 R. P. Brobst and G. Krauss, The Effect of Austenite Grain Size on Microcracking in Martensite of an Fe-1.22 C Alloy, *Met Trans*, Vol 5, 1974, p 457-462
3.40 M. G. Mendiratta, J. Sasser and G. Krauss, Effect of Dissolved Carbon on Microcracking in Martensite of an Fe-1.39 C Alloy, *Met Trans*, Vol 3, 1972, p 351-353
3.41 C. A. Apple, R. N. Caron and G. Krauss, Packet Microstructure in an Fe-0.2% C Martensite, *Met Trans*, Vol 5, 1974, p 593-599
3.42 G. R. Speich, Tempering of Low-Carbon Martensite, *Trans TMS-AIME*, Vol 245, 1969, p 2552-2564
3.43 T. Swarr and G. Krauss, The Effect of Structure on the Deformation of As-quenched and Tempered Martensite in an Fe-0.2% C Alloy, *Met Trans*, Vol 7A, 1976, p 41-48
3.44 R. F. Hehemann, Ferrous and Nonferrous Bainitic Structures, in *Metals Handbook*, 8th Ed., Vol 8, American Society for Metals, Metals Park, Oh., 1973, p 194-196

3.45 G. R. Srinivasan and C. M. Wayman, The Crystallography of the Bainite Transformation, *Acta Met*, Vol 16, 1969, p 621-636
3.46 R. F. Hehemann, K. R. Kinsman and H. I. Aaronson, A Debate on the Bainite Reaction, *Met Trans*, Vol 3, 1972, p 1077-1094
3.47 E. C. Bain and H. W. Paxton, *Alloying Elements in Steel*, American Society for Metals, Metals Park, Oh., 1966, p 36
3.48 J. W. Christian, in *Martensite Fundamentals and Technology,* E. R. Petty (Ed.), Longman, London, 1970, p 13

CHAPTER 4

Isothermal and Continuous Cooling Transformation Diagrams

This chapter describes the transformation diagrams that have been developed to define the progress of diffusion-controlled phase transformations of austenite to various mixtures of ferrite and cementite. Both isothermal and continuous cooling transformation diagrams are described, and references to atlases containing collections of these diagrams for a variety of steels are given. The availability of these diagrams makes possible the selection of steels and the design of heat treatments that will either produce desirable microstructures of ferrite and cementite or avoid diffusion-controlled transformations, and thereby produce martensitic microstructures of maximum hardness.

Isothermal Transformation Diagrams

Diagrams that define the transformation of austenite as a function of time at constant temperatures are referred to as isothermal transformation (IT) diagrams or time-temperature-transformation (TTT) diagrams. An IT diagram for 1080 steel has already been presented in Fig. 2.3 in connection with the description of the nucleation and growth kinetics of pearlite formation. The IT diagram for eutectoid steel with negligible alloy content is quite straightforward. Only pearlite forms above the nose of the IT diagram, and only bainite forms below the nose. The curves defining the beginning and end of pearlite or bainite formation are the major features of the diagram.

Steels with carbon content above or below the eutectoid composition and alloy steels have more complex transformation diagrams. Figure 4.1 shows schematic IT diagrams for eutectoid steel and a hypoeutectoid plain carbon steel containing nominally 0.5% carbon. Also shown is their relationship to the iron-carbon diagram. The beginning and ending curves for pearlite formation approach the Ae_1 temperature at very long transformation times and move to shorter times with decreasing transformation temperature for reasons discussed in Chapter 2. The IT diagram for the hypoeutectoid steel has an extra curve to mark the beginning of proeutectoid ferrite formation. As indicated in Fig. 4.1, the latter curve approaches the Ae_3 temperature for the 0.5% carbon steel with increasing transformation time. Hypoeutectoid steels with lower carbon contents would have higher Ac_3 temperatures and, therefore, expanded regions of proeutectoid ferrite coexistence with austenite. Similarly, hypereutectoid steels would have IT diagrams with curves for the beginning of proeutectoid cementite formation.

Figure 4.1 shows other differences between the IT diagrams for eutectoid and hypoeutectoid steels. One difference is in M_s temperatures: the lower the carbon content the higher the M_s temperature. Another difference is the acceleration of austenite transformation to proeutectoid ferrite with decreasing carbon content, as shown by the position of the nose of the hypoeutectoid steel at shorter times relative to that of the eutectoid steel. The dotted lines in Fig. 4.1(b) and (c) reflect experimental uncertainty in the exact positions of the beginning of transformation curves.

IT diagrams are usually produced by metallographic examination of series of specimens held for various times at various temperatures between Ae_3 or A_{cm} and M_s. More than a hundred specimens are often required to determine a complete IT diagram for a given steel (Ref 4.1). The procedure used is to heat the metallographic specimens in the single-phase austenite field for a sufficient time, usually 1 h, to produce a homogeneous austenite. The austenitizing treatment sets the austenite grain size and the extent of carbide solution. Both of the latter microstructural factors may influence the course of isothermal transformation of austenite, and therefore it is necessary to record the austenitizing temperature used to determine the IT. Once austenitizing is complete, a series of specimens is cooled rapidly, usually by immersion in a molten salt bath, to a given isothermal transformation temperature. The specimens are held for various times and then

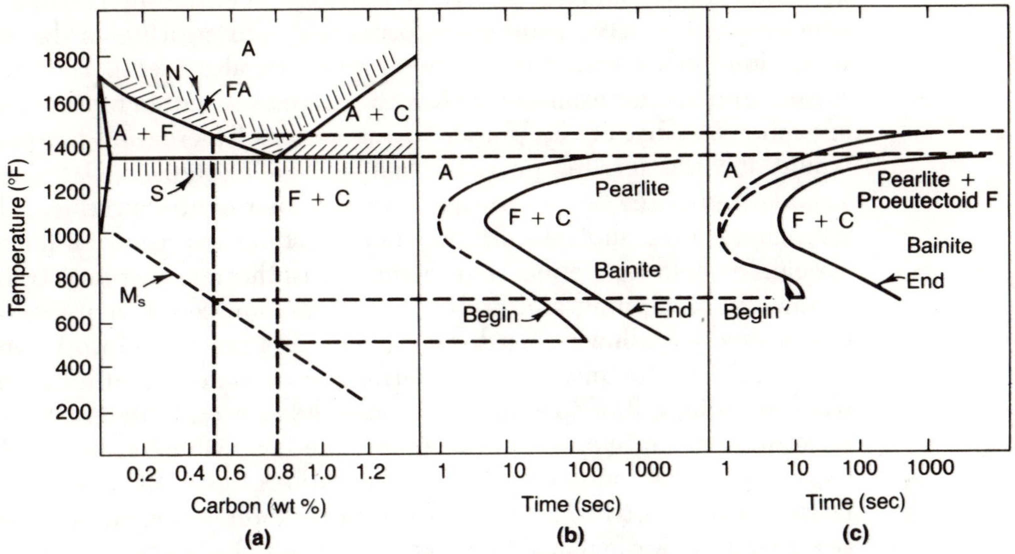

Fig. 4.1. Relationship to Fe-C diagram (a) of IT diagrams of eutectoid steel (b) and steel containing 0.5% carbon (c). The regions identified as N, FA and S in (a) are temperature ranges for normalizing, full annealing, and spheroidizing heat treatments, respectively, as discussed in Chapter 5. (Ref 4.14)

quenched to room temperature. The specimens held for the shortest times will transform completely to martensite on cooling because there is insufficient time at the hold temperature for any diffusion-controlled transformation. The austenite in specimens held for longer periods of time would transform to ferrite, cementite, pearlite and/or bainite, depending on temperature and the composition of the steel. The detection of the first small amounts of these phases in specimens largely transformed to martensite establishes the time for the beginning of transformation at a given temperature. With longer holding times at the transformation temperature, more and more of the austenite transforms to ferrite, cementite or mixtures of ferrite and cementite, and less of the specimen is martensitic after quenching to room temperature. Finally, after holding for a sufficiently long period of time at temperature, transformation of the austenite is complete prior to quenching, and the time for the end of transformation is established. When the process is repeated for a number of temperatures, the complete IT diagram is established.

Although metallographic examination of specimens isothermally held for various times is the most accurate method of determining IT diagrams (particularly with respect to differentiating regions of proeutectoid ferrite, cementite formation, and pearlite or bainite formation), other experimental techniques are also useful. Hardness measurements, for example, reflect the phases present in transformed specimens. A list of the phases in the order of increasing hardness would include ferrite, pearlite, bainite and martensite. Hardness would, therefore, be a maximum for microstructures produced by quenching after short isothermal holding times to a minimum for specimens held long enough for complete isothermal austenite transformation. Beginning and end of transformation could therefore be established by following hardness changes as a function of isothermal holding time. Dilatometry, an experimental technique that measures changes in length of specimens, has also been used to determine IT diagrams. The application of this technique is possible because of the expansion that accompanies the transformation of austenite to ferrite or ferrite-carbide mixtures, as discussed in Chapter 1. Dilatometry has been used by German investigators (Ref 4.2) for IT diagram determination. By cross checking with metallographic examination, dilatometry has been found to indicate the beginning of transformation after about 3% of the austenite has been transformed, as compared to the ability of microstructural examination to reveal the first 1% of austenite transformation. Figure 4.2 compares IT diagrams determined by dilatometry and metallography and shows the greater sensitivity of the latter technique for IT diagram determination.

Continuous Cooling Transformation Diagrams

Many of the heat treatments performed on steel are carried out by continuous cooling rather than by isothermal holding, and as a result diagrams that represent the transformation of austenite on cooling at various rates have been developed. The latter type of diagram for a given steel is referred to as a continuous cooling (CC) diagram or cooling transformation (CT) diagram. Generally, continuous cooling shifts the beginning of austenite transformation to lower temperatures and longer times. Figure 4.3 shows a derived (i.e., not experimentally determined) CT diagram for eutectoid steel and its relationship to the IT diagram (Ref 4.1 and 4.4). Also shown in the top part of Figure 4.3 is a Jominy specimen. The latter specimen is water quenched only at one end, and therefore the cooling rate is a maximum at that end and drops with increasing distance into the specimen. The cooling rates at various locations of a Jominy specimen have been measured by attachment of

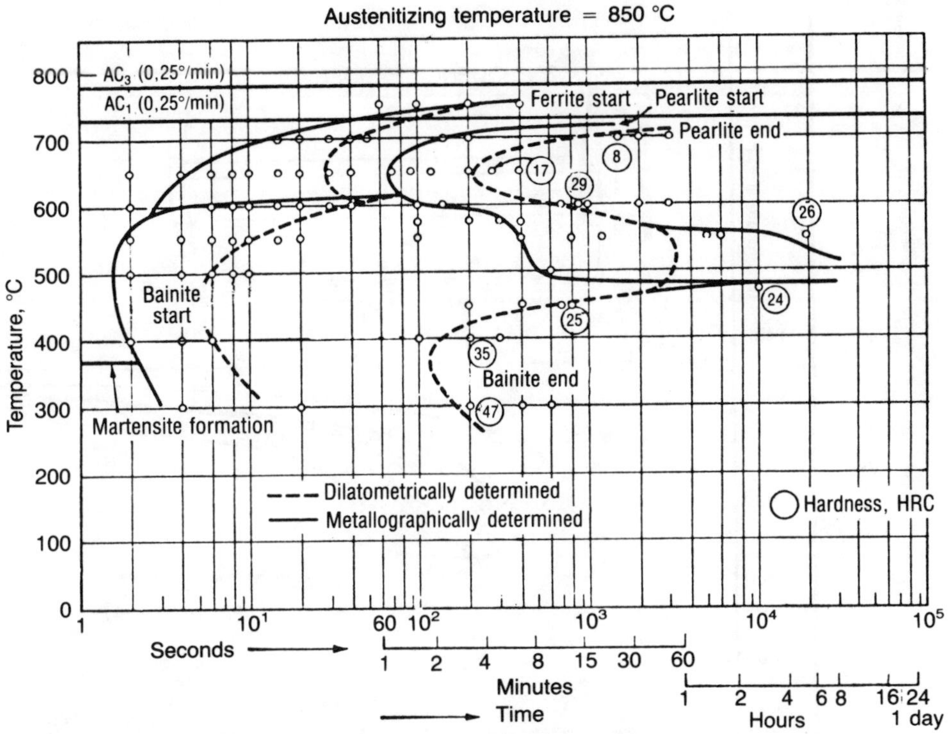

Fig. 4.2. Comparison of IT diagram for steel with German designation 42 CrMo 4 (0.38% C, 0.99% Cr and 0.16% Mo) determined by dilatometry (dashed lines) and metallography (continuous lines). (Ref 4.2)

thermocouples, and four of these cooling rates have been superimposed on the lower part of Fig. 4.3. With decreasing cooling rate or increasing distance from the quenched end of the Jominy specimen, the austenite transforms to microstructures containing increasingly greater quantities of pearlite. The decreased hardness associated with the replacement of martensite by pearlite with decreasing cooling rate is also shown in the top part of Fig. 4.3.

In general, especially for hardenable alloy steels, attempts to derive CT diagrams from IT diagrams without experimental verification have proven unsatisfactory (Ref 4.3). For example, the bainite transformation range is dominated by pearlite formation in eutectoid steel and has not been included in the derived curve of Fig. 4.3. The following list of CT characteristics with no IT counterparts has been published (Ref 4.4):

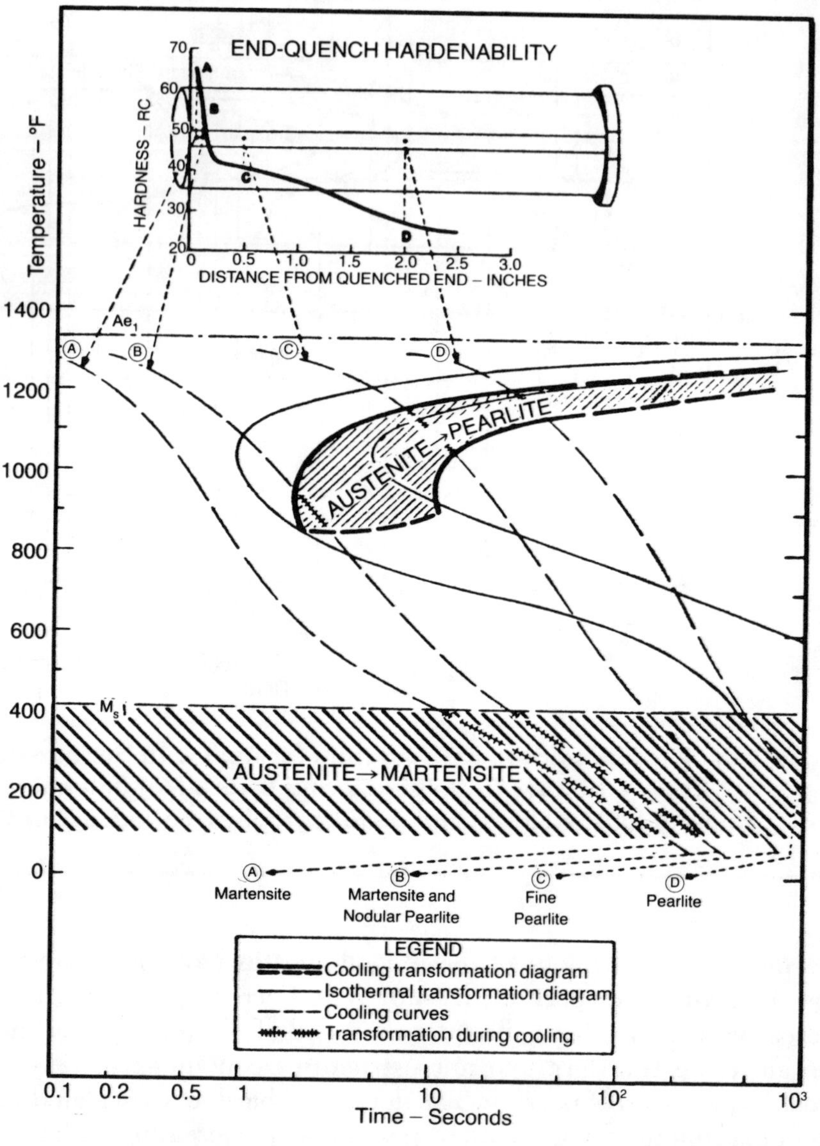

Fig. 4.3. Relationship of CT (heavy lines) and IT (light lines) diagrams of eutectoid steel. Also shown are Jominy end-quench specimen and four cooling rates from different positions on the specimen superimposed on the CT diagram. (Ref 4.4)

(a) the depression of the M_s temperature at slow cooling rates
(b) the tempering of martensite that takes place on cooling from the M_s temperature to about 204 °C (400 °F)
(c) the prevalence of bainite as a transformation product
(d) the extraordinary variety of microstructures encountered
(e) the unexpected occurrence of ferrite in a high-carbon steel such as AISI 52100

The following comments expand on the observations in the above list. The depression of the M_s temperature with decreasing cooling rate in a given steel is due to the rejection of carbon into austenite as ferritic or bainitic structures form on cooling. The untransformed austenite therefore has higher carbon concentration and a lower M_s temperature, as discussed in Chapter 3. The tempering of martensite on cooling is referred to as autotempering and is most common in low-carbon steels with high M_s temperatures. The latter situation results in the presence of martensite over a large temperature range on cooling. During this period of the quench, carbon has sufficient mobility to form the carbides characteristic of tempered martensite. Bainite formation, item (c), is promoted by certain alloying elements, in particular molybdenum, and by the more rapid cooling rates that favor shear transformation over diffusion-controlled transformation. The complexity of microstructures is due to the increasing fineness and intermixing of the austenite transformation products as transformation proceeds at successively lower temperatures on cooling. Finally, proeutectoid ferrite is sometimes observed in high-carbon steels where normally proeutectoid cementite would be expected because not all of the carbides may be dissolved during austenitizing. As a result, some of the carbon is tied up in carbide particles, and the austenite has a lower than expected carbon content approaching that of a hypoeutectoid steel.

In addition to the above differences between IT and CT diagrams, frequently there is a gap noted in CT diagrams. This gap represents a temperature range where apparently no transformation occurs on cooling and may be due to carbon enrichment of austenite on cooling as high-temperature ferrite forms and/or changes in incubation times for pearlite and bainite nucleation on cooling (Ref 4.3).

As a result of the differences between isothermal and continuous cooling transformation, CT diagrams are determined primarily by experiment, although there is still some interest in calculating CT diagrams from IT diagrams (Ref 4.3). The use of quenching dilatometers, in which changes in length and temperature with time of a standard specimen are simultaneously recorded, is now well established as the major approach to experimental determination of CT diagrams. The changes in specimen length due to the expansion associated with austenite transformation can therefore be related to points on a series of cooling curves. Metallographic examination of the transformed specimens then establishes the microstructure produced by a given cooling sequence. Experimental and instrumental details of the dilatometric approach are given in published atlases of CT diagrams (Ref 4.2, 4.5), and Eldis (Ref 4.6) has critically reviewed the relationship of dilatometry to the construction of CT diagrams.

CT diagrams for alloy steels are more complicated than that shown in Fig. 4.3 for eutectoid steel. Figures 4.4 and 4.5 show IT and CT diagrams for SAE 4140 steel and 42 CrMo 4 steel determined by U. S. Steel and Max-Planck Institut für Eisenforschung investigators, respectively. The steels are quite comparable in composition and contain nominally 0.4% carbon, 1% chromium and 0.2% molybdenum as the major alloy additions. The CT diagram in Fig. 4.4 was derived from the IT diagram and that in Fig. 4.5 was experimentally determined. In the case of the 4140-type steel there is relatively good agreement between the two methods of CT diagram determination, and the diagrams show the dominance of ferrite and bainite formation at intermediate rates of cooling.

Figures 4.6 and 4.7 are CT diagrams that show a number of effects of alloying on cooling transformation. Two steels, containing 1.4% nickel, 0.36% silicon and 0.85% manganese and differing only in molybdenum content, are compared. The diagrams have been selected from an atlas that systematically characterizes the effects of molybdenum, chromium, nickel and silicon on CT diagrams of 0.4% carbon steels (Ref 4.5). The microstructures resulting from selected cooling curves from Fig. 4.6 and 4.7 are shown in Fig. 4.8 and 4.9, respectively. Each cooling curve and microstructure is identified by the DPH hardness of the microstructure produced by that cooling sequence.

Figures 4.6 and 4.8 show that nickel depresses the Ac_3 and Ac_1 temperatures in accord with its role as an austenite stabilizer in steels and increases hardenability (i.e., the ability to form martensite on cooling) primarily by shifting the proeutectoid and pearlite transformation to longer time periods. Although the austenite-ferrite and austenite-pearlite regions are not differentiated in Fig. 4.6, the microstructures in Fig. 4.8 show that equiaxed proeutectoid ferrite and pearlite are the transformation products for continuous cooling that produces DPH hardnesses of 219, 210 and 185 [see Fig. 4.8(b), (c) and (d), respectively]. Kirkaldy (Ref 4.7) has attributed the improve-

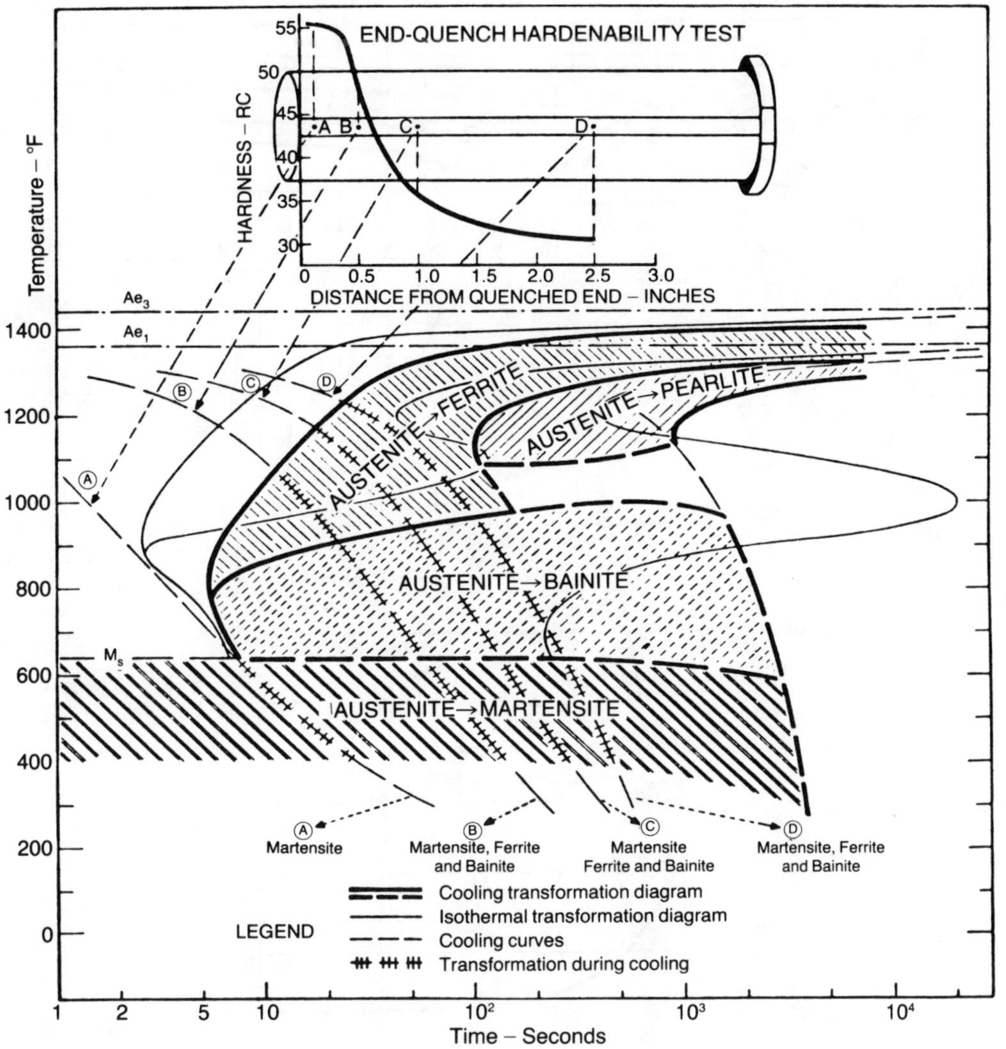

Fig. 4.4. CT diagram (heavy lines) for 4140 steel. Also shown are Jominy end-quench data and IT diagram (light lines). (Ref 4.4)

ment in hardenability due to elements such as nickel, copper and manganese to the lowering of the transformation temperatures and the attendant lower rates of diffusion.

Figures 4.7 and 4.9 show that the addition of about 0.5% molybdenum to the 1.4% nickel steel produces significant changes in cooling

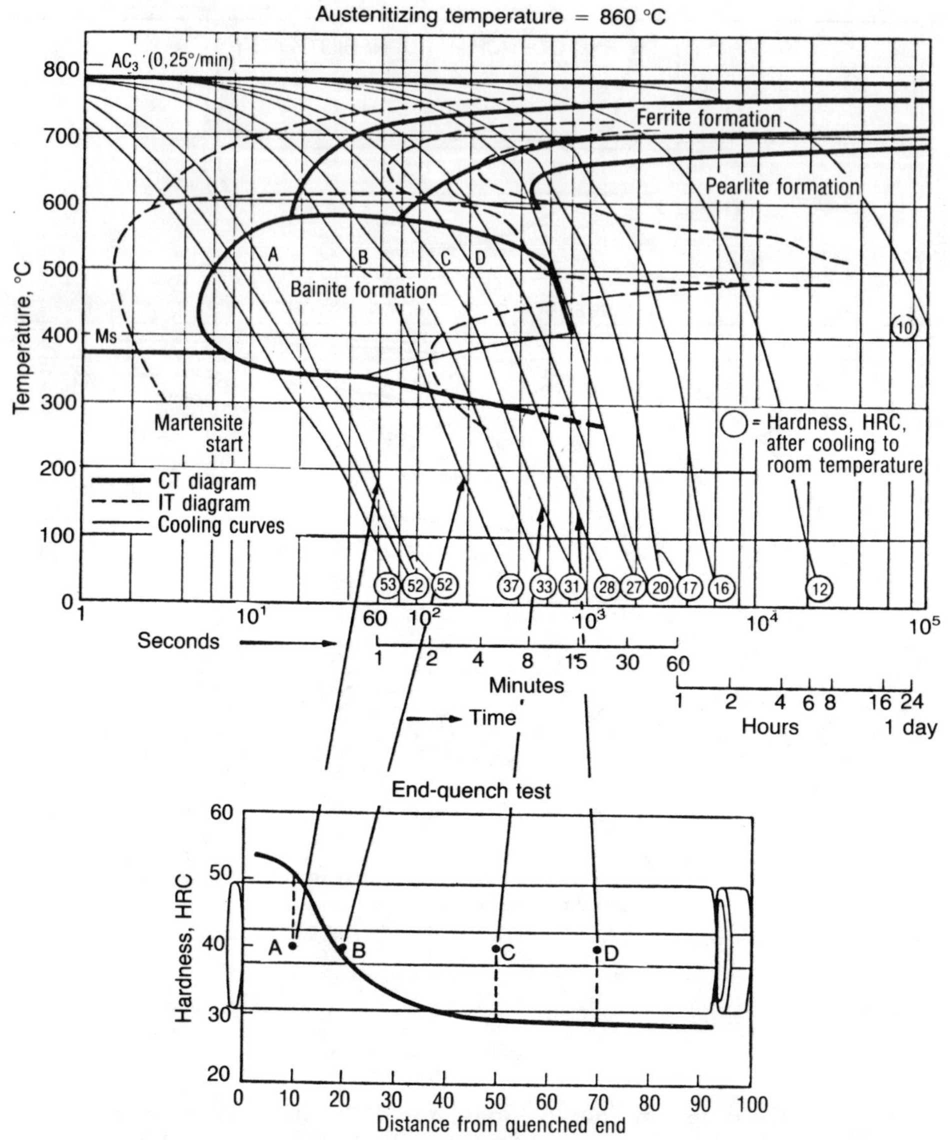

Fig. 4.5. Experimentally determined CT diagram (continuous lines) for steel with German designation 42 CrMo 4 (0.38% C, 0.99% Cr and 0.16% Mo) for comparison with that derived for a similar steel, 4140, as shown in Fig. 4.4. IT diagram is also shown (dashed lines). (Ref 4.2)

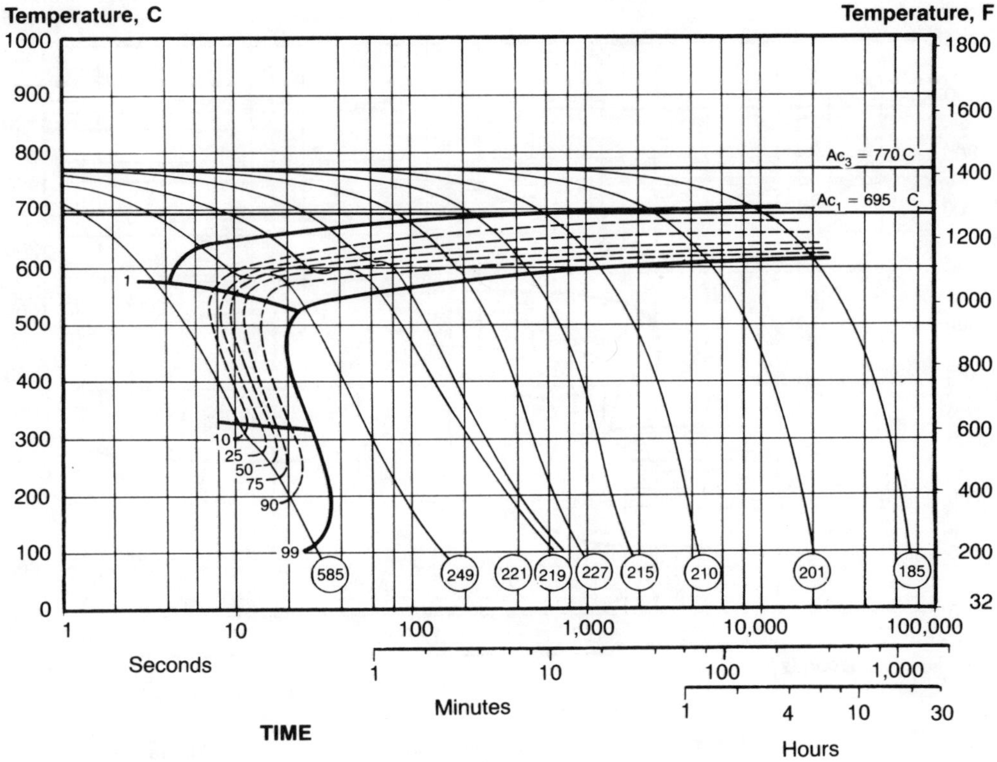

Fig. 4.6. CT diagram for a steel containing 0.37% C, 0.36% Si, 0.85% Mn, 1.44% Ni and 0.02% Mo. The steel was austenitized at 800 °C (1470 °F) for 20 min. The circled numbers correspond to DPH hardness of microstructures produced by cooling at the rates shown. (Ref 4.5)

transformation characteristics and microstructure. Hardenability is greatly improved, pearlite and equiaxed proeutectoid ferrite formation is severely retarded, and the bainite transformation becomes quite prominent. The gap that sometimes forms between two mechanisms of transformation is also apparent. The strong effect of molybdenum and similar ferrite stabilizers such as chromium and silicon has been attributed (Ref 4.7) to the fact that molybdenum must diffuse or partition during pearlite formation. Since molybdenum diffuses very sluggishly below Ae$_1$, the pearlite transformation is significantly retarded. Ferrite formation by a shear mechanism, on the other hand, requires no such partitioning of substitutional elements, and as a result the lower

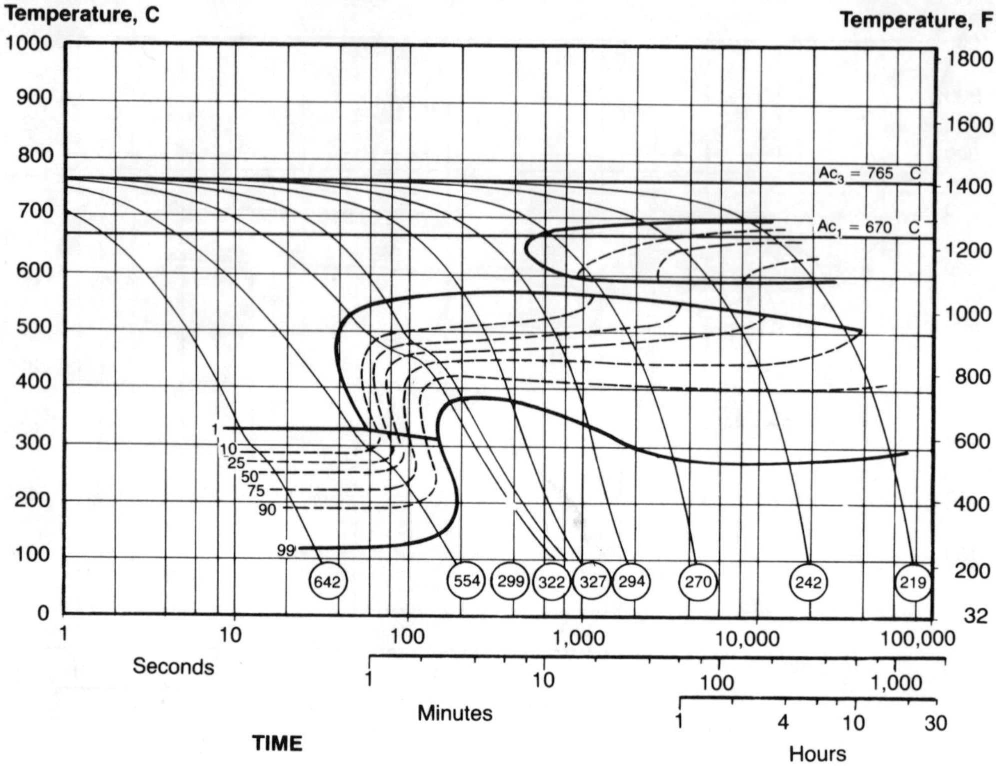

Fig. 4.7. CT diagram for a steel containing 0.37% C, 0.36% Si, 0.84% Mn, 1.40% Ni and 0.47% Mo. The steel was austenitized at 795 °C (1465 °F) for 70 min. The circled numbers correspond to DPH hardness of microstructures produced by cooling at the rates shown. (Ref 4.5)

nose for Widmanstätten ferrite and bainite (which is nucleated by ferrite) are prominent features of the CT diagram for the steel containing molybdenum. Of course, the excellent hardenability shown in Fig. 4.7 is due to the combination of both the nickel and molybdenum alloying effects.

Figure 4.6 through 4.9 show other examples of the characteristics of continuous cooling transformation discussed earlier. The M_s temperatures are slightly lowered once some bainite has formed and the microstructures, especially those produced in the Ni-Mo steel (see Fig. 4.9) are quite complex. Figures 4.6 and 4.7 also show the effects of the evolution of the heat of formation of the austenite transformation products on the cooling curves. This phenomenon is referred to as

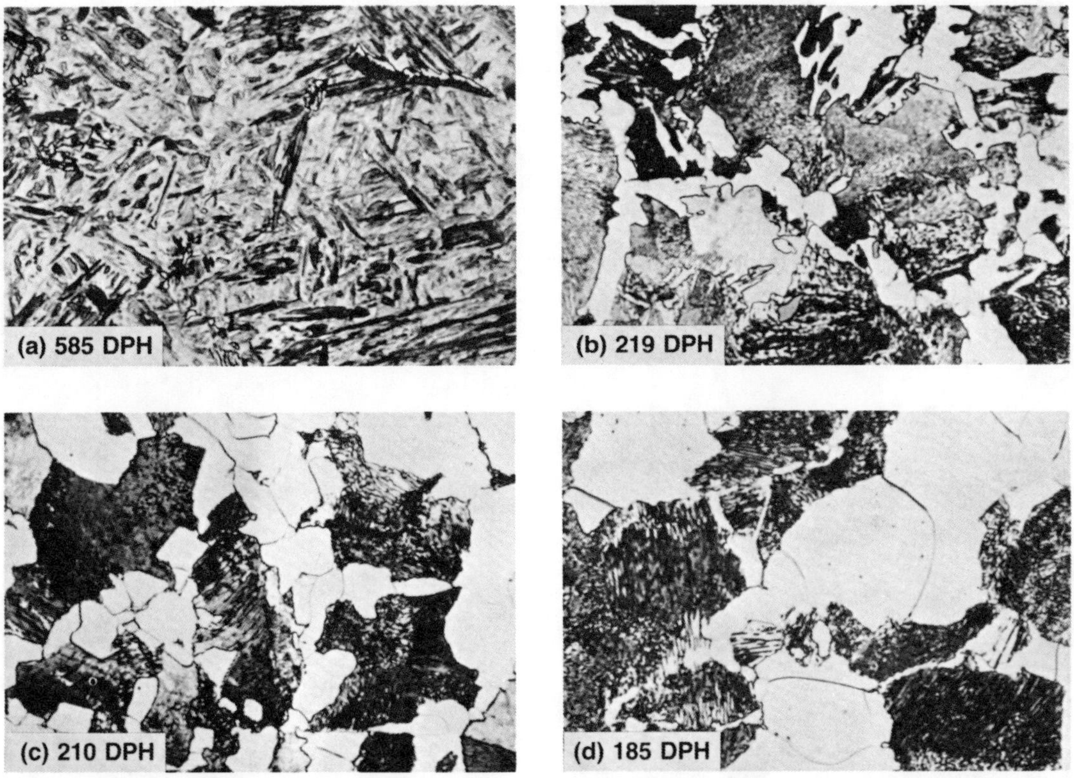

Fig. 4.8. Microstructures produced by cooling steel shown in Fig. 4.6 at four rates as identified by DPH hardness in Fig. 4.6. 2% nital etch. Magnification, 1000×; shown here at 75%. (Ref 4.5)

"recalescence" and causes changes in slopes of the cooling curves and sometimes even temperature increases on transformation. Figures 4.6 and 4.7 show that recalescence is most prominent at high and intermediate cooling rates; at slow cooling rates the specimen has sufficient time to dissipate to its surroundings the heat generated by transformation. Also, measurable recalescent effects lag the dilatometric detection of transformation and therefore are not suitable for accurate CT diagram determination.

Continuous Cooling Transformation and Bar Diameter

An atlas that presents continuous cooling transformation as a function of bar diameter rather than time has recently been prepared by the British Steel Corporation (Ref 4.8). This atlas is especially valuable to the heat treater because it permits an estimate of the microstructure

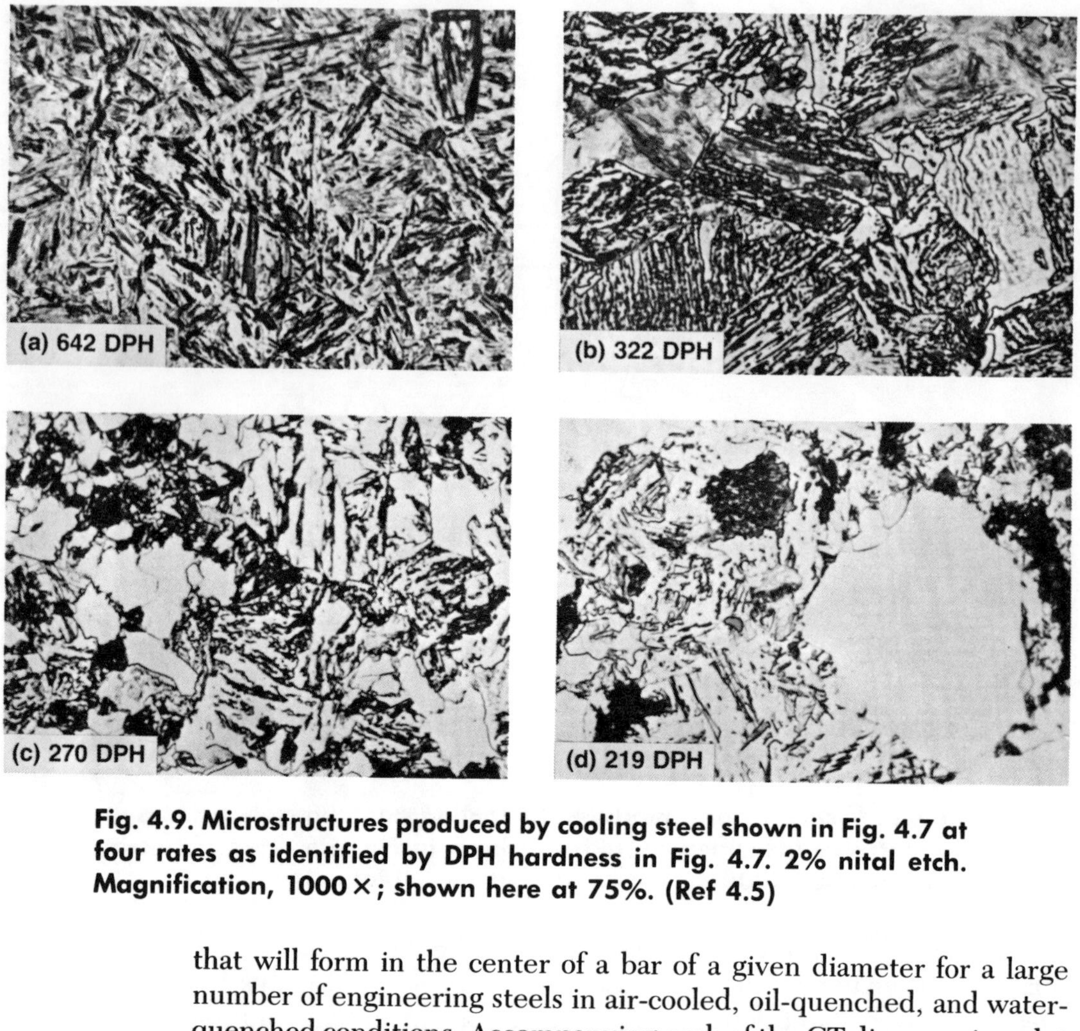

Fig. 4.9. Microstructures produced by cooling steel shown in Fig. 4.7 at four rates as identified by DPH hardness in Fig. 4.7. 2% nital etch. Magnification, 1000×; shown here at 75%. (Ref 4.5)

that will form in the center of a bar of a given diameter for a large number of engineering steels in air-cooled, oil-quenched, and water-quenched conditions. Accompanying each of the CT diagrams is a plot of hardness as related to the bar diameters in the as-cooled condition and sometimes the hardness after tempering. The bar diameter characterization is essentially a representation of hardenability, a subject that is developed more fully in Chapter 6 but is included here because of its direct relationship to continuous cooling transformations.

Figure 4.10 shows the CT diagram for a plain carbon steel containing 0.38% carbon, 0.20% silicon, and 0.70% manganese. The abscissa is plotted as bar diameter associated with air cooling, oil quenching, and water quenching. Vertical lines associated with a given diameter show the microstructures to be expected in the center of a bar of that diameter. For example, the vertical dashed line identified as "Air Cool" shows that a microstructure of ferrite, pearlite and a small amount of bainite is expected in a 10-mm diameter bar that has been air

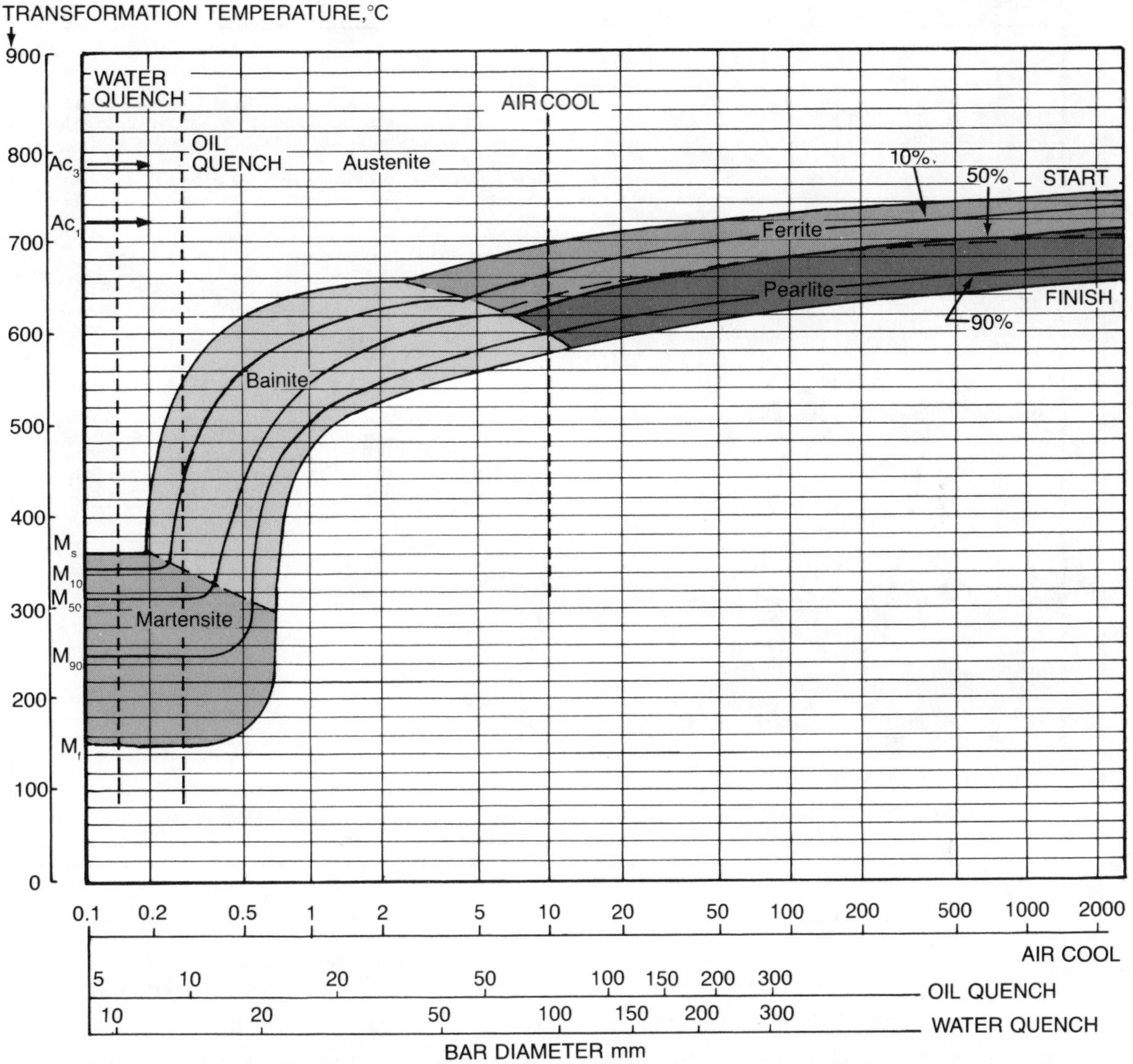

Fig. 4.10. CT diagram for plain carbon steel containing 0.38% C and 0.70% Mn. Transformation and microstructures are plotted as a function of bar diameter. (Ref 4.8)

cooled. Likewise, the vertical dashed lines marked "Water Quench" and "Oil Quench" indicate that martensite and bainite plus martensite, respectively, would be expected for 10-mm diameter bars quenched in the two different media. Figure 4.11 shows the CT diagram for a more highly alloyed 0.40% carbon steel. The diagram shows that a 10-mm bar of this steel, even if air cooled, would be entirely martensitic

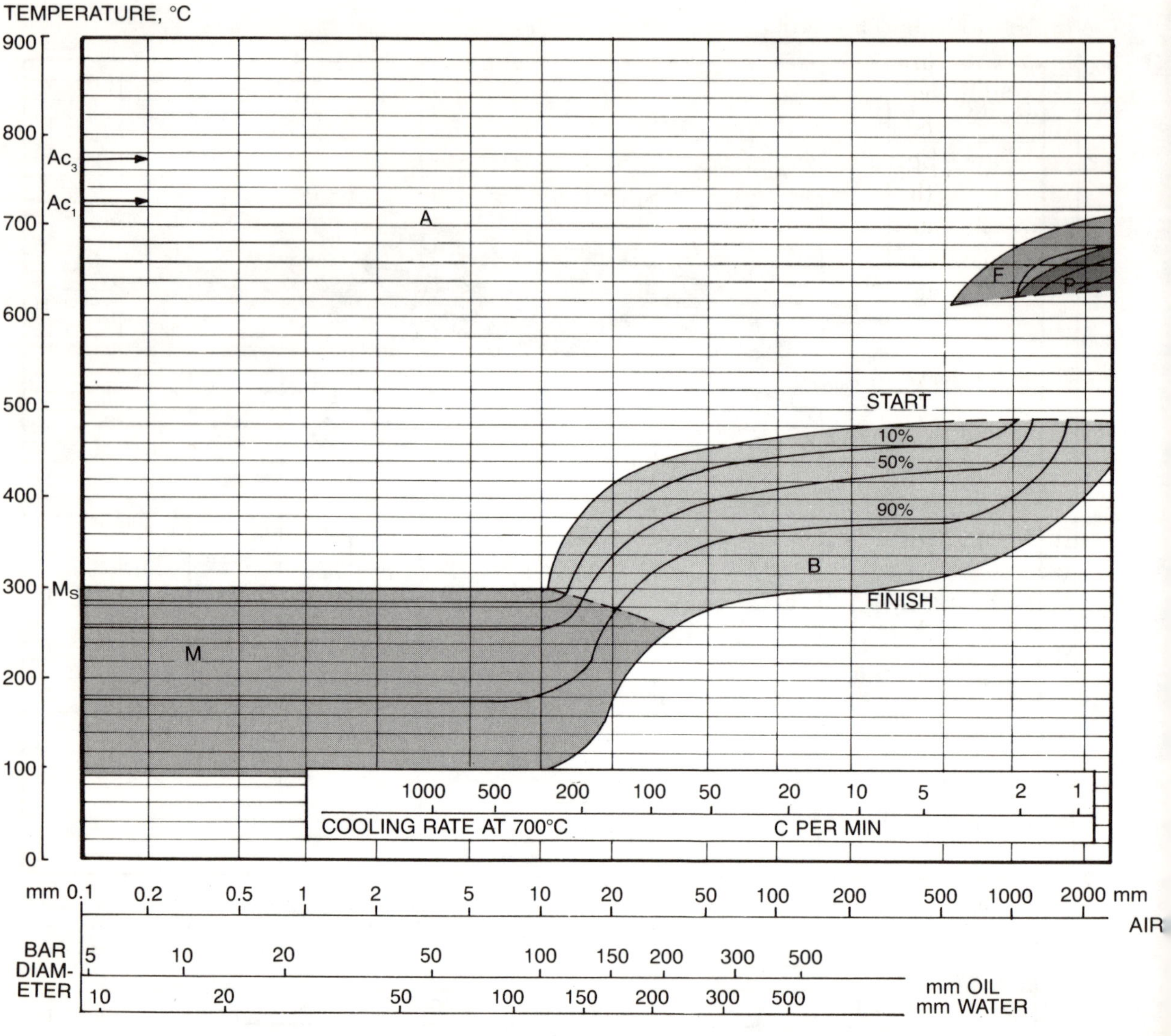

Fig. 4.11. CT diagram for an alloy steel with 0.40% C, 1.50% Ni, 1.20% Cr and 0.30% Mo, plotted as a function of bar diameter. Steel was austenitized at 850 °C (1562 °F); previous treatment: rolling, then softening at 650 °C (1202 °F). (Ref 4.8)

and that oil-quenched bars up to 100 mm in diameter would be fully hardened.

The CT diagrams relating microstructure to bar diameter, therefore, permit a direct assessment not only of the possibility of producing maximum hardness in a bar of a given diameter, but also of the ability to produce air-cooled or normalized structures with ferrite-pearlite microstructures of low hardness. It should be remembered that the microstructures are those that would be present at the center of the bars, and that microstructure and hardness gradients may exist between the center and surface of the bars because of cooling rate variations between those points. Some variation in microstructures from those represented in the CT diagrams must also be expected because of chemistry variations within specification limits of a given grade of steel, variations in austenitizing, and/or different degrees of agitation during water or oil quenching.

Summary

A wealth of information characterizing the isothermal (Ref 4.1, 4.4, 4.9, 4.10) and continuous cooling (Ref 4.2, 4.4, 4.5, 4.8, 4.10 to 4.13) transformation behavior of a large number of steels is now available in the form of atlases. There is a strong trend to the use of CT diagrams because they better represent the large number of heat treatments that are based directly on continuous cooling. The latter trend is in turn related to the ready availability of CT diagrams, a situation that has been significantly aided by the development of experimentally convenient and accurate dilatometric techniques for CT diagram determination.

References

4.1 *Atlas of Isothermal Transformation Diagrams*, 2nd Ed., United States Steel Corp., Pittsburgh, 1951

4.2 *Atlas zur Wärmebehandlung der Stähle*, Vol 1-4, Max-Planck-Institut für Eisenforschung, in cooperation with the Vereins Deutscher Eisenhüttenleute, Verlag Stahleisen M.B.H., Düsseldorf, 1954-1976

4.3 A. K. Cavanagh, *Met Trans A*, Vol 10A, 1979, p 129-132

4.4 *Atlas of Isothermal Transformation and Cooling Transformation Diagrams*, American Society for Metals, Metals Park, Oh., 1977

4.5 W. W. Cias, *Phase Transformation Kinetics and Hardenability of Medium-Carbon Alloy Steels*, Climax Molybdenum Co., Greenwich, Conn., 1972

4.6 G. T. Eldis, A Critical Review of Data Sources for Isothermal and Continuous Transformation Diagrams, in *Hardenability Concepts with Applications to Steel*, AIME, Warrendale, Pa., 1978, p 126-153

4.7 J. S. Kirkaldy, Prediction of Alloy Hardenability from Thermodynamic and Kinetic Data, *Met Trans*, Vol 4, 1973, p 2327-2333

4.8 M. Atkins, *Atlas of Continuous Cooling Transformation Diagrams for Engineering Steels*, British Steel Corp., Sheffield, 1977; revised U.S. edition published by American Society for Metals, Metals Park, Oh., 1980

4.9 *Supplement to Atlas of Isothermal Transformation Diagrams*, United States Steel Corp., Pittsburgh, 1953

4.10 A. Schrader and A. Rose, Structure of Steels, Vol 2, *De Ferri Metallographia*, Verlag Stahleisen MBH, Düsseldorf, 1966

4.11 *Transformation Diagrams of Steels Made in France*, Vol 1-4, I.R.S.I.D., St. Germaine-en-Laye, 1953-1960

4.12 M. Economopoulos, L. Harbraleen and N. Lambert, *Transformation Diagrams of Steels Made in Benelux Countries*, C.N.R.M., Brussels, 1967

4.13 A. A. Popov and L. E. Popova, *Isothermal and Thermokinetic Diagrams of the Breakdown of Supercooled Austenite*, Metalurgiya, Moscow, 1965

4.14 G. Krauss and J. F. Libsch, *Phase Diagrams in Ceramic, Glass and Metal Technology*, A. M. Alper (Ed.), Academic Press, New York, 1970

CHAPTER 5

Heat Treatments To Produce Ferrite and Pearlite

This chapter describes heat treatments such as full annealing, normalizing and spheroidizing that have been developed to produce uniformity in microstructure, improve ductility, reduce residual stresses, and/or improve machinability of steels. The microstructures that are produced by these heat treatments consist of various distributions of ferrite and cementite, and therefore are produced by relatively long holding times at temperature and slow cooling rates. The latter conditions permit the diffusion-controlled formation of desirable ferrite and cementite microstructures. The heat treatments are described in terms of the Fe-C diagram and the various transformation diagrams discussed in earlier chapters. Relationships for the mechanical properties of mixtures of ferrite and pearlite as influenced by steel composition are presented in the last section of this chapter.

Full Annealing and Homogenizing

The term annealing has been used in its broadest sense to refer to any heat treatment that has as its objective the development of a nonmartensitic microstructure of low hardness and high ductility. This understanding of annealing is much too broad, however, and a number of more specific annealing heat treatments have been developed and defined. Full annealing is a heat treatment accomplished by heating steels into the single-phase austenite field and slowly cooling, usually in a furnace, through the critical transformation ranges. When the term annealing is used without an adjective in reference to carbon steels, full annealing is the implied heat treatment practice (Ref 5.1).

Figure 5.1 shows the temperature ranges for several heat treatments involving austenitizing superimposed on the Fe-C diagram. As shown, the temperature for full annealing is a function of the carbon content of the steel, staying just above the A_3 temperature for hypoeutectoid steels and above the A_1 for hypereutectoid steels. The critical temperatures will vary somewhat with the alloy content of the steel, but the objective of heating into the single-phase austenite field for low- and medium-carbon steels and into the austenite-cementite field for high-carbon steels remains the same no matter what the steel composition.

The reason for heating the hypereutectoid steels in the two-phase field is to agglomerate or spheroidize the proeutectoid cementite. If such steels are heated above A_{cm}, proeutectoid cementite would form on slow cooling at the austenite grain boundaries, as discussed in Chapter 2. The resulting network of carbides on the austenitic grain boundaries provides an easy fracture path and renders the steel brittle to forming or service stresses. Figure 5.2(a) shows a carbide network developed in SAE 52100 steel, a high-carbon bearing steel containing nominally 1% carbon and 1.5% chromium. Figure 5.2(b) shows how fracture produced by impact loading has followed the carbide network along prior austenite grain boundaries in a microstructure similar to that shown in Fig. 5.2(a). In Fig. 5.2(b), the steel has been hardened by quenching from the austenite-cementite field and martensite coexists with the carbide network. A carbide network formed on slow cooling from above A_{cm} in a 52100 steel is shown in Fig. 5.3(a). Pearlite instead of martensite has formed within the austenite grains. The object of full annealing high-carbon steels in the austenite-carbide field, then, is to break up such continuous carbide networks by agglomeration into separated, spherical carbide particles. The driving force for this process is the reduction in austenite/cementite interface area and thus the reduction in interfacial energy that accompanies spheroidization. Figure 5.3(b) shows the partial spheroidization of a cementite network. Although the structure was formed during austenitizing for hardening, the austenitizing temperature ranges for hardening and full annealing are identical in high-carbon steels.

Not only is the temperature range of heating an important part of full annealing, but the slow cooling rate associated with full annealing is also a vital part of the process. Figure 5.4 compares schematic temperature-time schedules for full annealing and the normalizing

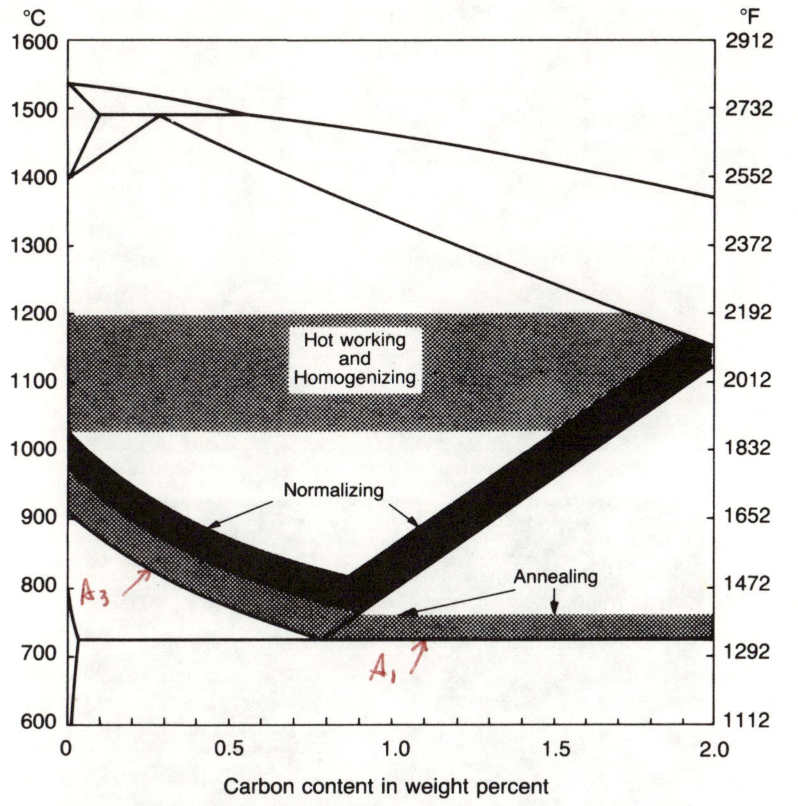

Fig. 5.1. Portion of the Fe-C diagram with temperature ranges for full annealing, normalizing, hot working and homogenizing indicated. (Courtesy of M. D. Geib, Colorado School of Mines, Golden, Colo.)

heat treatments discussed in the next section of this chapter. The cooling rates are superimposed on a schematic CT diagram for a hypoeutectoid steel. The slow cooling rates characteristic of furnace cooling insure that the austenite transforms first to proeutectoid ferrite and then to pearlite at temperatures approaching the equilibrium A_3 and A_1 temperatures. As a result the ferrite will be equiaxed and relatively coarse-grained, and the pearlite will have a coarse interlamellar spacing. The latter microstructural characteristics lower hardness and strength and increase ductility, the major objectives of the full annealing treatment. Once the austenite has fully transformed to ferrite and pearlite, the cooling rate could be increased to reduce the time of annealing and thereby improve productivity. A number of

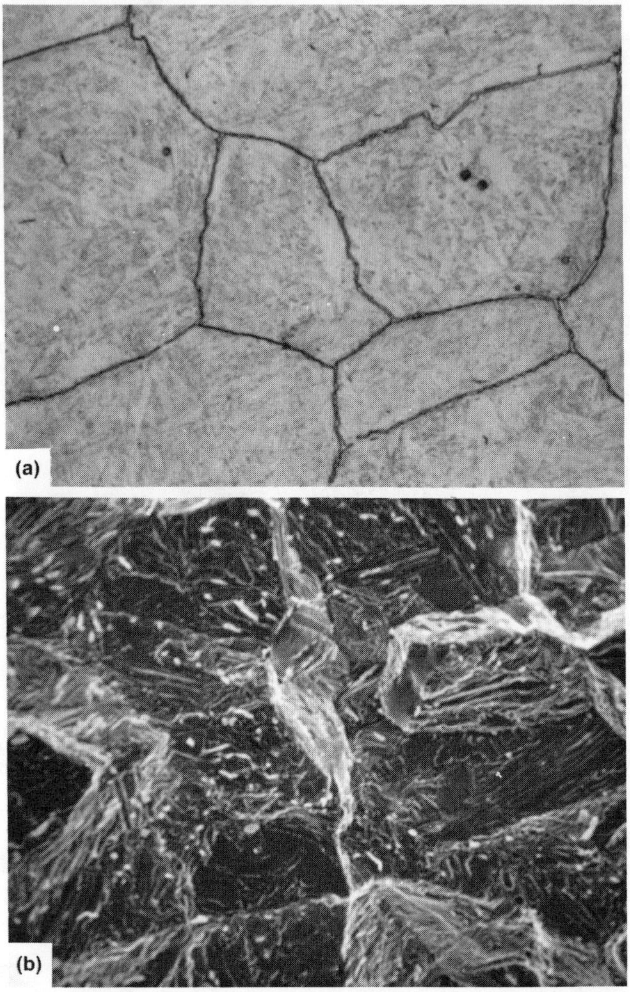

Fig. 5.2. (a) Carbide network at prior austenite grain boundaries in 52100 steel. Light micrograph. Nital etch. Magnification, 600×; shown here at 75%. (b) Fracture along grain boundary carbides in 52100 steel. Scanning electron micrograph. Magnification, 415×; shown here at 75%. (Courtesy of T. Ando, Colorado School of Mines, Golden, Colo.)

additional rules for developing optimum full annealing practices and properties are given in Ref 5.1.

Figure 5.1 also shows the temperature range for homogenizing, a type of annealing treatment usually performed in earlier stages of steel

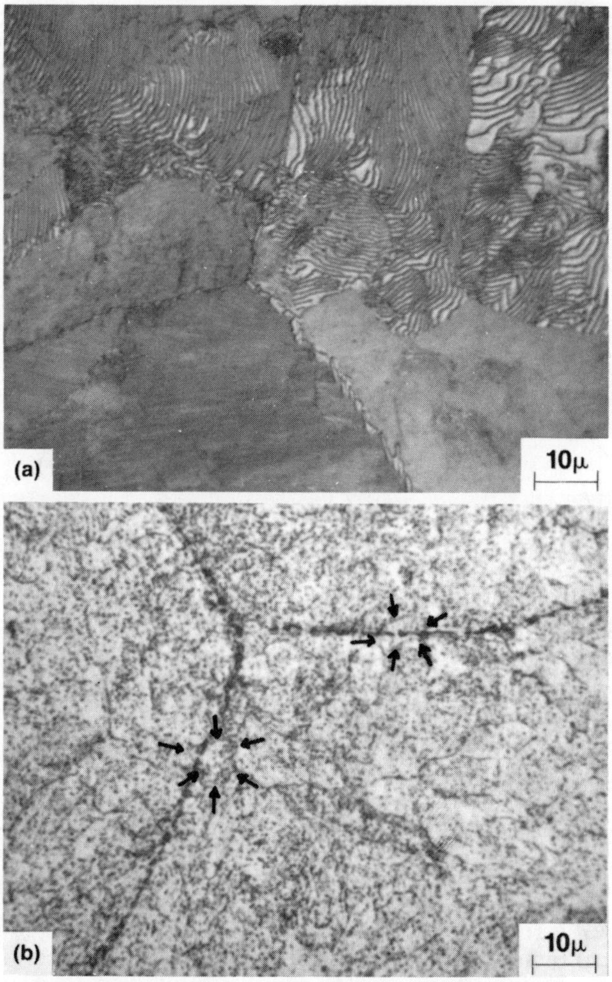

Fig. 5.3. (a) Proeutectoid cementite network in normalized 52100 steel. (b) Residual cementite network after austenitizing structure in (a) at 850 °C (1562 °F) for hardening. Very fine particles are from spheroidization of cementite in original pearlite matrix. Arrows point to fine austenite grains that have formed on austenitizing. (Ref 5.2)

processing prior to hot rolling or forging, working operations that are also performed in the same temperature range. Homogenizing is performed at high temperatures in the austenite phase field to speed the diffusion-controlled reduction of segregation or chemical concen-

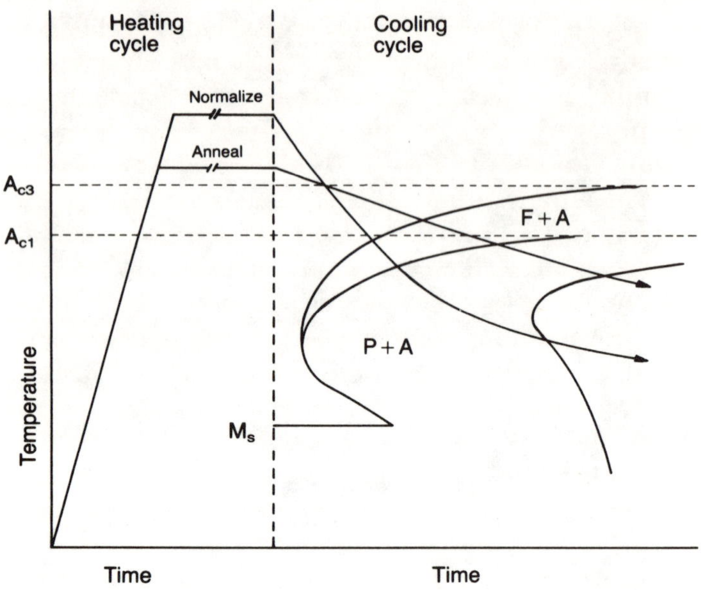

Fig. 5.4. Schematic time-temperature cycles for normalizing and full annealing. The slower cooling of annealing results in higher temperature transformation to ferrite and pearlite and coarser microstructures than does normalizing. (Courtesy of M. D. Geib, Colorado School of Mines, Golden, Colo.)

tration gradients that are produced by ingot solidification. In addition, second phases such as carbides are dissolved as fully as possible. The resulting uniformity or homogeneity of the austenite not only improves hot workability but also contributes to uniformity in the response of a steel to subsequent annealing or hardening operations.

Normalizing

Normalizing is a heat treatment which, similar to full annealing, produces a uniform microstructure of ferrite and pearlite. There are, however, several important differences between normalizing and annealing. Normalizing in hypoeutectoid steels is performed at temperatures somewhat higher than those used for annealing, while in hypereutectoid steels the heating temperature range is above the A_{cm} (see Fig. 5.1). In normalizing, heating is followed by air cooling, in contrast to the slower furnace cooling of full annealing.

The somewhat higher austenitizing temperatures used for normalizing, as compared to those used for annealing hypoeutectoid steels, in effect produce greater uniformity in austenitic structure and composition similar to a homogenizing treatment, although at a much lower temperature and for shorter times than those used for homogenizing. Another of the major objectives of normalizing is to refine the grain size that frequently becomes very coarse during hot working at high temperatures or that is present in as-solidified steel castings. As such hot worked or cast products are heated through the Ac_1 and Ac_3 temperatures, new austenite grains are nucleated and, if the austenitizing temperature is limited to the range shown in Fig. 5.1, a uniform fine-grained austenitic structure is produced. Exceeding the indicated temperature range might result in excessive austenitic grain size, as discussed in Chapter 7. Normalizing, then, produces a uniform, fine-grained austenite grain structure that in hypoeutectoid steels transforms to ferrite-pearlite microstructures on air cooling. The resulting microstructure may have good uniformity and desirable mechanical properties for a given application or may be reaustenitized for final hardening by quenching to martensite.

In hypereutectoid steels, normalizing is performed above the A_{cm} not only to refine austenitic grain size but also to dissolve carbides and carbide networks that may have developed during prior processing. The normalized structures that result respond more readily to the spheroidizing treatments for good machinability as described below and/or provide better response to a subsequent and final hardening heat treatment. There is the possibility that continuous carbide networks may develop on cooling from a normalizing temperature above A_{cm}, and that as a result a somewhat brittle normalized microstructure might develop. On subsequent austenitizing for hardening, however, the network carbides agglomerate or spheroidize somewhat (see Fig. 5.3(b), and fracture toughness is in fact improved relative to a microstructure without the partially spheroidized network (Ref 5.2).

The air-cooling step of a normalizing treatment produces subtle but significant differences in microstructures compared to those produced by full annealing. Figure 5.4 shows schematically that air cooling lowers the temperature range over which proeutectoid ferrite and pearlite form compared to the transformation range in full annealing. As a result, both the ferrite grain size and the pearlite interlamellar spacing are reduced compared to those in the same steel in the fully annealed condition. The finer microstructure of a normalized steel in turn has higher strength and hardness and slightly lower ductility than a fully annealed steel.

The actual mechanical properties of any normalized or annealed steel are determined by a number of factors, the most important being carbon content. The higher the carbon content, the more pearlite that forms and the higher the strength and hardness of the steel. Quantitative relationships for the contributions of carbon and other parameters to the mechanical properties of ferrite-pearlite steels are discussed in a later section of this chapter.

It is also important to realize that the air cooling associated with a normalizing heat treatment produces a range of cooling rates depending on the section size. Heavier sections air cool at much lower rates than do light sections because of the added time required for thermal conductivity to lower the temperature of central portions of the work piece. Two important consequences follow from the effect of section size on cooling rate. In very heavy sections, the surface may cool at significantly higher rates than the interior, thus producing residual stresses. In very light sections, especially in alloy hardenable steels, air cooling may actually be rapid enough to produce bainitic or martensitic microstructures instead of ferrite and pearlite. The British Steel Corporation atlas (Ref 5.3) which plots cooling transformation as a function of air-cooling section size (see Chapter 4) enables this effect to be evaluated. Other aspects of normalizing carbon steels are discussed in Ref 5.1.

Spheroidizing

The most ductile, softest condition of any steel is associated with a microstructure that consists of spherical carbide particles uniformly dispersed in a ferrite matrix. Figure 5.5 shows a spheroidized microstructure in an 0.66C-1Mn steel. The high ductility of such a microstructure is directly related to the continuous ductile ferrite matrix; pearlite (see Fig. 2.2) with its fine lamellar carbides separating the ferrite, more effectively hinders deformation and, therefore, increases hardness and lowers ductility compared to a spheroidized structure. The good ductility of spheroidized microstructures is extremely important for low- and medium-carbon steels that are cold formed, and the low hardness of spheroidized structures is important for high-carbon steels that undergo extensive machining prior to final hardening.

Spheroidized microstructures are the most stable microstructures found in steels and will form in any prior structure heated at temperatures high enough and times long enough to permit the diffusion-dependent development of the spherical carbide particles. As a result, there are many different heat treatment approaches for producing spheroidized microstructures. The slowest spheroidizing is associated

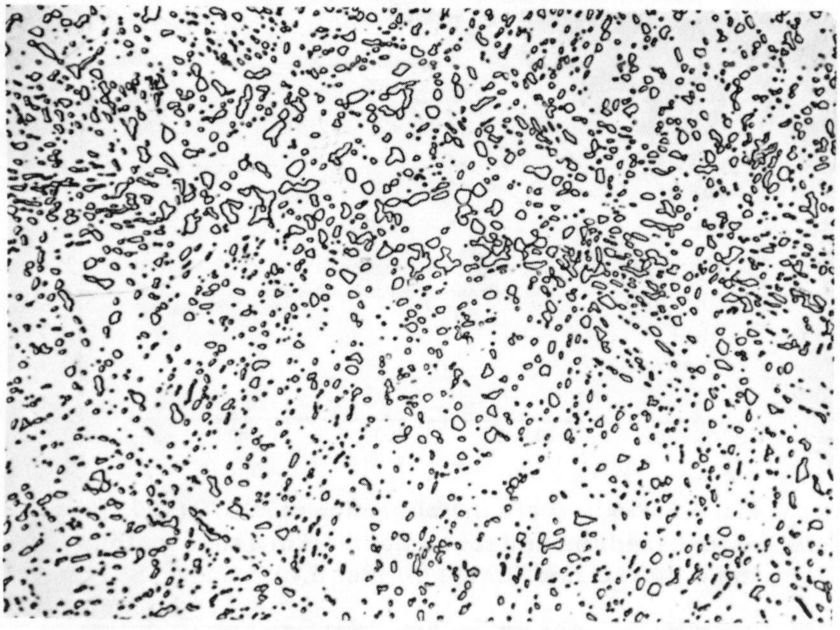

Fig. 5.5. Spheroidized microstructure in an Fe-0.66C-1Mn alloy formed by heating martensite at 704 °C (1300 °F) for 24 h. Picral etch. Magnification, 1000×. (Courtesy of A. R. Marder and A. Benscoter, Bethlehem Steel Corp., Bethlehem, Pa.)

with pearlitic microstructures, especially those with coarse interlamellar spacings. Figure 5.6 shows the percent of carbides that have spheroidized in fine to coarse pearlites produced by isothermally transforming an 0.74C-0.71Si steel between 700 and 580 °C (1292 and 1076 °F), followed by annealing at 700 °C (1292 °F) (Ref 5.4). Many hundreds of hours are required to spheroidize the pearlitic microstructures. Spheroidizing is more rapid if the carbides are initially in the form of discrete particles, as in bainite, and even more rapid if the starting structure is martensite. Spheroidizing of martensitic microstructures is most frequently performed on highly alloyed tool steels that form martensite on air cooling as shown schematically in Fig. 5.7.

Spheroidizing at rates much faster than those shown in Fig. 5.6 is accomplished by either complete or partial austenitizing, and then holding just below Ac_1, cooling very slowly through the Ac_1, or cycling above and below Ac_1 (Ref 5.1, 5.5). These temperature ranges for spheroidizing are shown in Fig. 5.8. It is important to limit the austenitizing temperature in order to retain a degree of heterogeneity in the

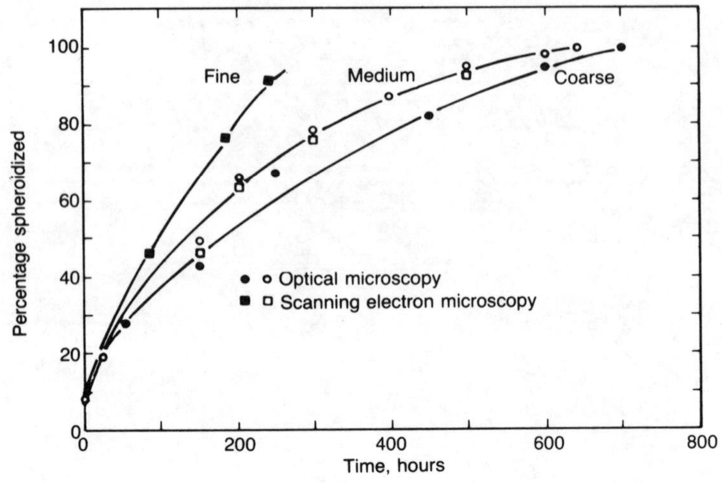

Fig. 5.6. Progress of spheroidization at 700 °C (1292 °F) of fine, medium and coarse pearlites in a steel containing 0.74% C and 0.71% Si. (Ref 5.4)

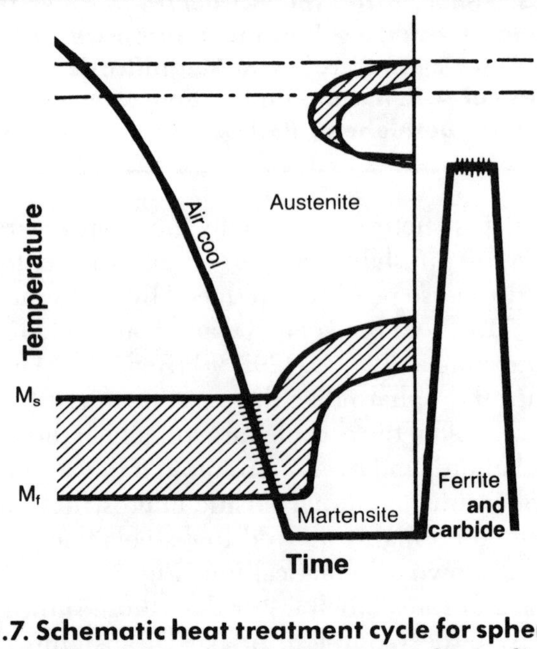

Fig. 5.7. Schematic heat treatment cycle for spheroidizing an air-hardening steel. Martensite forms first and then is tempered close to the Ac_1 to produce a spheroidized structure. (Ref 5.5)

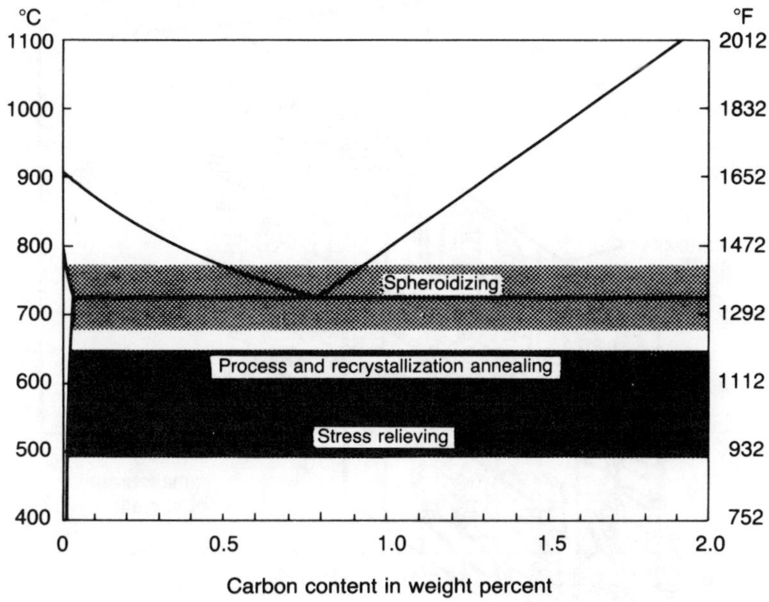

Fig. 5.8. Portion of the Fe-C diagram with temperature ranges for process annealing, recrystallization annealing, stress relieving and spheroidizing indicated. (Courtesy of M. D. Geib, Colorado School of Mines, Golden, Colo.)

austenite, especially since undissolved carbide particles appear to promote the transformation of the austenite to spheroidized microstructures. As noted earlier, homogenized austenite free of undissolved carbides as produced by normalizing or full annealing promotes the formation of pearlitic structures rather than spheroidized structures.

Spheroidized microstructures are stable because the ferrite is generally strain-free and because the spherical shape of the cementite particles is one of minimum interfacial area per unit volume of particle. Lamellar cementite particles, as present in pearlite, have a very large interfacial area per unit volume of particle and therefore high interfacial energy. In order to reduce the interfacial energy, cementite lamellae or plates break up into smaller particles that eventually assume spherical shapes. Figure 5.9 shows a representation of the breakup process of a single plate as determined by serial sectioning of a specimen annealed for 150 h at 700 °C (1292 °F) (Ref 5.4). Once the lamellae have broken up, the small spherical particles dissolve at the expense of the larger particles, again driven by the reduction in interfacial energy.

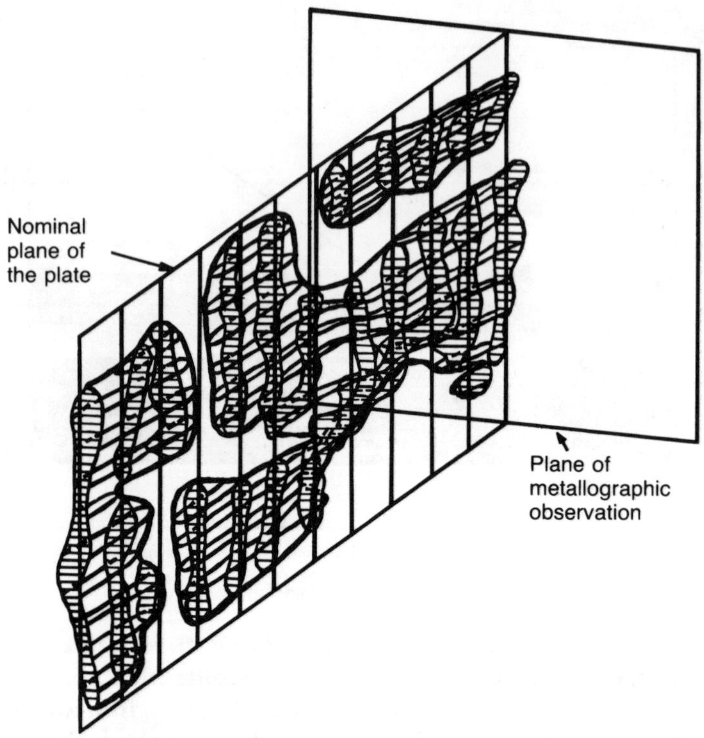

Fig. 5.9. Representation of a partially spheroidized cementite plate in coarse pearlite structure annealed for 150 h at 700 °C (1292 °F). (Ref 5.4)

The following equation (Ref 5.4, 5.7) describes the rate of coarsening of a spheroidized microstructure:

$$\frac{dr}{dr} = \frac{2\gamma\, V^{2}_{Fe_3C}\ X_c\ D_c^{eff}}{V_{Fe}\ RT_{r_1}}\left(\frac{1}{\bar{r}} - \frac{1}{r_1}\right) \qquad \text{(Eq 5.1)}$$

where γ is the interfacial energy; V_{Fe_3C} and V_{Fe} are the molar volumes of cementite and ferrite; X_c is the mole fraction of carbon in equilibrium with cementite in ferrite; D_c^{eff} is the effective carbon diffusion coefficient; R is the gas constant; T is the absolute temperature; r_1 is the radius of newly created particles; and $\bar{r}$ is the mean size of the already spheroidized particles. Equation 5.1 shows that the rate of spheroidization is directly related to the diffusion of carbon in ferrite and decreases as the average size of particles in a spheroidized microstructure increases. Alloying elements slow the rate of carbon diffusion in ferrite, and therefore the spheroidization process. Also if present, strong carbide-forming elements would have to diffuse for alloy carbide coarsening, and thus would greatly reduce the rate of spheroidization.

Process and Recrystallization Annealing

Process and recrystallization annealing are similar subcritical annealing treatments usually applied to restore ductility to cold worked steel products of a variety of shapes. Since these heat treatments are performed in the ferrite and cementite two-phase field of the Fe-C diagram (see Fig. 5.8), no phase transformation accompanies the microstructural changes produced by the treatments. Generally, the microstructure of the low- and medium-carbon steels prior to cold work is spheroidized or largely ferritic with small amounts of pearlite, both highly ductile microstructures. The ferrite in these microstructures is equiaxed and strain-free. Cold working deforms or work hardens the ferrite, tending to elongate the ferrite grains in the direction of working and introducing a high density of crystal imperfections within the grains. On heating, high strain energy of the deformed ferrite at first drives recovery, a mechanism by which some of the crystal imperfections are eliminated or rearranged into new configurations (Ref 5.8), and eventually drives recrystallization, a process where new, strain-free equiaxed grains nucleate and grow in the deformed ferrite (Ref 5.8). The end result of the subcritical annealing process is a restoration of the ductile, spheroidized microstructure which is again capable of undergoing significant cold deformation.

Figure 5.10 shows the effects of cold work and recrystallization annealing on a very low-carbon (0.003%) sheet steel (Ref 5.9). Because of the low carbon content no carbides are visible and the only second-phase particles present are oxide and sulfide inclusions. Figure 5.10(a) shows the deformed microstructure produced by a 60% reduction in sheet thickness by cold rolling. The elongated grain structure and etching effects associated with the deformation within the grains are clearly visible. Figure 5.10(b) shows the same steel after about 80% recrystallization has been produced by annealing at 538 °C (1000 °F) for 2 h. The recrystallized ferrite grains are quite fine and equiaxed, and etching shows no evidence of deformation.

A third stage of the annealing process, grain growth or coarsening has not yet developed in the example shown in Fig. 5.10(b). Prolonged heating beyond the point of recrystallization would cause such grain

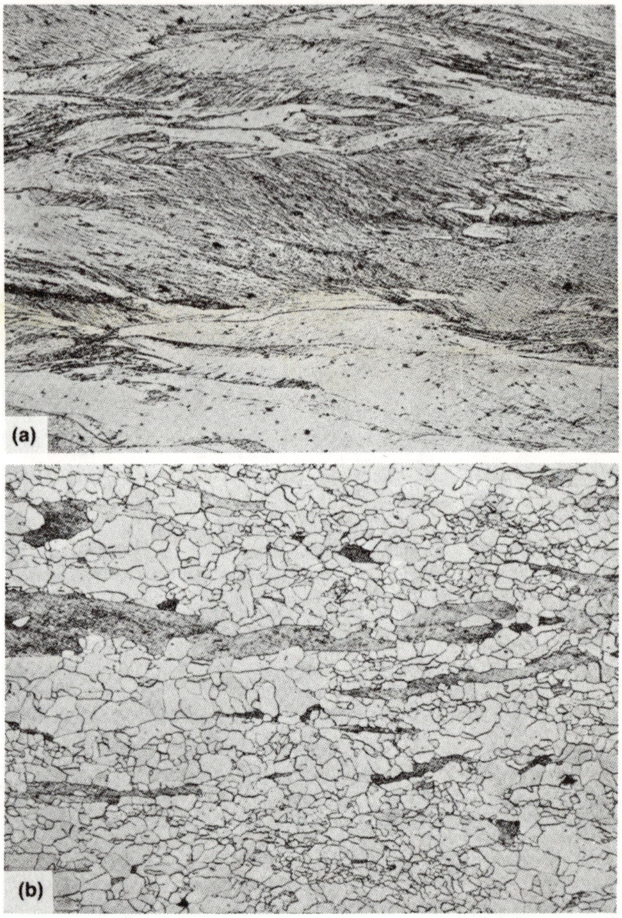

Fig. 5.10. (a) Microstructure of an Fe-0.003C alloy cold rolled 60%. (b) Microstructure of an Fe-0.003C alloy after annealing at 538 °C (1000 °F) for 2 h. About 80% of structure has recrystallized to fine equiaxed grains. Light micrographs. Nital etch. Magnification, 100×; shown here at 75%. (Courtesy of D. A. Witmer, Bethlehem Steel Corp., Bethlehem, Pa.)

growth to occur, a process driven by the reduction in grain boundary or interfacial energy made possible by an increase in grain size. Reference 5.8 describes the mechanisms of recovery and recrystallization stages of annealing in detail.

Stress Relieving

A number of thermal and mechanical processes produce residual stresses that might be detrimental to the performance of fabricated steel parts or assemblies. The residual stresses may cause distortion, cracking during heat treatment or processing, or failure below design stresses in service. One source of residual stresses is the cooling of heavy sections after austenitizing. Even during air cooling, the surface of a heavy section may transform to ferrite and cementite well before the center. When the center eventually transforms, the volume expansion associated with the ferrite formation is restrained by the cooler, already transformed surface. As a result the center is compressed and the surface put in tension. Quenching to form martensite produces a similar but even more severe residual stress problem, even in smaller sections, and is one reason why hardenable steels are alloyed to permit martensite formation at lower cooling rates. Martensitic steels, as discussed in Chapter 8, are invariably tempered, a process that reduces residual stresses and increases ductility and toughness. Machining and cold work also may introduce residual stresses in steel, due to differences in the amount of deformation between the surface and interior regions of a part. Welding is another process which produces residual tensile stresses. As the weld metal solidifies and contracts, it is restrained by the adjacent base metal. As a result, stress-relieving treatments are frequently specified for welded assemblies.

Residual stresses are reduced or eliminated by subcritical heat treatments performed at temperatures either somewhat below or overlapping those used for process or recrystallization annealing (see Fig. 5.8). Heating to and cooling from the stress-relief temperature must be done slowly, especially in heavy sections or large welded assemblies, in order to avoid introducing new thermal stresses and possible cracking during the stress-relief treatment itself (Ref 5.10).

The objective of stress relieving is not to produce major changes of mechanical properties by recrystallization as do the subcritical annealing treatments discussed in the previous section. Rather the stress relief is accomplished by recovery mechanisms that precede recrystallization, a situation that is aided by the difference in the kinetics between the two mechanisms (Ref 5.8). Recovery starts almost immediately on heating and reaching temperature. The rate of recovery is very high initially and decreases with increasing time at temperature. Recrystallization, on the other hand, requires an incubation period and starts very slowly. Therefore it is possible to relieve residual stresses without changing mechanical properties significantly. For example, a recent investigation (Ref 5.11) of stress relief in cold extruded mild steel bars shows that residual stresses are almost completely relieved without any hardness decrease after heating at 500 °C (932 °F) for 1 h. The latter result shows that the strengthening produced by cold working can be used without the harmful effects of residual surface tensile stresses present in the as-deformed condition. Other aspects of stress relieving are discussed in Ref 5.12.

Mechanical Properties of Ferrite-Pearlite Microstructures

As discussed earlier, the objective of normalizing and full-annealing heat treatments of carbon steels is to produce microstructures consisting of ferrite and pearlite. Some relationships that show the effect of various parameters on the mechanical properties of steels containing ferrite and pearlite are described here.

Figure 5.11 shows a set of mechanical properties for ferrite-pearlite microstructures as a function of steel carbon content. Yield and tensile strengths increase and reduction of area, a measure of ductility, decreases with increasing carbon content because of the increase in pearlite content. The microstructures vary from essentially 100% ferrite to 100% pearlite as carbon is increased to 0.8%, the eutectoid carbon content. The divergence of the yield and ultimate strength curves with increasing carbon content indicates that pearlite increases the work hardening rate.

In addition to the mechanical properties that characterize the strength and ductility of steels, toughness or the energy absorbed during fracture is also of considerable engineering importance. Ferritic steels are unique in that they show a transition from ductile to brittle fracture when broken at successively lower temperatures. Generally the transition for a given steel and microstructure is determined by breaking a series of V-notched bars by impact loading at temperatures above and below room temperature. The specimens and machine for the testing are standardized in what is known as the Charpy impact test (Ref 5.15). The ductile fracture typical of higher temperatures proceeds by the growth of microvoids around carbides and/or inclusion particles, a fracture process that requires large amounts of shear or plastic deformation and, therefore, absorbs considerable energy. In contrast, the low-temperature brittle fracture of ferrite proceeds by cleavage between {100} planes of the ferrite grains. Little plastic deformation accompanies cleavage, and therefore little energy is absorbed during this type of fracture.

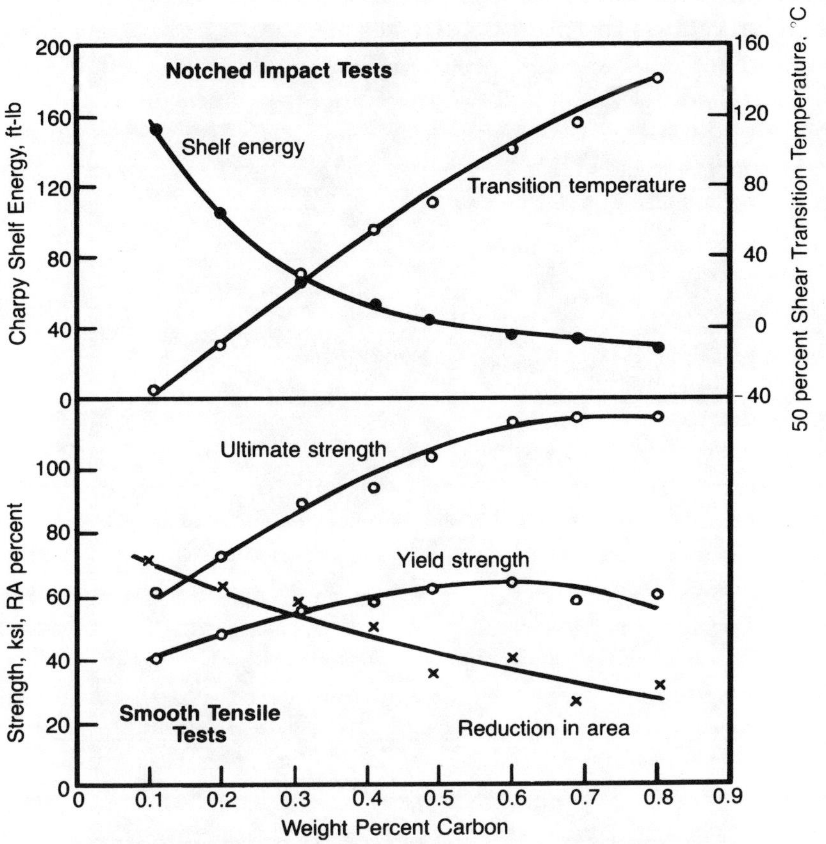

Fig. 5.11. Mechanical properties of ferrite-pearlite microstructures as a function of carbon content. (Ref 5.13)

Figure 5.12(a) shows a mixture of cleavage and ductile fracture on the overload fracture surface of a mild steel. Particles are associated with some of the large conical holes in the fracture surface, and clusters of fine dimples characteristic of ductile fracture are visible in the upper portions of the micrograph. Figure 5.12(b) shows primarily cleavage fracture and is characterized by very flat fracture facets. Cleavage steps (arrows) from one set of cleavage planes to another in a given grain are also a characteristic feature of cleavage fracture.

When energy absorbed during impact fracture is plotted as a function of testing temperature, a transition curve results. Figure 5.13 shows a family of transition curves for steels containing from 0.11% to 0.80% carbon. The energy absorbed by ductile fracture is known as the

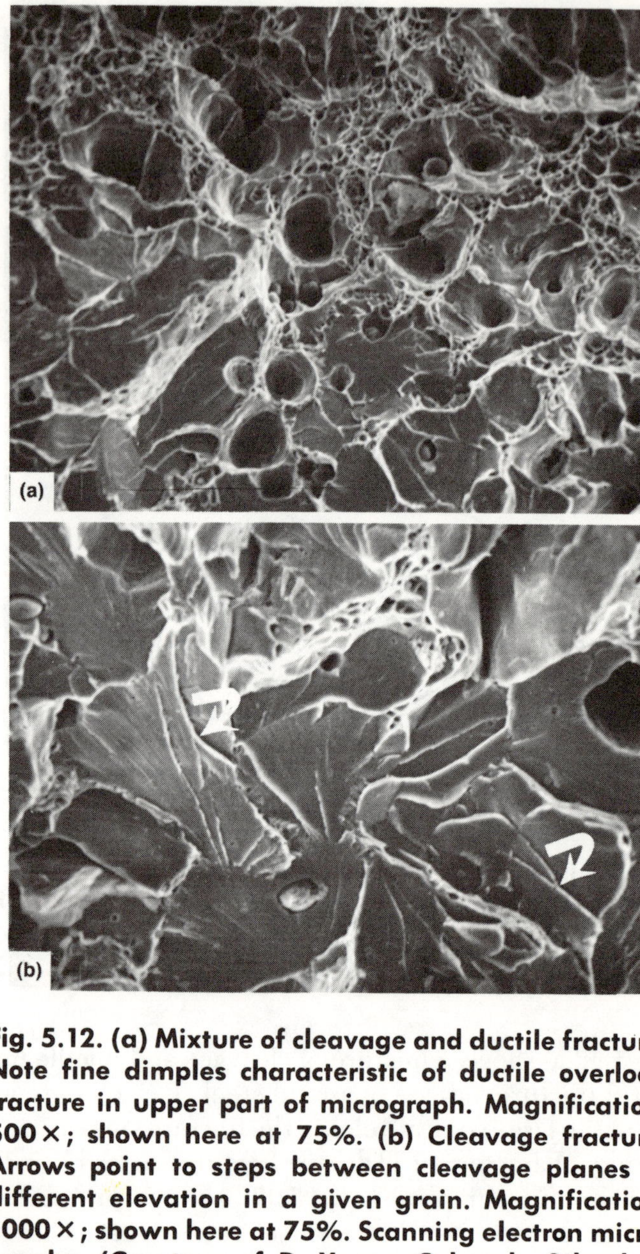

Fig. 5.12. (a) Mixture of cleavage and ductile fracture. Note fine dimples characteristic of ductile overload fracture in upper part of micrograph. Magnification, 500×; shown here at 75%. (b) Cleavage fracture. Arrows point to steps between cleavage planes of different elevation in a given grain. Magnification, 1000×; shown here at 75%. Scanning electron micrographs. (Courtesy of D. Yaney, Colorado School of Mines, Golden, Colo.)

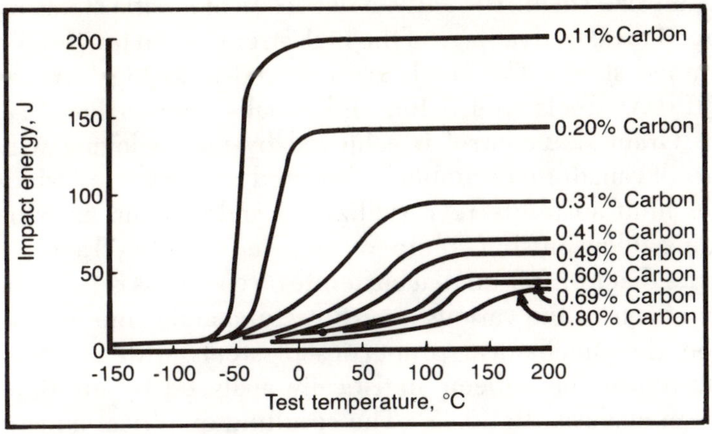

Fig. 5.13. Change in impact transition curves with increasing pearlite content in normalized carbon steels. (Ref 5.18)

"shelf energy" because it reaches a plateau or is essentially constant as a function of temperature. Figure 5.13 shows that increasing carbon content lowers shelf energy, and that, therefore, increasing amounts of pearlite adversely affect the ductile fracture toughness. The transition temperature marking the transition between ductile and brittle fracture is also adversely affected by increasing carbon content. Figure 5.13 shows that high-carbon steels with large amounts of pearlite have high transition temperatures and therefore will fail in a brittle manner even well above room temperature. Low-carbon steels, on the other hand, have subzero transition temperatures and are quite tough at room temperature. Changes in shelf energy and transition temperature (taken as the testing temperature where fracture is 50% cleavage and 50% shear) as a function of carbon content are summarized in the upper part of Fig. 5.11.

At any given carbon content, the mechanical properties and toughness of a steel may be significantly affected not only by pearlite content but also by ferrite grain size and chemical composition. The refinement of ferritic grain size, for example, increases both strength and toughness. Good steel mill and heat treatment practice is therefore directed to producing as fine a ferrite grain size as possible for critical applications. Alloying, low finishing temperatures for hot rolling, and low austenitizing temperatures for normalizing are all techniques used to keep grain size small. Recently, a new class of ferritic steels has been developed to take advantage of the high strength and toughness of very fine-grained steels. The steels are referred to as high-strength low-alloy (HSLA) steels (Ref 5.16), and usually contain less than 0.2% carbon. Grain size control is achieved by microalloying with small amounts of vanadium or niobium that produce very fine carbides. The carbides limit austenite recrystallization and/or grain growth during hot rolling at low finishing temperatures (see Chapter 7) and as a result the ferrite that forms from that austenite on cooling is remarkably fine.

The effects of the various microstructural and composition parameters on the mechanical properties of steels with ferrite-pearlite microstructures have been statistically analyzed by multiple linear regression analysis (Ref 5.17). The resulting empirical equations are limited to steels containing less than 0.25% carbon and therefore to microstructures that are largely ferritic. The following selected equations for yield strength and impact transition temperature illustrate the effects of the various parameters:

$$\text{Yield strength (MPa)} (\pm 31 \text{ MPa}) =$$

$$K + 37(\%\text{Mn}) + 83(\%\text{ Si}) + 2918 (N_f) + 15.1(d^{-1/2}) \quad \text{(Eq 5.2)}$$

$$\text{Impact transition temperature (°C)} (\pm 30\ \text{°C}) =$$

$$-19 + 44(\%\text{Si}) + 700N_f - 11.5(d^{-1/2}) + 2.2(\%\text{ pearlite}) \quad \text{(Eq 5.3)}$$

In these equations, K is 88 MPa for air-cooled steel and 62 MPa for furnace-cooled steel (Ref 5.20); N_f is the free nitrogen dissolved in the ferrite lattice (i.e., not combined as a stable nitride); and d is the mean linear intercept in polygonal ferrite (mm). The beneficial effect of the ferrite grain size on both strength and toughness is apparent from the equations. The effect of manganese and silicon is to increase both yield and tensile strength by solid solution strengthening of the ferrite. The manganese and silicon replace iron on the bcc lattice of ferrite, and are said to dissolve substitutionally. Nitrogen, however, dissolves interstitially and the equations show that it not only is a very potent strengthener of mild steels but also significantly promotes brittle cleavage fracture. The equations also show that the colonies of pearlite in low-carbon steels have no statistically important effect on yield strength. However, other equations show that pearlite does increase tensile strength, an effect attributed to the decrease in yield extension produced by larger pearlite content (Ref 5.17). Figure 5.11 shows a similar greater effect of pearlite on ultimate strength than on yield strength, especially beyond the 0.25% carbon for which Eq 5.2 and Eq 5.3 hold.

In addition to the largely ferritic microstructures of mild steels with carbon contents less than 0.25%, ferrite-pearlite microstructures in medium-carbon steels (Ref 5.21) and fully pearlitic microstructures of eutectoid steels (Ref 5.22) have also been analyzed to correlate various microstructural parameters with mechanical properties. Whereas low-carbon steels are used for structural applications where relatively low strength but good ductility and toughness are required for formability and service conditions, the eutectoid steels are used where high hardness and wear resistance are of major concern. Railroad rails are an example of an important application where the latter properties are important, and the need for improved rail steels has led to the development of the following equations for yield strength and toughness of pearlitic steels (Ref 5.22):

$$\text{Yield strength (MPa)} = 2.18(S^{-1/2}) - 0.40(P^{-1/2}) - 2.88(d^{-1/2}) + 52.30 \qquad \text{(Eq 5.4)}$$

$$\text{Transition temperature (°C)} \; -0.83(P^{-1/2}) - 2.98(d^{-1/2}) + 217.84 \qquad \text{(Eq 5.5)}$$

where S is the pearlite interlamellar spacing, P is the pearlite colony size, and d is the austenite grain size.

Equations 5.5 and 5.6 show that the most important parameter affecting yield strength is the interlamellar spacing while the transition temperature is affected most strongly by the austenite grain size from which the pearlite develops. The transition temperature for fully pearlitic steels is invariably above room temperature, and therefore cleavage on the {100} planes of the ferrite in pearlite is the characteristic fracture mode. The size of the cleavage facets is a strong function of but always smaller than the austenite grain size and appears to be related to common ferrite orientations in several adjoining pearlite colonies (Ref 5.22). Air cooling or normalizing of eutectoid rail steels containing only manganese and silicon produces fine interlamellar pearlite spacings, generally on the order of 2000 Å (200 nm) (Ref 5.24). Alloying of rail steels with vanadium, chromium and molybdenum produces even finer pearlite interlamellar spacings (Ref 5.25 - 5.27).

References

5.1 Heat Treating of Carbon and Low-Alloy Steels, in *Metals Handbook*, Vol 2, 8th Ed., American Society for Metals, Oh., 1964, p 1-10

5.2 K. Nakazawa and G. Krauss, Martensite and Fracture in 52100 Steel, *Met Trans A*, Vol 9A, 1978, p 681-689

5.3 M. Atkins, *Atlas of Continuous Cooling Transformation Diagrams for Engineering Steels*, British Steel Corp., Sheffield, 1977

5.4 S. Chattopadhyay and C. M. Sellars, Quantitative Measurements of Pearlite Spheroidization, *Metallography*, Vol 10, 1977, p 89-105

5.5 E. C. Rollason, *Fundamental Aspects of Molybdenum on Transformation of Steel*, Climax Molybdenum Co., London

5.6 P. Payson, W. L. Hodapp and J. Leeder, The Spheroidizing of Steel by Isothermal Transformation, *Trans ASM*, Vol 28, 1940, p 306-332

5.7 R. L. Fullman, Measurement of Particle Sizes in Opaque Bodies, *Trans AIME*, Vol 197, 1953, p 447-452

5.8 P. G. Shewmon, *Transformations in Metals*, McGraw-Hill, New York, 1969

5.9 D. Witmer and G. Krauss, Effect of Thermal History on the Recrystallization Behavior of Low Carbon 0.305 Mn Steels Containing Oxygen and Sulfur, *Trans ASM*, Vol 62, 1969, p 447-456

5.10 A. Vinckier and A. Dhooge, Reheat Cracking in Welded Structures During Stress Relief Heat Treatments, *J Heat Treating*, Vol 1, 1979, p 72-80

5.11 M. B. Adeyemi, R. A. Stark and G. F. Modlen, Isothermal Stress Relief of Cold Extruded Mild Steel Rods, *Proceedings of the Metals Society, Heat Treatment '79*, Birmingham, England, 22-24 May 1979

5.12 K. E. Thelning, *Steel and Its Heat Treatment*, Bofors Handbook, 1075, Butterworths, London, 1975

5.13 A. R. Rosenfield, G. T. Hahn and J. D. Embury, Fracture of Steels Containing Pearlite, *Met Trans*, Vol 3, 1972, p 2797-2804

5.14 K. W. Burns and F. B. Pickering, Deformation and Fracture of Ferrite-Pearlite Structures, *J Iron Steel Inst*, Vol 202, 1964, p 899-906

5.15 G. E. Dieter, Jr., *Mechanical Metallurgy*, McGraw-Hill, New York, 1961, p 371-375

5.16 *Microalloying 75*, Union Carbide Corp., 1977, distributed by American Society for Metals, Metals Park, Oh.

5.17 F. B. Pickering, The Effect of Composition and Microstructure on Ductility and Toughness, in *Toward Improved Ductility and Toughness*, Climax Molybdenum Development Co., Japan, 1971, p 9-31

5.18 F. B. Pickering, The Optimization of Microstructures in Steel and Their Relationship to Mechanical Properties, in *Hardenability Concepts with Applications to Steel*, D. V. Doane and J. S. Kirkaldy (Eds.), AIME, Warrendale, Pa., 1978, p 179-228

5.19 R. W. K. Honeycombe and F. B. Pickering, Ferrite and Bainite in Alloy Steels, *Met Trans*, Vol 3, 1972, p 1099-1112

5.20 T. Gladman, D. Dulieu and I. D. McIvor, Structure-Property Relationships in High-Strength Microalloyed Steels, in *Microalloying 75*, Union Carbide Corp., 1977, distributed by American Society for Metals, Metals Park, Oh.; p 32-54

5.21 T. Gladman, I. D. McIvor and F. B. Pickering, Some Aspects of the Structure-Property Relationships in High-Carbon Ferrite-Pearlite Steels, *JISI*, Vol 210, 1972, p 916-930
5.22 J. M. Hyzak and I. M. Bernstein, The Role of Microstructure on the Strength and Toughness of Fully Pearlitic Steels, *Met Trans*, Vol 74, 1976, p 1217-1224
5.23 Y-J. Park and I. M. Bernstein, Mechanism of Cleavage Fracture in Fully Pearlitic 1080 Rail Steel, in *Rail Steels — Developments, Processing and Use*, STP 644, ASTM, 1978, p 287-302
5.24 D. E. Sonon, J. V. Pellegrino and J. M. Wandrisco, A Metallurgical Examination of Control-Cooled, Carbon-Steel Rails with Service-Developed Defects, in *Rail Steels — Developments, Processing and Use*, STP 644, ASTM, 1978, p 99-117
5.25 G. K. Bouse, I. M. Bernstein and D. H. Stone' in *Rail Steels — Developments, Processing and Use*, STP-644, ASTM, 1978, p 145-166
5.26 S. Marich and P. Curcio, in *Rail Steels — Developments, Processing and Use*, STP 644, ASTM, 1978, p 167-210
5.27 Y. E. Smith and F. B. Fletcher, in *Rail Steels — Developments, Processing and Use*, STP 644, ASTM, 1978, p 212-232

CHAPTER 6

Hardness and Hardenability

A martensitic microstructure is the hardest microstructure that can be produced in any carbon steel, but it can be produced only if the transformation of austenite to mixtures of ferrite and cementite is avoided. This chapter first describes the relationship of martensite hardness to carbon content, and then discusses other factors that determine whether or not that hardness can be achieved throughout a given part fabricated from a given steel. The term hardenability is used to describe both the ease of martensite formation and the technology that relates section size, cooling rates and composition to the hardening of steel. This chapter emphasizes principles and the classical approach to hardenability as developed by Grossmann, Bain and their contemporaries and then describes more recent approaches to the continually developing technology of hardenability.

Hardness and Carbon Content

The maximum hardness that can be produced in any given carbon steel is that associated with a fully martensitic microstructure. Figure 6.1 shows the much higher hardness of martensite relative to that of ferrite-pearlite or spheroidized microstructures for the entire range of carbon content usually found in steels. The high hardness and associated high strength, fatigue resistance and wear resistance are the prime reasons for the quenching heat treatments that produce martensite. Almost all martensite is tempered, and depending on the amount of tempering, hardness in a given quench and tempered steel may vary

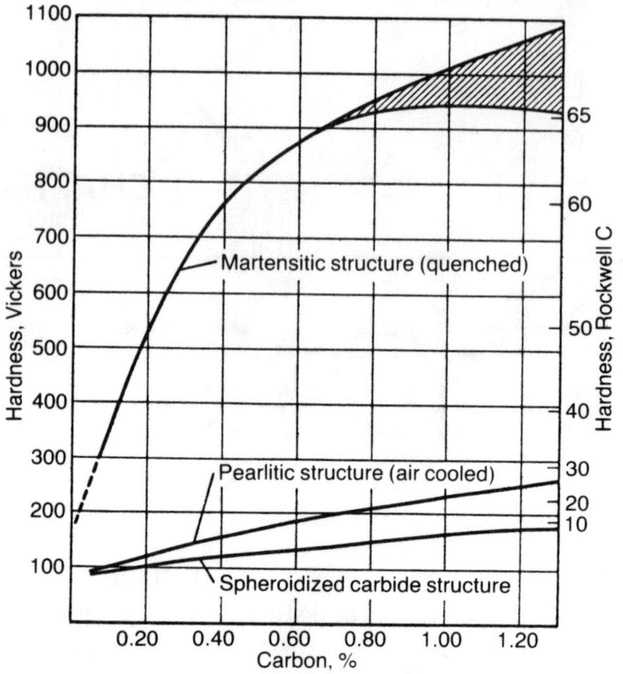

Fig. 6.1. Hardness as a function of carbon content for martensitic, ferrite-pearlite and spheroidized microstructures in steels. Cross-hatched area shows effect of retained austenite. (Ref 6.1)

from close to the maximum shown for martensite to the minimum associated with the spheroidized carbide structure. Heat treatments to form martensite are generally applied to steels containing more than 0.3% carbon. In these steels the gains in hardness are most substantial. Also, steels containing less than 0.3% carbon tend to be difficult to harden and are usually used with the ferrite-pearlite microstructures produced by the heat treatments described in Chapter 5. Rockwell C readings below 20 are not considered valid and are included in Fig. 6.1 only for comparative purposes.

Figure 6.2 is a summary plot of many investigations of the hardness of martensitic microstructures as a function of the carbon content of steels and Fe-C alloys, and shows the range of hardness that may develop in largely martensitic microstructures in steels of a given carbon content. Special care was taken in all of the investigations to insure that no proeutectoid phases or mixtures of ferrite and cementite formed. However, the martensitic microstructures may have contained various amounts of retained austenite because M_f drops below

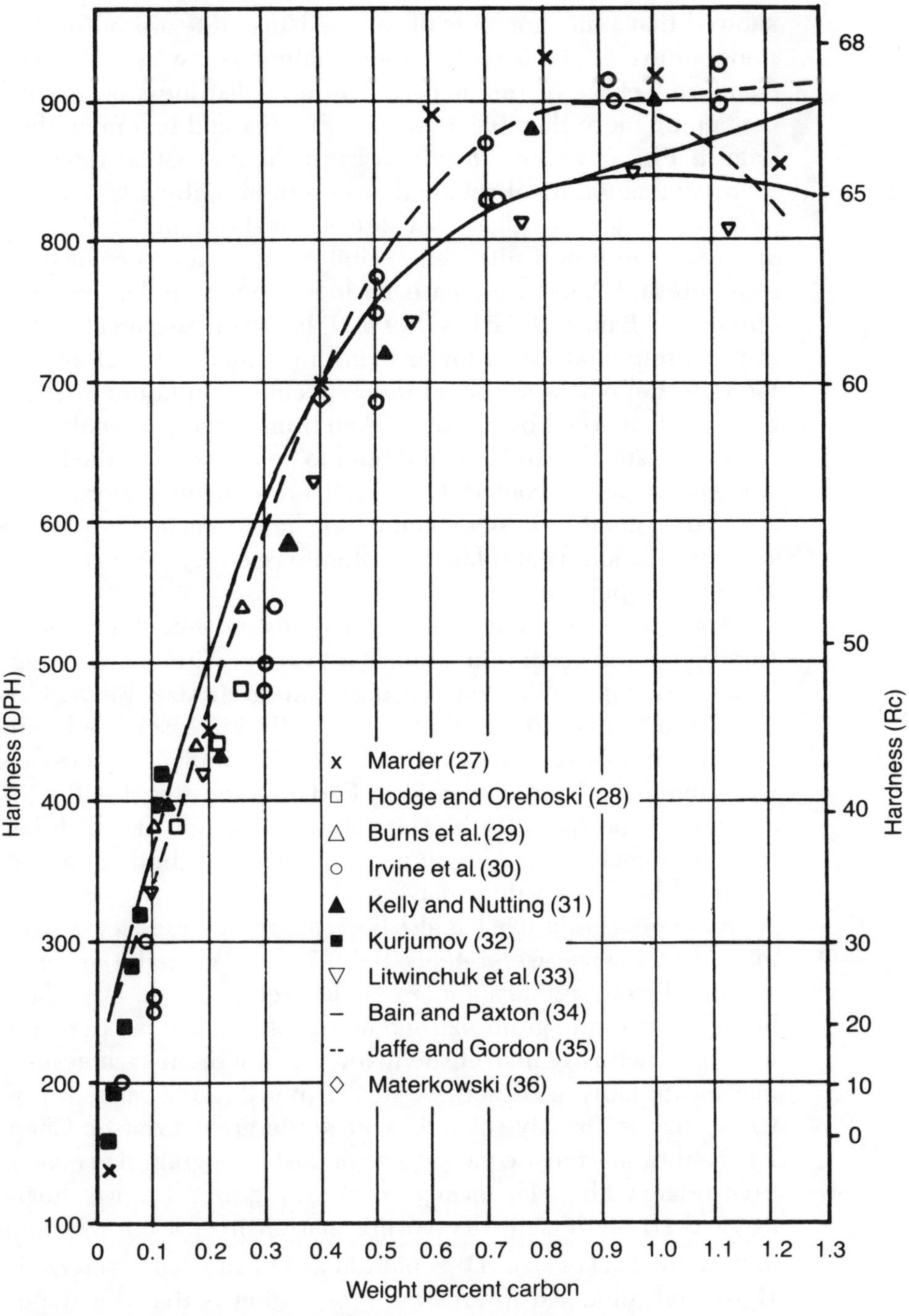

Fig. 6.2. Summary of hardness of martensite as a function of carbon content in Fe-C alloys and steels. (Ref 6.2)

room temperature even in low-carbon steels. For example, Fig. 3.8 showed that small amounts of retained austenite are present at room temperature in steels with carbon content as low as 0.3%. The most significant effect of retained austenite on hardness occurs in steels containing more than 0.7% carbon; Fig. 6.1 and several of the sets of data in Fig. 6.2 show the decrease in hardness that develops with increasing amounts of retained austenite in high-carbon steels.

Some of the investigators whose data is shown in Fig. 6.2 quenched specimens in liquid nitrogen (−196 °C) in order to reduce retained austenite and thereby increase hardness. For example, the continuous curve after Bain and Paxton (Ref 6.1), based on as-quenched hardness at room temperature, is lower than the dashed curve after Jaffee and Gordon (Ref 6.3) who cooled their specimens in liquid nitrogen. The data points marked by x's were taken from specimens cooled in liquid helium (−269 °C) (Ref 6.4) and tend to be higher than the hardness of steels not as deeply cooled. The effectiveness of the subzero treatments is of course greatest in steels containing more than 0.4% carbon, where significant amounts of retained austenite (see Fig. 3.8) may be present at room temperature.

Apart from differences in retained austenite content, some of the variation in the maximum hardness of various carbon levels might also be due to aging or differences in austenitic grain size. Figure 6.3 shows that room temperature aging significantly increases the hardness of martensitic Fe-Ni-C alloys (Ref 6.5, 6.6). Similar hardness changes with time have been observed in Fe-C martensites (Ref 6.4); thus, if attention is not paid to the time after quenching at which hardness measurements are made, some variation contributing to scatter in reported hardness values may occur.

Austenitic grain size has also been observed to affect the strength of martensite in low-carbon steels (Ref 6.7, 6.8). When the austenite grain size is reduced, significant increases in strength occur. The relationship between austenite grain size and martensite structure is a result of the unique structure of martensite in low- and medium-carbon steels. The martensite laths, as described in Chapter 3, are arranged in packets whose size is directly related to austenite grain size (see Chapter 7). Thus either martensite packet size or austenite grain size may be used to correlate with mechanical properties. Figure 6.4 shows the increase in yield strength with decreasing martensite packet size in an Fe-0.2C alloy. Packet size (D) is plotted as $D^{-1/2}$ in what is referred to as a Hall-Petch plot. An interesting observation is that the slope of the Fe-0.2C martensite curve is steeper than that of lath martensite in an Fe-Mn alloy without carbon. This observation was explained by the

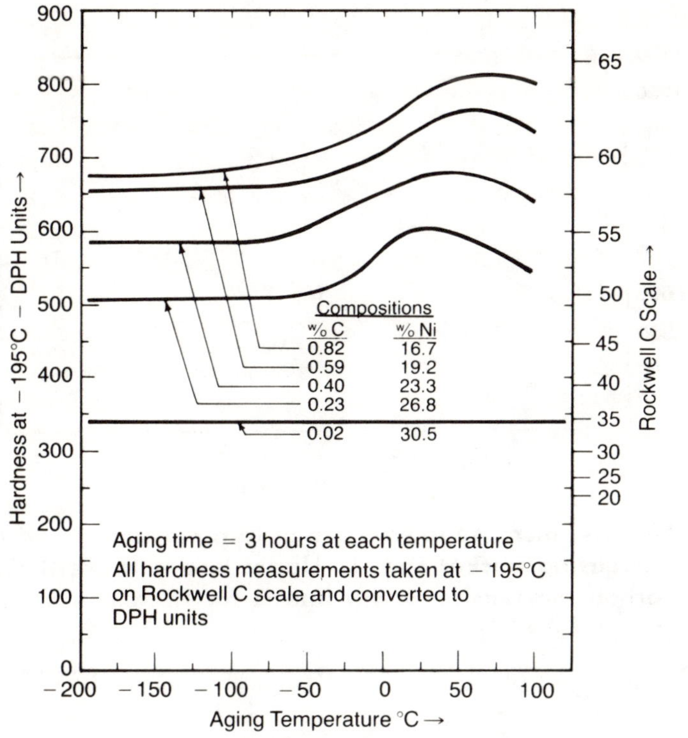

Fig. 6.3. Hardness of Fe-Ni-C alloy martensites at − 195 °C (320 °F) after aging for 3 h at the temperatures shown. (Ref 6.5)

segregation of carbon atoms to packet boundaries where they make the initial yielding process more difficult; the more so the finer the packet size (Ref 6.8).

Martensite Strength

The reason for the very high hardness of carbon-containing martensites has long intrigued metallurgists. Cohen in the 1962 Howe Memorial Lecture (Ref 6.5) followed the historical development of the theories of martensite strength in steels, and emphasized the important role that carbon atoms trapped in the octahedral interstitial sites play in the strengthening of martensite. Figure 6.5 shows a schematic representation of the displacement of the iron atoms due to carbon

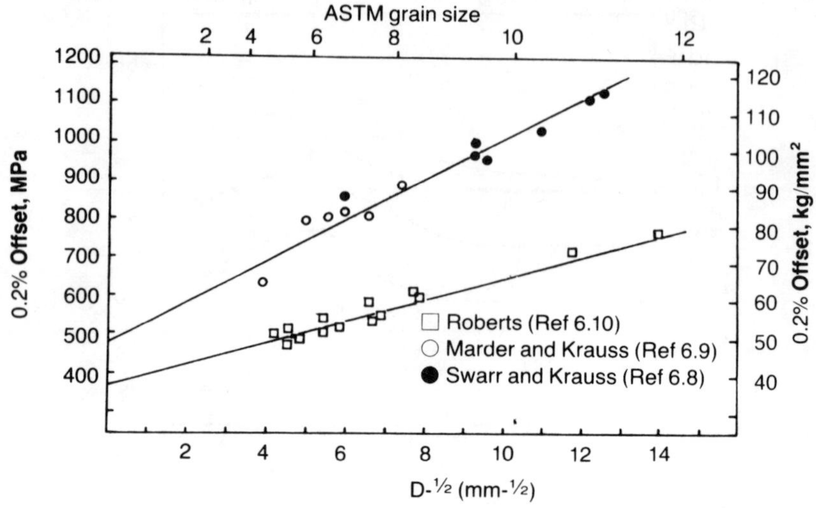

Fig. 6.4. Increase in strength of lath martensites with decreasing packet size, D. Upper line is for Fe-0.2% carbon martensite; lower line is for Fe-Mn martensite. (Ref 6.8)

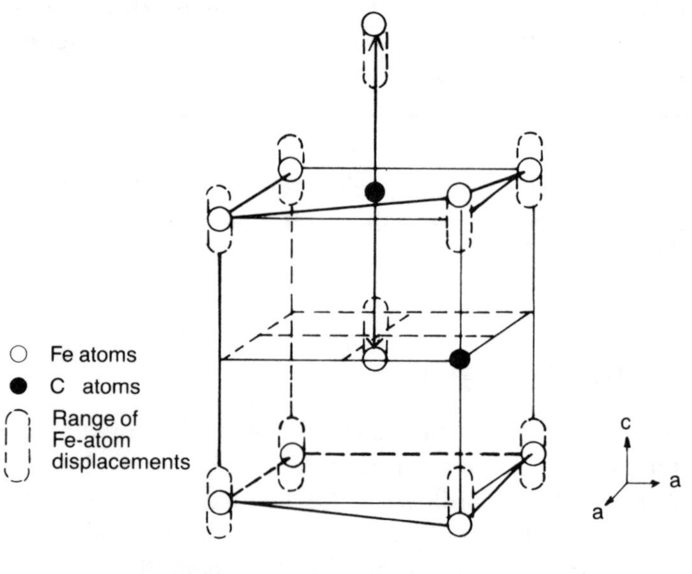

Fig. 6.5. Iron atom displacements due to carbon atoms in martensite. (Ref 6.5)

atoms in the body-centered tetragonal lattice of martensite. This distortion of the iron lattice makes the movement of dislocations very difficult and is considered to be a major cause of the high strength of martensite.

In addition to the solid solution strengthening by carbon, the substructure of martensite also contributes to strength. Chapter 3 showed that martensitic transformation was unique in that it introduced a high density of dislocations and/or fine twins into a martensite lath or plate. The contribution to the strength of martensite by the substructure is relatively constant as a function of carbon content and, except at low carbon concentrations, does not make nearly as great a contribution as does the carbon solid solution strengthening. The following equation (Ref 6.11, 6.12) for the 0.2% offset yield strength of martensite ($\sigma_{0.2}$), determined from a series of Fe-Ni-C alloys with subzero M_s temperatures, permits a quantitative assessment of the carbon and substructure contributions:

$$\sigma_{0.2}\,(\text{MPa}) = 461 + 13.1 \times 10^5\,(\text{w/o C})^{1/2} \qquad \text{(Eq 6.1)}$$

The second term shows the strong effect of carbon and that the strengthening of martensite follows a square-root dependency with carbon content, a functional relationship that correlates well with the initial rapid increase in strength with carbon content and the more gradual strength increases at higher carbon contents. The first term includes the strengthening contribution of 20% nickel (20 000 psi or 138 MPa), the friction stress or stress to move dislocations in pure bcc iron (10 000 psi or 69 MPa) and the contribution of the martensitic substructure (37 000 psi or 255 MPa). Equation 6.1 holds for unaged martensite, a result possible because of the low M_s temperatures of the Fe-Ni-C alloys. The martensite, therefore, was formed and mechanically tested at low temperatures where aging was minimal.

Carbon steels, especially those of low carbon content, have high M_s temperatures and undergo considerable carbon atom rearrangement during quenching before reaching room temperature, a process referred to as autotempering. The carbon atoms segregate to the dislocation fine structure and/or lath and packet boundaries (Ref 6.13). One result of the segregation, a very high dependency of strength on packet size, has already been mentioned (Ref 6.8). Despite the effects of carbon atom segregation, the yield strength of low-carbon martensite still follows a square root dependency on carbon content as shown in the following equation (Ref 6.14):

$$\sigma_{0.2}\,(\text{MPa}) = 413 + 17.2 \times 10^5\,(\text{w/o C})^{1/2} \qquad \text{(Eq 6.2)}$$

This equation was determined from low-carbon Fe-C alloys containing up to 0.2% carbon and also fits data for martensitic steels containing 0.08 to 0.24% carbon and 0.4 to 0.5% manganese (Ref 6.15). Again, the first term includes all of the structural contributions to strength, including the austenitic grain size or packet size (in this case austenite grain size was roughly constant, between ASTM 7 and 9), lath size and dislocation fine structure.

Definitions of Hardenability

The above discussion shows that the maximum hardness of any steel is associated with a fully martensitic structure. This microstructure, however, can only be produced if the diffusion-dependent transformation of austenite can be suppressed by sufficiently rapid cooling. There are a number of factors that affect cooling rates throughout a given part and the response of a given steel to those cooling rates. Thus, the formation of martensite and high hardness may vary considerably throughout a given cross section or between identical cross sections fabricated from different steels. The subject of hardenability deals with the latter variations.

Hardenability is defined as the "susceptibility to hardening by rapid cooling" (Ref 6.16), or as "the property, in ferrous alloys, that determines the depth and distribution of hardness produced by quenching" (Ref 6.17). Both of these definitions emphasize hardness. As discussed above, the source of hardening is the formation and presence of martensite, and therefore a third definition of hardenability, "the capacity of a steel to transform partially or completely from austenite to some percentage of martensite at a given depth when cooled under some given conditions", more accurately describes the physical process underlying hardening. Siebert, Doane and Breen in their comprehensive book on hardenability prefer the latter structural definition (Ref 6.18).

Hardness Distribution

An experimental approach that demonstrates the striking effect of various factors on hardenability is the quenching of series of round bars of various diameters. The bars are completely austenitized, quenched and tempered. Hardness readings are then taken along diameters of the bar cross sections in order to show the distribution of hardness as a function of distance from the surface to the center of the bar.

Figures 6.6 and 6.7 show the results of water quenching bars of

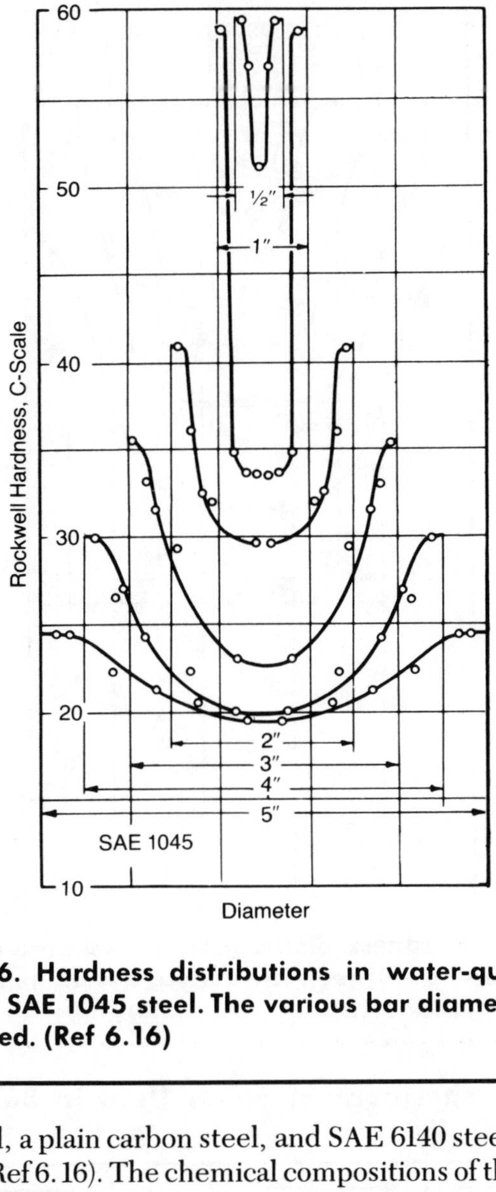

Fig. 6.6. Hardness distributions in water-quenched bars of SAE 1045 steel. The various bar diameters are indicated. (Ref 6.16)

SAE 1045 steel, a plain carbon steel, and SAE 6140 steel, an alloy steel, respectively (Ref 6.16). The chemical compositions of the two steels are given in Table 6.1:

Plain carbon and alloy steels are classified by the Society of Automotive Engineers (SAE) and the American Iron and Steel Institute (AISI) and are manufactured to various ranges of compositions (Ref 6.19). For example, the AISI-SAE specifications for steel designated as 1045 permit carbon in the range of 0.42 to 0.50% and manganese in the range of 0.60 to 0.90%. It is, therefore, important to state the exact composition of a heat of steel (as in Table 6.1) for the most accurate

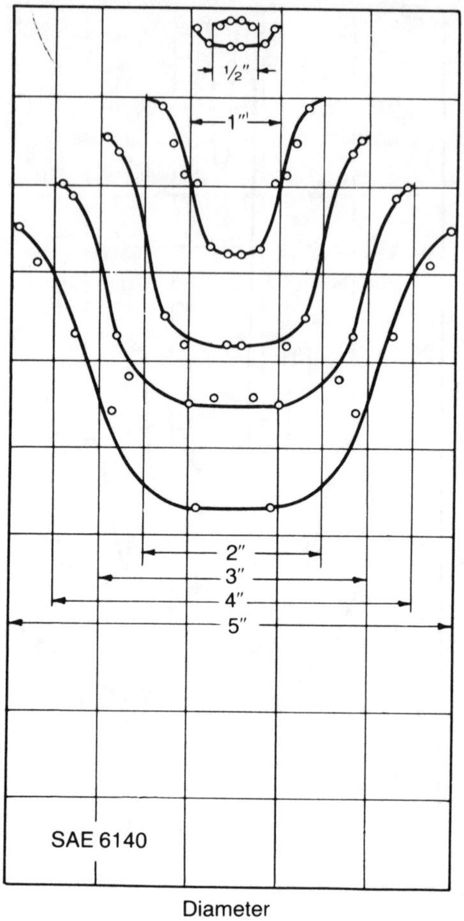

Fig. 6.7. Hardness distribution in water-quenched bars of SAE 6140 steel. The various bar diameters are indicated. (Ref 6.16)

Table 6.1. Compositions of Steels Used in Bar Quenching Experiments

Steel	Composition, % — C	Mn	P	S	Si	Cr	V
SAE 1045.....	0.48	0.60	0.022	0.016	0.17	..	..
SAE 6140.....	0.42	0.73	0.027	0.023	0.25	0.94	0.17

interpretation of the response to hardening.

Figure 6.6 shows that the maximum hardness in the SAE 1045 steel can be achieved only on the surface of bars with small diameters. Even in a 0.5-in. (12.7-mm) diameter bar, the hardness in the interior drops significantly. With increasing bar diameter, the surface hardness of the SAE 1045 steel drops significantly and the center hardness continues to decrease. The alloy steel, SAE 6140, on the other hand, develops higher hardness than the SAE 1045 steel at all bar diameters (see Fig. 6.7) but nevertheless still shows large variations in hardness from the surface to the center of the bars, especially in the larger sizes.

Figures 6.6 and 6.7 show the effects of bar diameter and alloy content on hardness distribution of water-quenched rounds. A third factor that influences hardness distribution is the rate of quenching. Figures 6.8 and 6.9 show the results of oil quenching on the hardness distribution in round bars of various diameters for the SAE 1045 and 6140 steels, respectively. Oil is a much less severe quenching medium than water, and so the cooling rates of oil-quenched bars are appreciably lower than those of water-quenched bars. Figure 6.8 shows that the hardening response of the SAE 1045 steel to oil quenching is very low. Even in the 0.5-in. (12.7-mm) diameter bar the surface hardness is well below the hardness expected from a fully martensitic structure of a 0.48% carbon steel (see Fig. 6.1 and 6.2). It is apparent, therefore, that the slower cooling associated with oil quenching has not been able to prevent the diffusion-controlled transformation to ferrite and/or pearlite in the SAE 1045 steel. The SAE 6140 steel, however, hardens well in the small bar sizes (see Fig. 6.9) and only in the larger sizes does the hardness distribution fall off significantly.

Comparison of Fig. 6.6 through 6.9 shows that the alloy steel, SAE 6140, is much more hardenable than the plain carbon SAE 1045 steel. SAE 6140 is, therefore, said to have a higher hardenability than the SAE 1045 steel. The plain carbon steel can be hardened but only in small sections and/or with very severe quenches. Fundamentally, the alloying elements in the SAE 6140 steel increase the time required for austenite to decompose to ferrite and/or ferrite-cementite mixtures, and thereby make it possible to form martensite at lower cooling rates. The effects of alloying elements on the diffusion-controlled decomposition of austenite in many steels are summarized in the IT and CT diagrams contained in the atlases described in Chapter 4.

Factors Affecting Cooling Rates

Two important factors influence cooling rates or the rates at which heat can be removed from a steel part. One is the ability of the heat to diffuse from the interior to the surface of the steel specimen, and the other is the ability of the quenching medium to remove heat from the surface of the part.

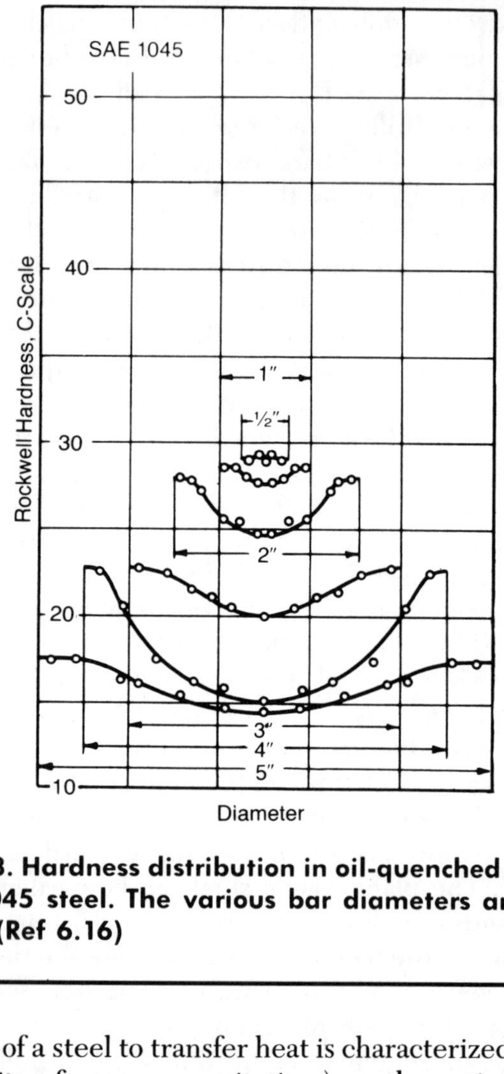

Fig. 6.8. Hardness distribution in oil-quenched bars of SAE 1045 steel. The various bar diameters are indicated. (Ref 6.16)

The ability of a steel to transfer heat is characterized by its thermal diffusivity (units of area per unit time) or the ratio of its thermal conductivity to the volume specific heat. The thermal diffusivity of austenitic transformation products increases with decreasing temperature, and plots of thermal diffusivity and conductivity for various structures as a function of temperature are reproduced in Ref 6.18. For a given quenching medium, the thermal diffusivity determines the temperature distribution as a function of position at any given time in the quenching process. For example, Fig. 6.10 shows cooling rates as a function of position in a quenched 1-in. diameter bar. The slower cooling rates at positions removed from the surface of the bar permit more time for diffusion-controlled transformations, and it is this type of

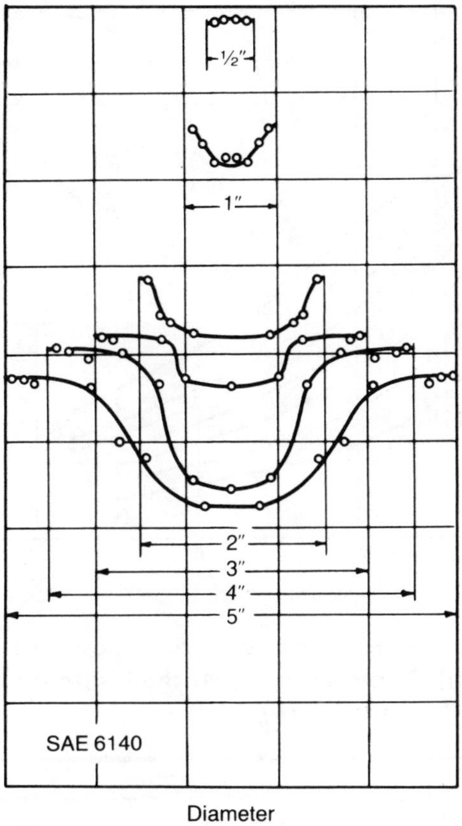

Fig. 6.9. Hardness distributions in oil-quenched bars of SAE 6140 steel. The various bar diameters are indicated. (Ref 6.16)

cooling behavior that results in the low center hardness of the bars shown in Fig. 6.6 through 6.9, especially in the larger sizes. Practically, however, there is little control of thermal properties possible in steels, and the most important control of cooling rates is performed by proper selection of quenching media.

The transfer of heat at the interface of a steel part and a quenching medium is a complex process that depends primarily on the emissivity of the steel (or the rate at which the surface of the steel radiates heat) and convection currents within the quenching medium that remove heat from the interface. The complexity of the process is illustrated in Fig. 6.11, a curve obtained by measuring temperature as a function of time in the center of a 0.5-in. (12.7-mm) diameter bar of steel during

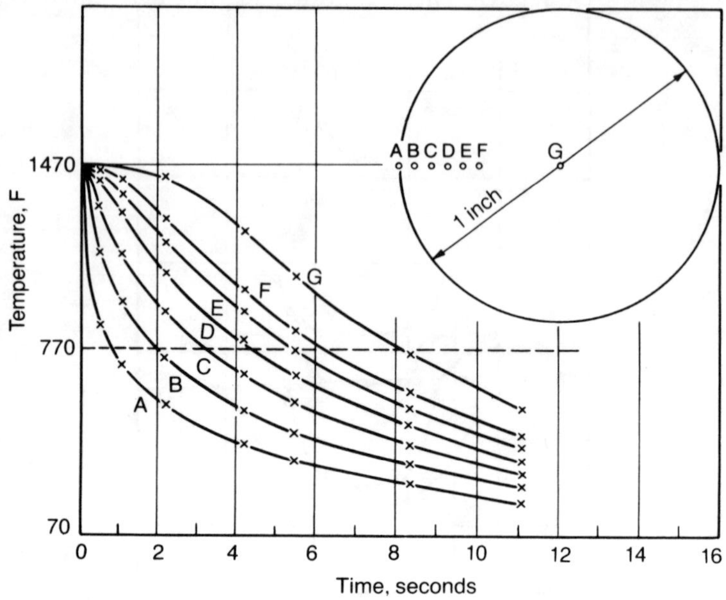

Fig. 6.10. Cooling curves at various positions in a 1-in. (25.4-mm) diameter bar quenched with a severity of quench H=4. (Ref 6.16)

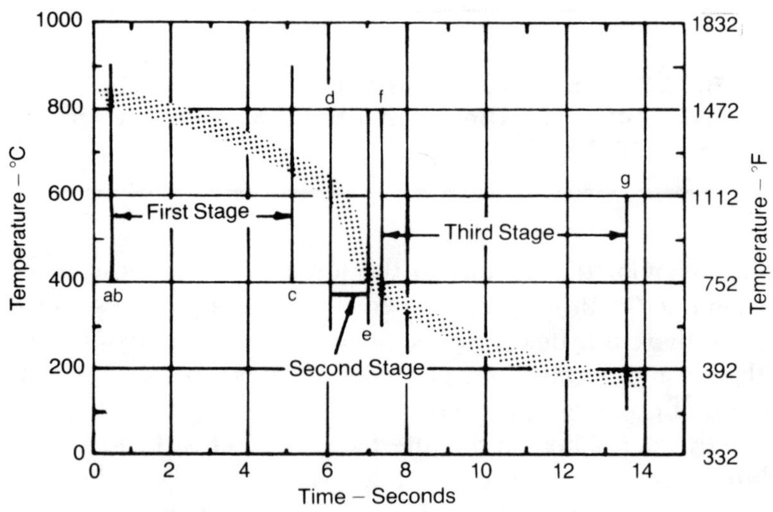

Fig. 6.11. Stages of cooling during water-quench. (Ref 6.24)

water quenching (Ref 6.20). Three stages of cooling are shown. The first stage is associated with the development of a layer of water vapor or steam immediately adjacent to the surface of the steel. The steam insulates the surface and produces a low cooling rate. In the second stage the vapor blanket breaks down and water comes in contact with the steel. The water vaporizes, but bubbles away, thereby continually bringing more water in contact with the surface. Cooling is quite rapid in this stage. When the surface temperature of the steel drops below the boiling point, vaporization stops and cooling is controlled by convection and conduction at the fluid-metal interface. The latter or third stage is characterized by relatively slow rates of cooling.

Understanding of the cooling process has important practical consequences. For example, if the low cooling rate of the first stage results in ferrite or pearlite, efforts should be made to increase the cooling rate in this stage. Agitation of the part or the quenchant or the use of brine solutions for quenching are effective in reducing the duration of the first stage of cooling.

Severity of Quench

The effectiveness of a given quenching medium is ranked by a parameter referred to as its "severity of quench". This measure of cooling or quenching power is identified by the letter "H", and is determined experimentally by quenching a series of round bars of a given steel. Figure 6.12 shows schematically the results of oil and water quenching bars of SAE 3140 steel (Ref 6.16). SAE 3140 is a nickel-chromium alloy steel containing nominally 0.40% carbon. The cross-hatched areas represent the unhardened areas of the various bars, assuming that less than 50% martensite represents an unhardened microstructure. The larger the bar diameter (D), the greater the unhardened diameter (D_u). Figure 6.13 plots the results of Fig. 6.12 as D_u/D versus D for both the oil- and water-quenched series. The steeper curve is associated with the oil quench, a result of the reduced ability of oil quenching to produce hardening in heavier sections. When the curves of Fig. 6.13 are matched to one of the large number of calculated curves that are characteristic of a wide range of quench severities (see Fig. 6.14), H can be determined. The matching can be performed by plotting D_u/D versus D as in Fig. 6.13 on transparent paper and finding the best correspondence to a D_u/D versus HD curve in Fig. 6.14. When the HD values are divided by corresponding D values, the H

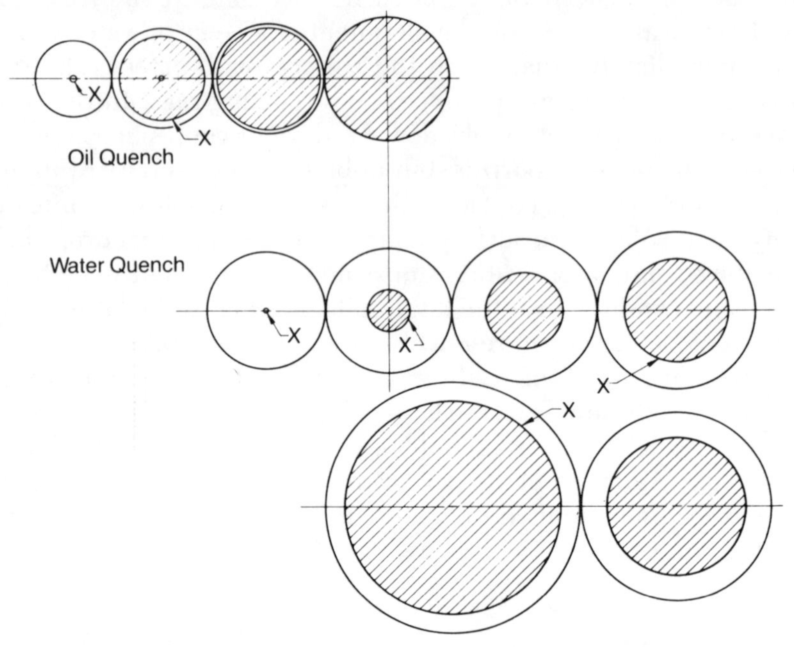

Fig. 6.12. Schematic representation of extent of hardening in oil-quenched and water-quenched bars of SAE 3140 steel of various diameters. The cross-hatched areas represent the unhardened core. (Ref 6.16)

value is obtained. For example, point A in Fig. 6.13, corresponding to a bar diameter of 1.83 in. (4.65 cm) would fall on an HD value of 2.6 in Fig. 6.14 when the curves are matched. Then H = 2.6/1.83 = 1.4 for the water quench of this example.

Table 6.2 lists H values for a number of commonly used quenches. The increase in severity of quench from air, H = 0.02, through brine quenching, H = 2, is shown. Also, the very strong effect of agitation or circulation on increasing the severity of quench in any given quenching medium is apparent. Another useful ranking of quenching media relative to water is shown in Table 6.3. This table not only ranks the cooling media, but also shows the large number of media available for cooling at various rates.

Quantitative Hardenability

Up to this point a number of important aspects of hardenability have been described. High hardness is related to martensite forma-

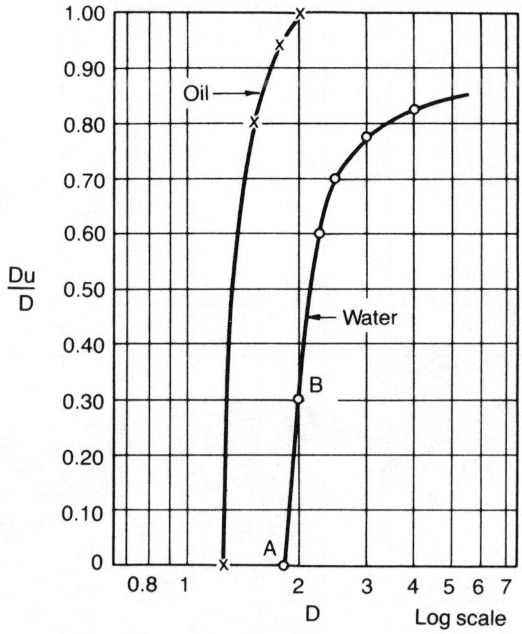

Fig. 6.13. Ratio of unhardened to hardened diameters as a function of bar diameter for oil- and water-quenched bars of 3140 steel. (Ref 6.16)

Table 6.2. Severity of Quench (H) for Various Quenching Media (Ref 6.16, 6.21)

	Air	Oil	Water	Brine
No circulation of fluid or agitation of piece	0.02	0.25 to 0.30	0.9 to 1.0	2
Mild circulation (or agitation)	...	0.30 to 0.35	1.0 to 1.1	2 to 2.2
Moderate circulation	...	0.35 to 0.40	1.2 to 1.3	...
Good circulation	...	0.4 to 0.5	1.4 to 1.5	...
Strong circulation	0.05	0.5 to 0.8	1.6 to 2.0	...
Violent circulation	...	0.8 to 1.1	4	5

tion, which in turn is dependent upon cooling rate. Cooling rate is affected by both the specimen size and severity of quench. However, there still remain the questions as to how hardenability is evaluated as a function of steel composition and how the effect of the large number of quenching media on hardness distribution can be evaluated without the time-consuming approach of quenching a series of round bars in the various quenching media. The first and now classical approach to

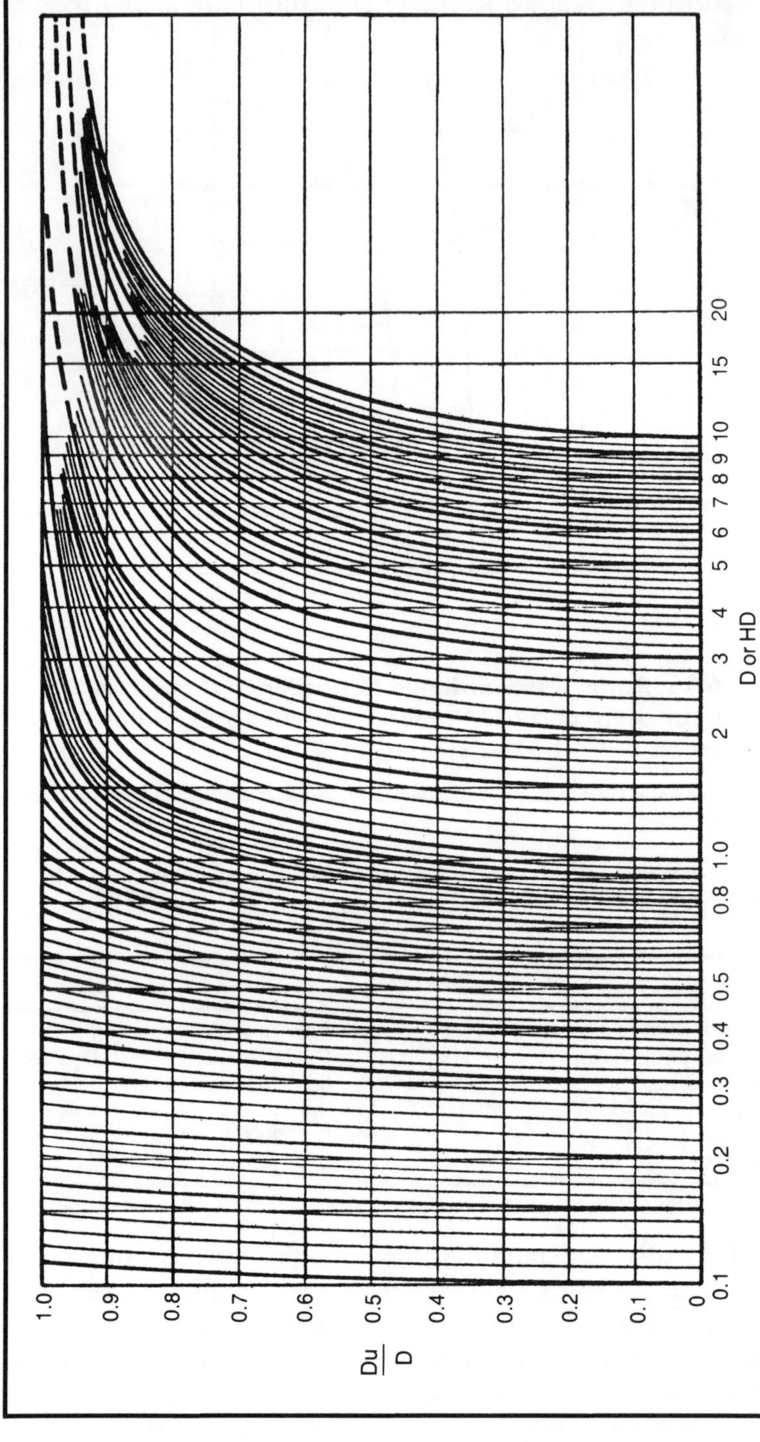

Fig. 6.14. Curves of D_u/D vs. D or HD for estimating severity of quench (H) for quenching baths. (Ref 6.16)

Table 6.3. Relative Cooling Rates in Different Quenching Media Ref (6.16, 6.22)

Quenching medium	Cooling rate (a) from 717 to 550 °C (1328 to 1022 °F) relative to that for water at 18 °C (65 °F)	Quenching medium	Cooling rate (a) from 717 to 550 °C (1328 to 1022 °F) relative to that for water at 18 °C (65 °F)
Aqueous solution, 10% LiCl	2.07	Oil 20204	0.20
Aqueous solution, 10% NaOH	2.06	Oil, Lupex light	0.18
Aqueous solution, 10% NaCl	1.96	Water at 122 F	0.17
Aqueous solution, 10% Na_2CO_3	1.38	Oil 25441	0.16
Aqueous solution, 10% H_2SO_4	1.22	Oil 14530	0.14
Water at 32 F	1.06	Emulsion of 10% oil in water	0.11
Water at 65 F	1.00	Copper plates	0.10
Aqueous solution, 10% H_3PO_4	0.99	Soap water	0.077
Mercury	0.78	Iron plates	0.061
$Sn_{30}Cd_{70}$ at 356 F	0.77	Carbon tetrachloride	0.055
Water at 77 F	0.72	Hydrogen	0.050
Rape seed oil	0.30	Water at 166 F	0.047
Trial oil No. 6	0.27	Water at 212 F	0.044
Oil P20	0.23	Liquid air	0.039
Oil 12455	0.22	Air	0.028
Glycerin	0.20	Vacuum	0.011

(a) Determined by quenching a 4-mm nichrome ball, which when quenched from 860 °C (1580 °F) into water at 18 °C (65 °F) cooled at the rate of 1810 °C (3260 °F) per second over the range 717 to 550 °C (1328 to 1022 °F). This cooling rate in water at 18 °C (65 °F) is rated as 1.00 in the table, and the rates in the other media are compared with it. (Ref 6.22)

these questions, described below, was developed in the 1930s and 1940s by Grossmann and Bain (Ref 6.16) and their many colleagues and contemporaries.

The Grossmann and Bain approach to hardenability *is* based on the definition of two parameters: the critical size and the ideal size. The critical size is the largest size of a bar quenched in a given medium which contains no unhardened core after quenching. An important aspect of this definition is that the hardness that separates the hardened from the unhardened core of a bar is associated with a microstructure assumed to contain only 50% martensite. This assumption underlies all of the graphical information associated with the Grossmann-Bain approach to hardenability. The reason for selecting the 50% martensite criterion for the critical diameter is shown in Fig. 6.15. Etching differences between the hardened surface of a bar and the unhardened center are most clearly developed close to the 50%

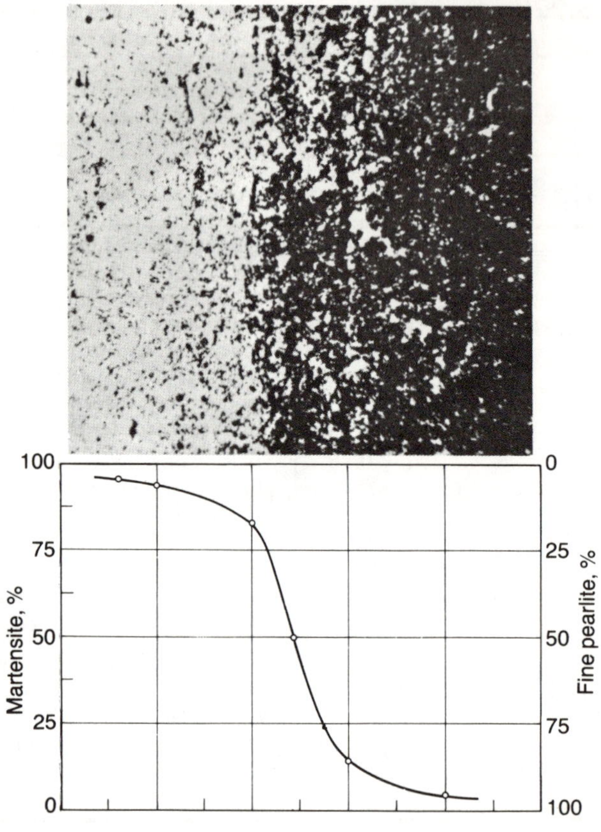

Fig. 6.15. Transition from martensitic to pearlitic microstructure between hardened and unhardened zones of a quenched steel. (Ref 6.16)

martensite-50% pearlite zone in a bar. Likewise when a quenched bar is broken, the same 50% martensite zone correlates well with a transition from very smooth or faceted intergranular fracture (now known to be related to austenitizing and the presence of impurities such as phosphorus) associated with a predominantly martensitic structure to a rough, transgranular surface associated with ductile fracture of the softer nonmartensitic transformation products of austenite. Therefore both etching and fracture observations, frequently on a macroscopic scale, could be readily used to evaluate depth of hardening at the 50% martensite level. Detection of martensite at levels above 50% in the microstructure would be much more difficult.

Not only does the etching and fracture response of a quenched bar change abruptly at the 50% martensite level, but hardness also changes rapidly as bar diameter increases through those associated with 50% martensite. Figure 6.16 shows center hardness as function of bar diameter for the chromium-nickel SAE 3140 steel quenched in oil and water. Each quenching medium produces a different critical diameter associated with the rapid changes in hardness with bar diameter close to Rockwell C 50. Judging the position of 50% martensite from hardness changes with bar diameter can be difficult, as for example in the water quenched data of Fig. 6.16. Therefore, the probable hardness associated with a 50% martensite structure, similar to those given for fully martensitic microstructures in Fig. 6.1 and 6.2, was determined as a function of carbon content. Figure 6.17 shows such a plot based on data from plain carbon steels. Alloy steels are expected to show somewhat higher hardness values, as is the case for the SAE 3140 steel. From Fig. 6.17 the hardness at 50% martensite for a 0.40% carbon steel would be expected to be Rockwell C 40, but Fig. 6.16 shows that the critical diameter was selected at Rockwell C 50. A possible explanation for this discrepancy may be the presence of large amounts of bainite having relatively high hardness together with 50% martensite in alloy steels, whereas ferrite and pearlite of relatively lower hardness might coexist with 50% martensite in plain carbon steels.

In summary of the above discussion, the critical size or diameter of a steel of given composition is directly related to a given quenching medium. The higher the quench severity the greater the critical size. The ideal size, on the other hand, is defined as the size of bar hardened to 50% martensite by a theoretically perfect quench in which it is assumed the surface of the bar cools instantly to the temperature of the quenching medium. The ideal size is a true measure of the hardenability associated with a given steel composition, and it can also be used to determine the critical size of the steels quenched in media of different quench severities.

Figures 6.18 and 6.19 show plots of critical size (D) versus ideal size (D_I) for various quench severities (H). The straight line identified by a quench severity of infinity shows that the critical size equals the ideal size for a theoretically perfect quench. However, as quench severity decreases, Fig. 6.18 and 6.19 show that the critical size for a given D_I decreases. Thus the concept of the ideal size permits a rapid estimate of the bar size that will harden to the 50% martensite level in quenches over the entire range of severities. Similar curves between critical plate thickness, ideal plate thickness and quench severity have also been developed (Ref 6.16).

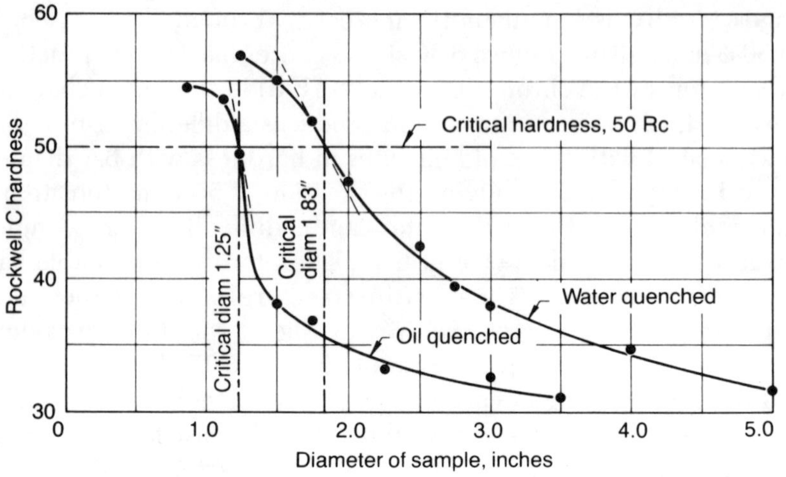

Fig. 6.16. Hardness at the center of water- and oil-quenched bars of SAE 3140 steel of various diameters. (Ref 6.16)

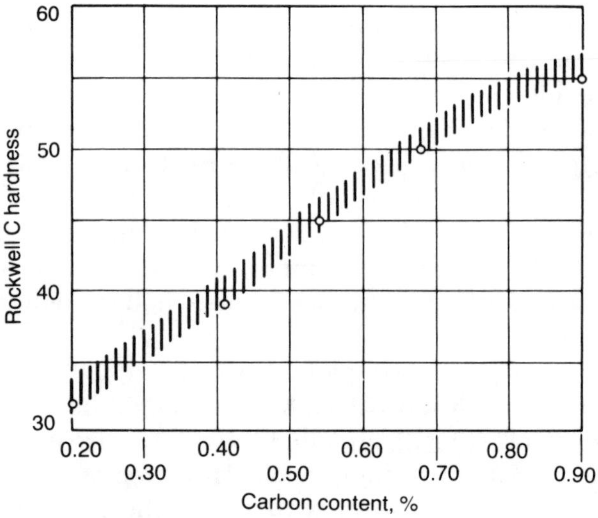

Fig. 6.17. Hardness as a function of carbon content for quenched structures that contain 50% martensite. (Ref 6.16)

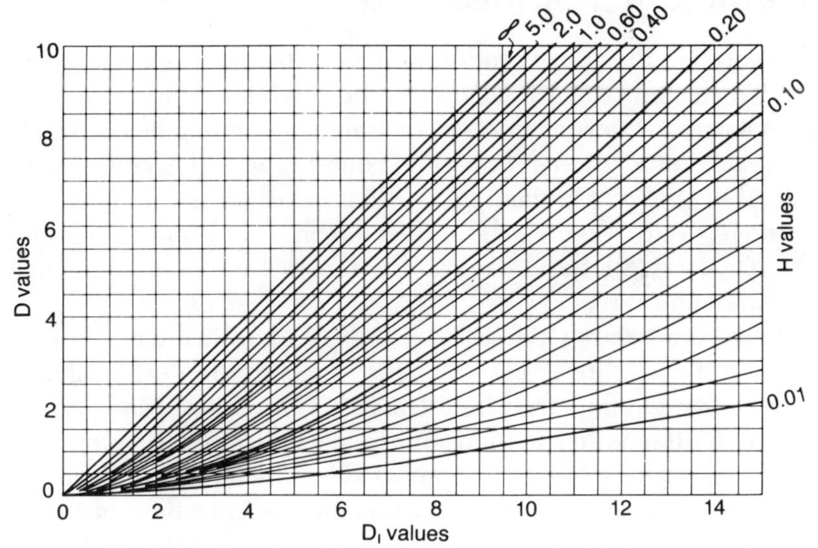

Fig. 6.18. Relationship between actual critical size (D), ideal critical size (D_I) and severity of quench (H). (Ref 6.16)

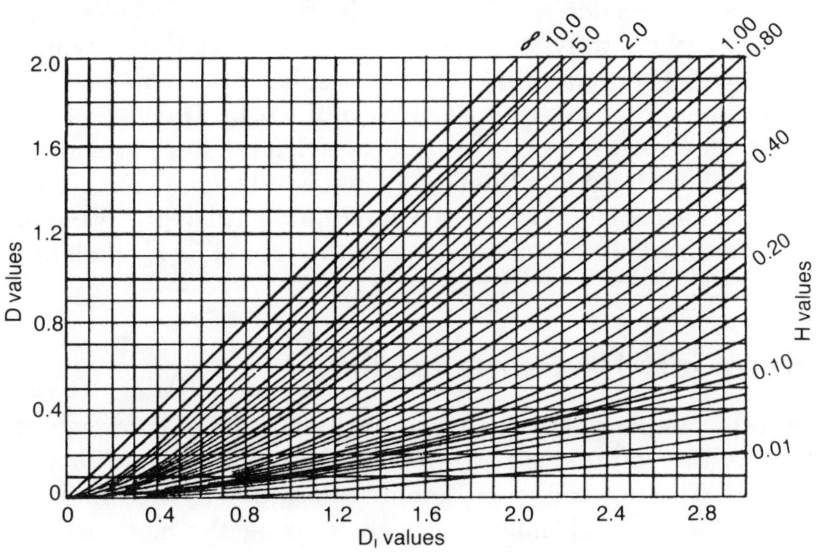

Fig. 6.19. Relationships similar to those shown in Fig. 6.18 but at a larger scale. (Ref 6.16)

Determination of Ideal Size

As noted in the above section, the ideal diameter is a true measure of the hardenability of a steel and can be used to compare the hardening response of different steels to the same quenching medium. Three factors, austenitic grain size, carbon content, and alloy content, affect the ideal diameter. Fundamentally, an increase in any of these factors reduces the rate at which the diffusion-controlled transformations of austenite occur and thereby makes martensite formation more likely at a given cooling rate.

Figure 6.20 shows the relationship of ideal diameter to carbon content and austenite grain size. This plot is used to establish a base hardenability, D_I, for a steel based on its carbon content and grain size. The base hardenability is then multiplied by factors as given in Fig. 6.21 for the various concentrations of alloying elements. As an example, Table 6.4 shows multiplying factors for concentrations of elements in a nickel-chromium steel containing 0.5% carbon (Ref 6.16). If the steel has an austenitic grain size of No. 7, then the base ideal diameter from Fig. 6.19 is 0.24 in. (6.1 mm). After multiplying by the factors in Table 6.4, an ideal diameter of 2.4 in. (61 mm) is obtained for the steel. The multiplying factors have been reviewed and revised over the years, and the reader is referred to Ref 6.18 for a recent complete compilation of multiplying factors for the common alloying elements.

Table 6.5 lists ranges of D_I for a number of commercial steels. Compositions of these steels are given in Ref 6.19. The letter H at the end of the SAE-AISI designation indicates that the steels are produced to specified hardenability limits. The range in D_I for a given steel is a result of the acceptable ranges of composition for that grade and other factors such as grain size and the concentrations of residual elements.

Jominy Test for Hardenability

Another important approach to the evaluation of hardenability is the use of the end-quench test developed by Jominy and Boegehold (Ref 6.23). The test is now commonly referred to as the Jominy test, and has the great advantage of characterizing the hardenability of a given steel from a single specimen rather than from a series of round bars.

Figure 6.22 shows the shape and dimensions of a Jominy specimen and the fixture for supporting the specimen in a quenching system. The specimen is cooled at one end by a column of water, thus the entire specimen experiences a range of cooling rates between those associ-

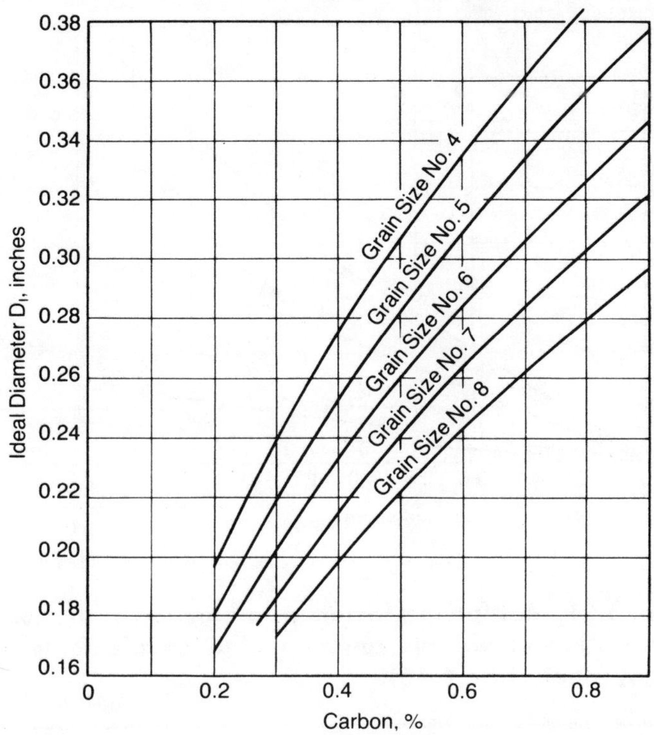

Fig. 6.20. Hardenability, expressed as ideal critical size, as a function of austenite grain size and carbon content of iron-carbon alloys. (Ref 6.16)

Table 6.4. Composition and Multipliers for a Ni-Cr Steel

	Carbon	Manganese	Element Silicon	Nickel	Chromium
Concentration....	0.50%	0.80%	0.25%	1.00%	0.28%
Multiplier	0.24	3.7	1.2	1.4	1.6

ated with water and air cooling. After quenching, parallel flats are ground on opposite sides of the specimen, and hardness readings are taken every 1/16 in. from the quenched end, and plotted as shown in Fig. 6.23. Hardenability differences between different grades of steels can be readily compared if Jominy curves are available. For example, Fig. 6.24 shows hardenability differences between different grades of alloy steels containing 0.5% carbon. Higher hardness persists to greater distances from the quenched end in the more hardenable steels.

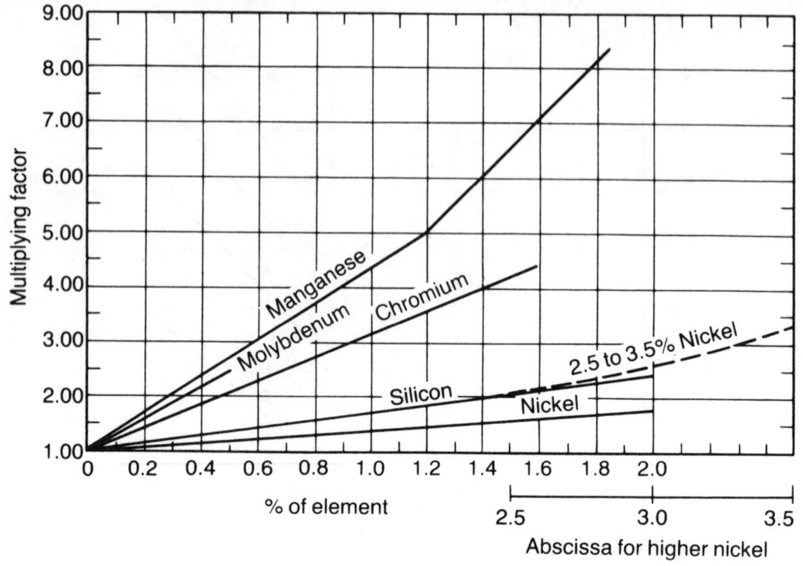

Fig. 6.21. Multiplying factors as a function of the concentration of various common alloying elements in alloy steels. (Ref 6.16)

The Jominy test method is now standardized in specifications of the American Society for Testing and Materials (ASTM Method A 255) and the Society of Automotive Engineers (SAE Standard J406). Figure 6.25 shows the method of presentation of the end-quench data for a single heat of AISI 8650 steel (Ref 6.25). For any grade of steel a hardenability band (see Fig. 6.26) develops because of the small variations in composition allowable in the grade. The SAE/AISI steels designated by the letter H (H-steels) are guaranteed to meet established hardenabilities.

A very important feature of the Jominy test is that each position of the specimen corresponds to a well-known cooling rate. The top scale of Fig. 6.25 shows approximate cooling rates corresponding to positions on the Jominy specimen. As developed previously, it is the cooling rate that determines the amount of martensite, and therefore the degree of hardness, that develops at a given point in a steel specimen. Therefore, if cooling rates as a function of position in parts of various geometries are known, it is possible to use Jominy curves to plot hardness profiles in the parts. Such correlations of cooling rate as a function of position in various sizes of bars and plates quenched in various media are available (Ref 6.25). Figure 6.27 shows equivalent

Table 6.5. Hardenabilities (Stated as a Range of D_I Values) for Various Steels (Ref 6.16)

Steel	D_I	Steel	D_I	Steel	D_I
1045	0.9 to 1.3	4135 H	2.5 to 3.3	8625 H	1.6 to 2.4
1090	1.2 to 1.6	4140 H	3.1 to 4.7	8627 H	1.7 to 2.7
1320 H	1.4 to 2.5	4317 H	1.7 to 2.4	8630 H	2.1 to 2.8
1330 H	1.9 to 2.7	4320 H	1.8 to 2.6	8632 H	2.2 to 2.9
1335 H	2.0 to 2.8	4340 H	4.6 to 6.0	8635 H	2.4 to 3.4
1340 H	2.3 to 3.2	X4620 H	1.4 to 2.2	8637 H	2.6 to 3.6
2330 H	2.3 to 3.2	4620 H	1.5 to 2.2	8640 H	2.7 to 3.7
2345	2.5 to 3.2	4621 H	1.9 to 2.6	8641 H	2.7 to 3.7
2512 H	1.5 to 2.5	4640 H	2.6 to 3.4	8642 H	2.8 to 3.9
2515 H	1.8 to 2.9	4812 H	1.7 to 2.7	8645 H	3.1 to 4.1
2517 H	2.0 to 3.0	4815 H	1.8 to 2.8	8647 H	3.0 to 4.1
3120 H	1.5 to 2.3	4817 H	2.2 to 2.9	8650 H	3.3 to 4.5
3130 H	2.0 to 2.8	4820 H	2.2 to 3.2	8720 H	1.8 to 2.4
3135 H	2.2 to 3.1	5120 H	1.2 to 1.9	8735 H	2.7 to 3.6
3140 H	2.6 to 3.4	5130 H	2.1 to 2.9	8740 H	2.7 to 3.7
3340	8.0 to 10.0	5132 H	2.2 to 2.9	8742 H	3.0 to 4.0
4032 H	1.6 to 2.2	5135 H	2.2 to 2.9	8745 H	3.2 to 4.3
4037 H	1.7 to 2.4	5140 H	2.2 to 3.1	8747 H	3.5 to 4.6
4042 H	1.7 to 2.4	5145 H	2.3 to 3.5	8750 H	3.8 to 4.9
4047 H	1.8 to 2.7	5150 H	2.5 to 3.7	9260 H	2.0 to 3.3
4047 H	1.7 to 2.4	5152 H	3.3 to 4.7	9261 H	2.6 to 3.7
4053 H	2.1 to 2.9	5160 H	2.8 to 4.0	9262 H	2.8 to 4.2
4063 H	2.2 to 3.5	6150 H	2.8 to 3.9	9437 H	2.4 to 3.7
4068 H	2.3 to 3.6	8617 H	1.3 to 2.3	9440 H	2.4 to 3.8
4130 H	1.8 to 2.6	8620 H	1.6 to 2.3	9442 H	2.8 to 4.2
4132 H	1.8 to 2.5	8622 H	1.6 to 2.3	9445 H	2.8 to 4.4

cooling rates for four positions in round bars quenched in water and oil. As bar diameter increases, the cooling rates at the surface and interior points decrease (see top scale of Fig. 6.27). The cooling rates correspond to equivalent distances from the quenched end (see bottom scale of Fig. 6.27), and those distances can be used to determine the hardness distribution in the rounds from appropriate Jominy curves.

The use of the Jominy data as described above is a highly accurate method of selecting steels of just the right hardenability for a given required hardness distribution. A steel can be selected that will not only satisfy the hardness requirements but also have just the right alloy content, therefore permitting selection at minimum cost from the many steels that might have sufficient or even excess hardenability for the application. On the other hand, alloy steels that can be hardened by moderate quenching may be selected to replace leaner steels in which the severe quenching required to obtain high hardness causes quench cracking.

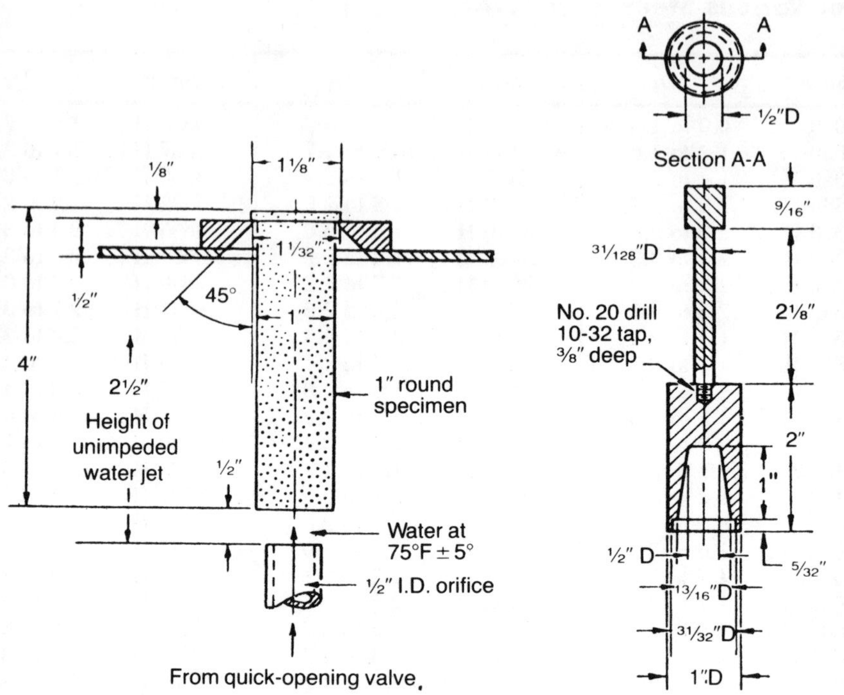

Fig. 6.22. Jominy-Boegehold specimen for end-quench test for hardenability. (Ref 6.16)

Recent Developments

The technology associated with hardenability is continually developing. A measure of this activity is the recent publication of two volumes, *Hardenability Concepts with Applications to Steel* (Ref 6.26) and *The Hardenability of Steels — Concepts, Metallurgical Influences and Industrial Applications* (Ref 6.18). A detailed review of these two volumes is published in Ref 6.27. The principles of hardenability and much of the hardenability testing as described in the preceding sections of this chapter remain the same, but the present emphasis is on developing more reliable and systematic hardenability data that can be used in rapid computer prediction of hardenability and the selec-

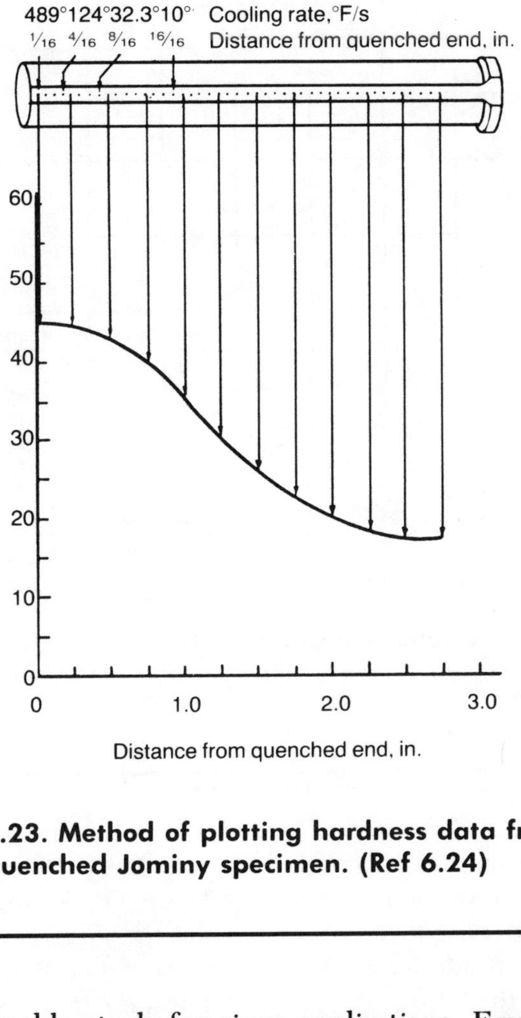

Fig. 6.23. Method of plotting hardness data from an end-quenched Jominy specimen. (Ref 6.24)

tion of hardenable steels for given applications. Emphasis has also been placed on evaluating the hardenability of shallow hardening low-carbon steels (Ref 6.28), high-carbon steels (Ref 6.29) and the boron steels (Ref 6.30, 6.31), all of which did not receive a great deal of attention in the early days when the hardenability of medium-carbon steels was of greatest importance.

Several computerized on-line systems for hardenability calculations have been developed and are presently in use. The International Harvester Company has developed a system identified as CHAT, an acronym for Computer Harmonized Application Tailored process. This

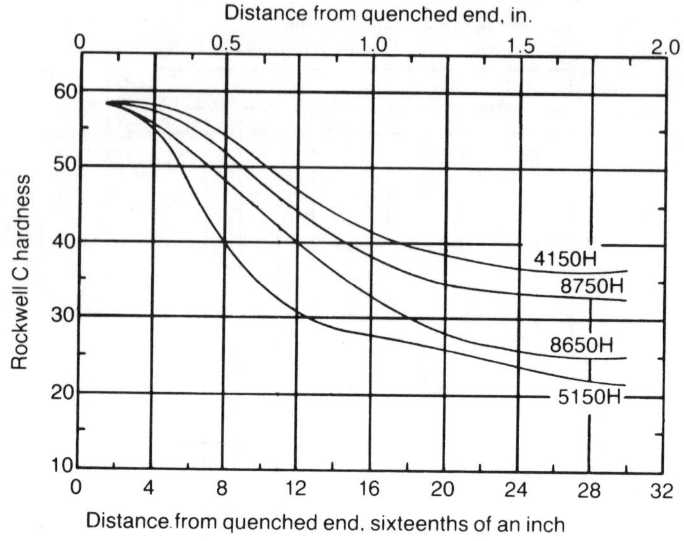

Fig. 6.24. Results of end-quench tests for four different grades of alloy steels, all containing 0.5% carbon. (Ref 6.24)

system is a two-part system in which the first part determines the hardenability requirements in terms of D_I for a given application, and the second part is used to design or select the most economical steel with composition adequate for the requirements. Another system, developed by the Creusot-Loire Corporation in France, predicts structures produced at various cooling rates by the use of CT diagrams and then calculates hardness, yield and tensile strengths expected from martensitic, bainitic, and/or ferrite-pearlite microstructures in the as-quenched and as-tempered conditions. Another system, the Minitech Alloy Steel Information System, calculates the inflection point of Jominy curves as well as entire Jominy curves for many types of steels. The reader is referred to Ref 6.26 for detailed articles about the development and applications of these and other systems for hardenability calculations.

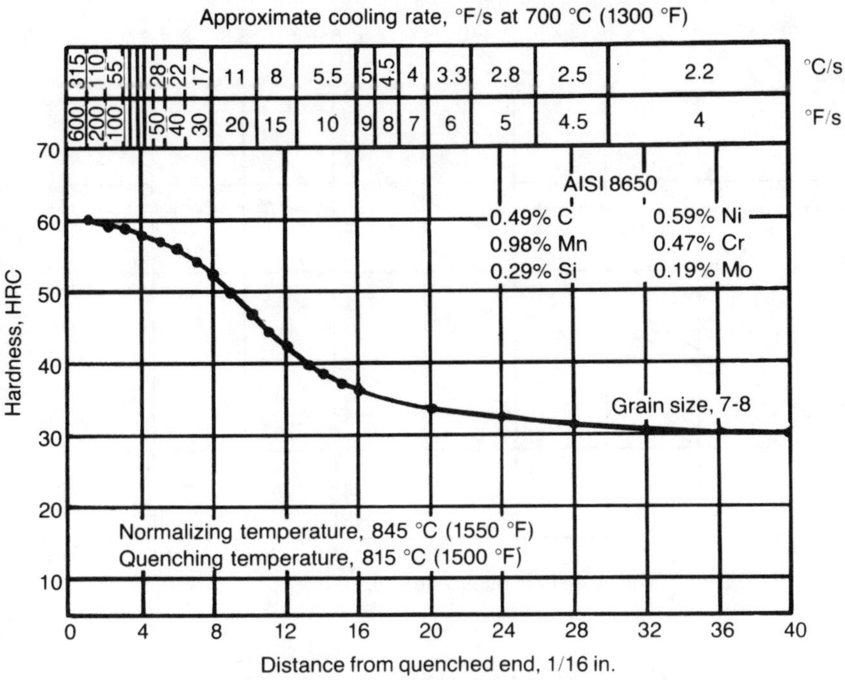

Fig. 6.25. Method for presenting end-quench hardenability data. Data presented here are for AISI 8650 steel. Note relationship of cooling rate (top) to distance from the quenched end. (Ref 6.25)

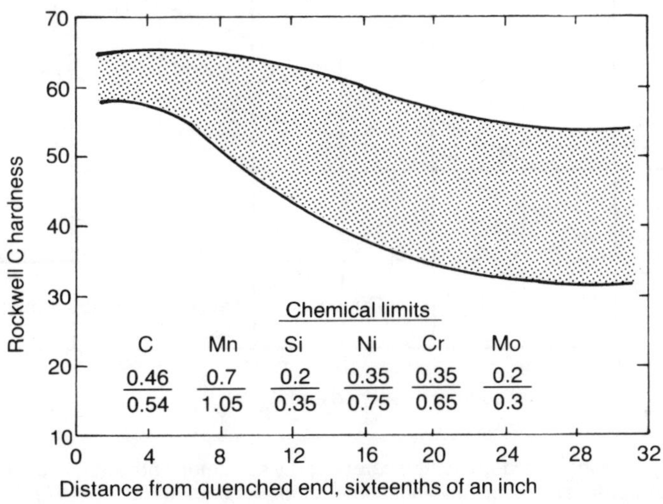

Fig. 6.26. Hardenability band for 8750H steel. (Ref 6.24)

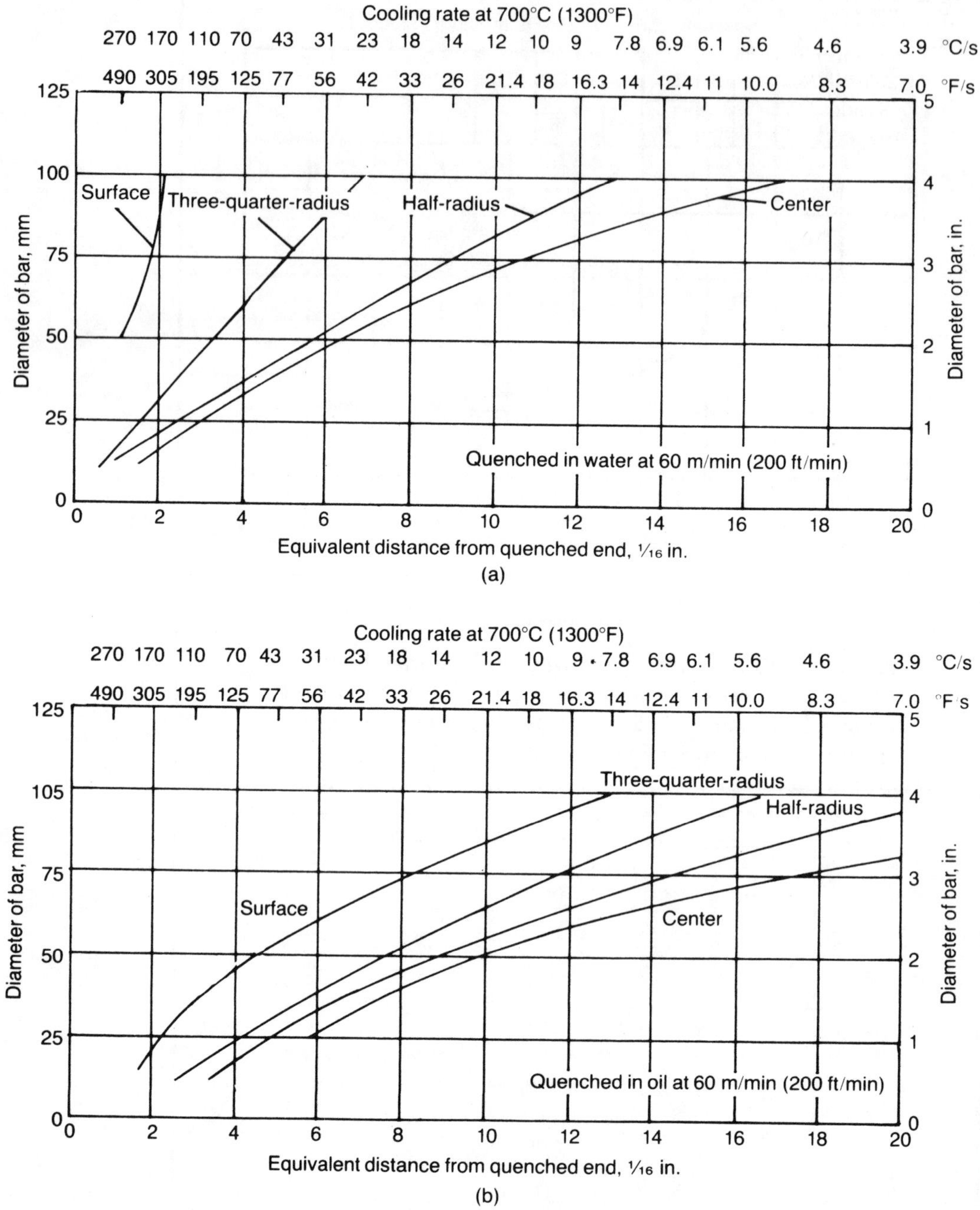

Correlation of equivalent cooling rates in the end-quenched hardenability specimen and quenched round bars free from scale. Data for surface hardness are for "mild agitation"; other data are for 60 m/min (200 ft/min).

Fig. 6.27. Equivalent cooling rates for round bars quenched in water and oil (Ref 6.25).

References

6.1 E. C. Bain and H. W. Paxton, *Alloying Elements in Steel*, 2nd Ed., American Society for Metals, Metals Park, Oh., 1961

6.2 G. Krauss, Martensitic Transformation, Structure and Properties in Hardenable Steels, in *Hardenability Concepts with Applications to Steel*, D. V. Doane and J. S. Kirkaldy (Eds.), AIME, Warrendale, Pa., 1978, p 229-248

6.3 L. D. Jaffee and E. Gordon, Temperability of Steels, *Trans ASM*, Vol 49, 1957, p 359-369

6.4 A. R. Marder, The Morphology and Strength of Iron-Carbon Martensite, Ph.D. Dissertation, Lehigh University, Bethlehem, Pa., 1968

6.5 M. Cohen, The Strengthening of Steel, *Trans TMS-AIME*, Vol 224, 1962, p 638-657

6.6 P. G. Winchell and M. Cohen, The Strength of Martensite, *Trans ASM*, Vol 55, 1962, p 347-361

6.7 T. E. Swarr and G. Krauss, Boundaries and the Strength of Low Carbon Ferrous Martensites, in *Grain Boundaries in Engineering Materials*, Claitor's Publishing Division, Baton Rouge, La., 1975, p 127-138

6.8 T. E. Swarr and G. Krauss, The Effect of Structure on the Deformation of As-Quenched and Tempered Martensite in an Fe-0.2% C Alloy, *Met Trans A*, Vol 7A, 1976, p 41-48

6.9 A. R. Marder and G. Krauss, The Effect of Morphology on the Strength of Lath Martensite, *Proceedings of Second International Conference on the Strength of Metals and Alloys*, Vol III, American Society for Metals, Metals Park, Oh., 1970, p 822-823

6.10 M. J. Roberts, Effect of Transformation Substructure on the Strength and Toughness of Fe-Mn Alloys, *Met Trans*, Vol 1, 1970, p 3287-3294

6.11 M. Cohen, Strengthening Mechanisms in Steel, *Trans J.I.M.*, Vol 9, 1968, Supplement

6.12 M. J. Roberts and W. S. Owen, Solid Solution Hardening and Thermally Activated Deformation in Iron-Nickel-Carbon Martensites, *JISI*, Vol 206, 1968, p 375-384

6.13 G. R. Speich, Tempering of Low-Carbon Martensite, *Trans TMS-AIME*, Vol 245, 1969, p 2553-2564

6.14 G. R. Speich and H. Warlimont, Yield Strength and Transformation Substructure of Low-Carbon Martensite, *JISI*, Vol 206, 1968, p 385-392

6.15 W. H. McFarland, Mechanical Properties of Low-Carbon Alloy-Free Martensite, *Trans TMS-AIME*, Vol 233, 1967, p 2028-2035

6.16 M. A. Grossmann and E. C. Bain, *Principles of Heat Treatment*, 5th Ed., American Society for Metals, Metals Park, Oh., 1964

6.17 Definition Relating to Metals and Metalworking, *Metals Handbook*, Vol 1, 8th Ed., American Society for Metals, Metals Park, Oh., 1961, p 20

6.18 C. A. Siebert, D. V. Doane and D. H. Breen, *The Hardenability of Steels — Concepts, Metallurgical Influences, and Industrial Applications*, American Society for Metals, Metals Park, Oh., 1977

6.19 Classification and Designation of Carbon and Alloy Steels, in *Metals Handbook*, Vol 1, 9th Ed., American Society for Metals, Metals Park, Oh., 1978, p 117-143

6.20 N. B. Pilling and T. D. Lynch, Cooling Properties of Technical Quenching Liquids, *Trans AIME*, Vol 62, 1920, p 665

6.21 M. A. Grossmann and M. Asimov, Hardenability and Quenching, *Iron Age*, Vol 145, 1940, p 25-29, 39-45
6.22 F. Wever, *Archiv für das Eisenhüttenwesen*, Vol 5, 1936-37, p 367
6.23 W. E. Jominy and A. L. Boegehold, *Trans ASM*, Vol 26, 1938, p 574
6.24 G. F. Melloy, *Hardness and Hardenability*, P. D. Harvey (Ed.), Metals Engineering Institute, Metals Park, Oh., 1977
6.25 C. F. Jatczak, Hardenability of Carbon and Alloy Steels, in *Metals Handbook*, Vol 1, 9th Ed., American Society for Metals, Metals Park, Oh., 1978, p 471-526
6.26 D. V. Doane and J. S. Kirkaldy (Eds.), *Hardenability Concepts with Applications to Steel*, AIME, Warrendale, Pa., 1978
6.27 D. V. Doane, Application of Hardenability Concepts in Heat Treatment of Steel, *J Heat Treating*, Vol 1, 1979, p 5-30
6.28 R. A. Grange, Estimating the Hardenability of Carbon Steels, *Met Trans*, Vol 4, 1973, p 2231-2244
6.29 C. F. Jatczak, Hardenability in High Carbon Steels, *Met Trans*, Vol 4, 1973, p 2267-2277
6.30 P. Maitrepierre, D. Thivellier, J. Rofes-Vernis, D. Rousseau, and R. Tricot, Microstructure and Hardenability of Low-Alloy Boron-Containing Steels, in *Hardenability Concepts with Applications to Steel*, AIME, Warrendale, Pa., 1978, p 421-447
6.31 B. M. Kapadia, Prediction of the Boron Hardenability Effect in Steel — A Comprehensive Review, in *Hardenability Concepts with Applications to Steel*, AIME, Warrendale, Pa., 1978, p 448-482

CHAPTER 7

Austenite in Steels

The formation of austenite and the control of austenite grain size are vital aspects of many of the annealing and hardening heat treatments described in earlier chapters. Both the transformation behavior of austenite and the mechanical properties of the microstructures formed from austenite are influenced by austenite grain size. This chapter discusses the formation of austenite from various types of starting microstructures and shows how austenite grain size may be revealed in various types of transformed structures. Finally, the control of austenite grain size by particle dispersions and hot rolling practice is described.

Austenite and Properties

The austenite in carbon steels, although it is stable only above the Ae_1 temperature, strongly influences the transformation and deformation behavior of heat treated steels. With respect to transformation, austenite grain boundaries are preferred sites for the nucleation of proeutectoid phases and pearlite. Therefore, if the austenite grain size in a steel is coarse, fewer nucleation sites are available and diffusion-controlled transformation of the austenite is retarded. As a result, hardenability is increased. The effect of austenite grain size on hardenability has already been discussed in Chapter 6, and the effect of austenitic grain size on the base hardenability (D_I) is shown in Fig. 6.20. Austenite grain size also affects martensite transformation kinetics by its effect on M_s temperatures. In Fe-Ni and Fe-Ni-C alloys decreasing austenite grain size significantly lowers M_s (Ref 7.1). The latter result is attributed to the higher strength of fine-grained austenite which in turn increases the shear resistance of the austenite-to-martensite transformation. Similar effects of grain size on M_s may occur in carbon steels.

The preferred nucleation and growth of proeutectoid phases and pearlite at austenitic grain boundaries establishes a direct relationship between austenite grain size and the grain size of the transformation products: the finer the austenite grain size the finer the grain size of the ferrite-cementite products. In normalized or annealed low-carbon steels, where the microstructure is predominantly ferritic, not only the strength but also the toughness of the steel is increased with decreasing grain size. Equations showing the effect of ferrite grain size on the mechanical properties of ferritic steels are given in Chapter 5. The reduction of austenite grain size in low-carbon steels, therefore, offers the possibility of significant improvement in the properties of ferritic microstructures and, in fact, plays a very important part in the recent development of low-carbon high-strength low-alloy (HSLA) steels as discussed later in this chapter.

In hardened steels, formation of martensite in fine-grained austenite is preferred because of the improved mechanical properties that result. The increase in yield or flow strength of low-carbon lath martensite with decreasing austenite grain size has already been described in the discussion of martensite hardness in Chapter 6. Lath martensite is unique in that the martensite units form in parallel arrays called packets (see Chapter 3) which subdivide the parent austenite grain. Each packet is effectively a grain because most of the laths have the same orientation. Figure 7.1 shows that packet size is directly related to austenitic grain size, and helps explain why the strength of lath martensite may be related to either prior austenite grain size or packet size (Ref 7.2-7.5).

The impact toughness of hardened steels, as determined by Charpy impact testing, improves with decreasing austenitic grain size. The reasons for this effect of austenite grain size are complex and may be in part related to the segregation of impurity atoms to austenite grain boundaries during austenitizing (Ref 7.6). In steels austenitized at high temperatures, and therefore with coarse austenitic grain size, fracture of Charpy impact specimens frequently occurs along prior austenite grain boundaries. Such intergranular fracture is quite brittle and often reflects the effects of grain boundary segregation. Analytical techniques such as Auger electron spectroscopy verify the presence of elements such as phosphorus in very thin layers on the grain boundary fracture surfaces of quench and tempered steels (Ref 7.6). Since tempering in certain temperature ranges aggravates intergran-

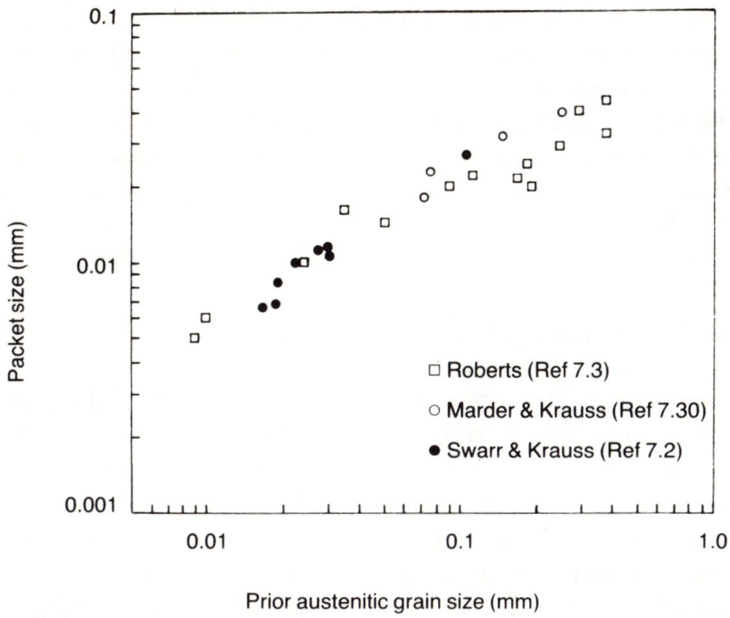

Fig. 7.1. Relationship of martensite packet size to austenite grain size. Roberts (Ref 7.3) data are for Fe-Mn alloys; other data are for an Fe-0.2C alloy. (Ref 7.2)

ular fracture, a more complete discussion of mechanisms of embrittlement in hardened steel is deferred to the chapter on tempering (see Chapter 8).

However, when toughness of martensitic steels is evaluated by the propagation of a very sharp crack that has been fatigued into a specimen, a testing technique referred to as fracture toughness testing, coarse-grained specimens appear to show better fracture toughness than fine-grained specimens (Ref 7.7). This apparent anomaly is resolved by understanding the differences between fracture toughness and Charpy V-notch impact toughness testing and their relationship to the microstructure of hardened steel (Ref 7.8). The probability that the sharp crack used to obtain fracture toughness lies away from grain boundaries and within the martensite and retained austenite of a hardened steel is greater for a coarse-grained specimen than for a fine-grained specimen. Therefore, in coarse-grained specimens, the detrimental effects of grain boundary embrittlement are initially minimized when the sharp crack begins to propagate, while the beneficial matrix effects of homogenization, retained austenite, and the solution of second-phase particles due to high-temperature austenitizing are enhanced. In Charpy V-notch impact testing, however, the large extent of the stress field ahead of the notch immediately causes stress to be applied to embrittled grain boundaries, and even during fracture toughness testing, the propagating crack will soon find and follow any embrittled grain boundaries (Ref 7.9). The best metallurgical and heat treatment practice, then, is to avoid high-temperature austenitizing that produces coarse grains and the attendant high susceptibility to intergranular fracture of hardened steels.

Austenite Formation

The manner in which austenite forms in a given steel depends very much on the microstructure present prior to heating for austenitizing. This section describes austenite formation from microstructures consisting of pearlite, ferrite, spheroidized cementite and martensite.

Figure 7.2 shows the development of austenite in the pearlitic microstructure of a eutectoid steel, from the work of Vilella as published in Ref 7.10. A series of specimens were heated into the austenite field, held for the times shown and quenched. Areas of austenite formation are visible as white patches within the lamellar pearlitic structure. The austenite, of course, transformed to martensite on quenching, but the etching differences between the newly formed martensite and the preexisting pearlite allow clear delineation of the extent of the austenite formation. In addition to the gradual development of the austenite, Fig. 7.2 shows that not all of the cementite is dissolved as the austenite grows into the pearlite. Some of the cementite persists in the form of spheroidized particles (the small dark spots in the white areas), and only dissolves with longer holding times at temperature. More recent work (Ref 7.11) confirms the persistence of cementite in the austenite after initial austenite formation.

When the amount of austenite formed in Fig. 7.2 is plotted as a function of austenitizing time, the curve shown in Fig. 7.3 results. The austenite formation requires some incubation or time for the first nuclei to form and then proceeds at a more rapid rate as more nuclei develop and grow. The time dependence is expected because of the carbon diffusion required to produce an austenite containing 0.8% carbon from the low-carbon (0.02%) ferrite and high-carbon (6.67%) cementite from the pearlite. At higher temperatures the diffusion rate increases, and austenite forms more rapidly. Figure 7.4 shows the acceleration of austenite formation in a pearlitic 0.80% carbon steel when the austenitizing temperature was raised from 730 to 751 °C (1346 to 1385 °F) (Ref 7.12).

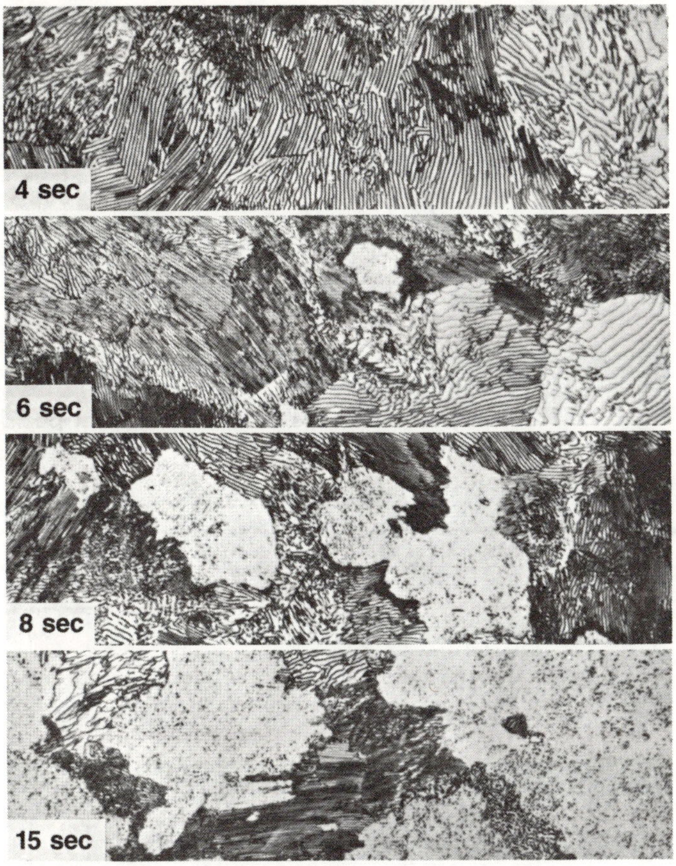

Fig. 7.2. Formation of austenite (light patches) from pearlite as a function of time. (Ref 7.10)

In microstructures consisting initially of ferrite and spheroidized cementite particles, austenite forms first at the interface between the carbides and the ferrite. Ferrite and cementite, therefore, combine to form austenite, exactly the reverse of the reaction that forms pearlite (see Eq 2.1). Figure 7.5 shows the development of austenite at carbide particles in a spheroidized low-carbon steel (Ref 7.10). The cementite particles are soon enveloped by the austenite, and subsequent austenite formation depends on carbon diffusion through the austenite as the carbides dissolve. Figure 7.6 shows that the latter process leads to a slower rate of austenite formation compared to that in a pearlitic steel (see Fig. 7.3) where the closely spaced ferrite and cementite lamellae reduce the diffusion distances for austenite formation. Judd and Paxton

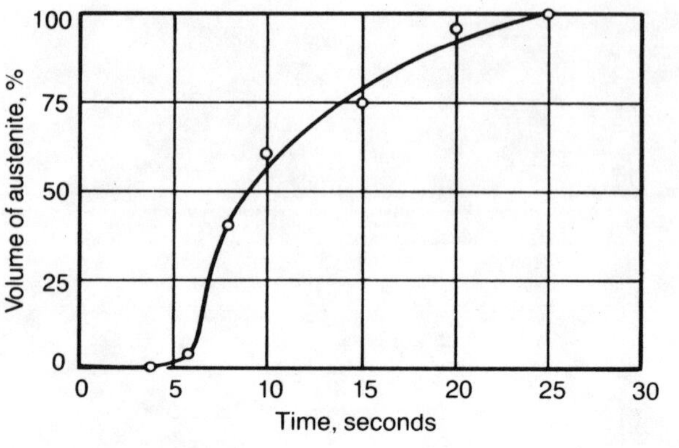

Fig. 7.3. Volume percent austenite formed from pearlite in a eutectoid steel as a function of time at a constant austenitizing temperature. (Ref 7.10)

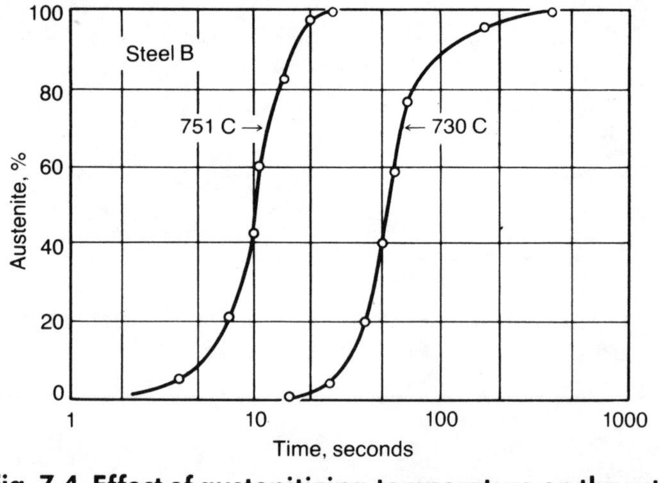

Fig. 7.4. Effect of austenitizing temperature on the rate of austenite formation from pearlite in eutectoid steel. (Ref 7.10)

(Ref 7.13) and Speich and Szirmae (Ref 7.11) show similar effects associated with austenite formation in spheroidite.

Figure 7.7 shows schematically the preferred nucleation sites for austenite formation in three different microstructures (Ref 7.11). In ferritic microstructures, nucleation of austenite occurs primarily at ferrite grain boundary faces and edges. In spheroidized microstruc-

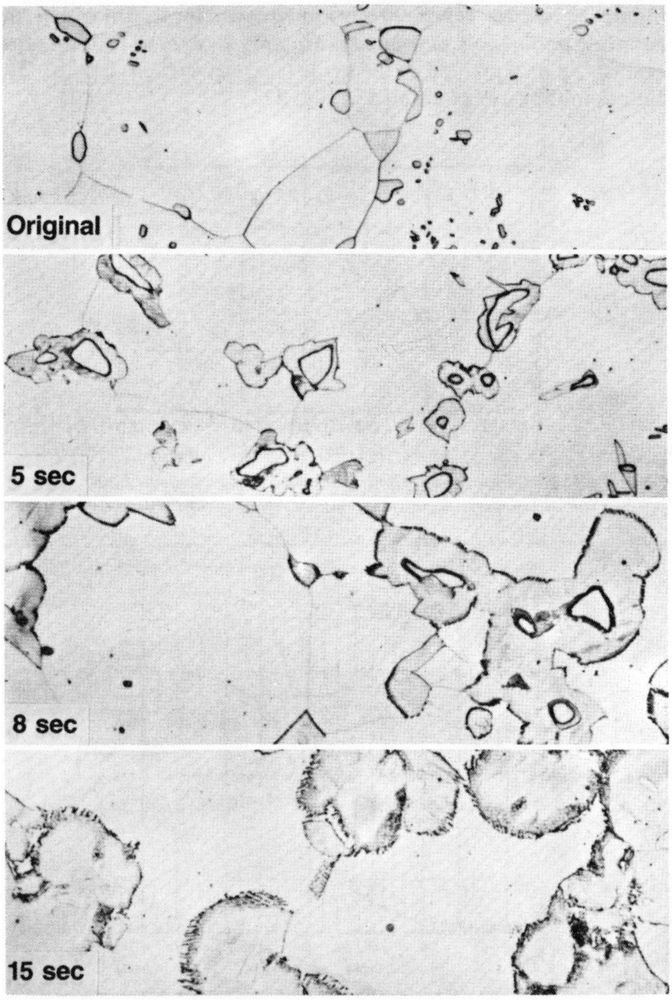

Fig. 7.5. Formation of austenite around cementite particles of a spheroidized steel. (Ref 7.10)

tures, austenite nucleates at cementite particles associated with ferrite grain boundaries. In pearlite, austenite nucleates primarily at intersections of pearlite colonies but also at the interfaces of ferrite and cementite lamellae within a colony.

The formation of austenite in martensitic microstructures of carbon steels produces two morphologies of austenite crystals (Ref 7.13-7.15). One morphology is equiaxed and tends to form at prior austenite grain

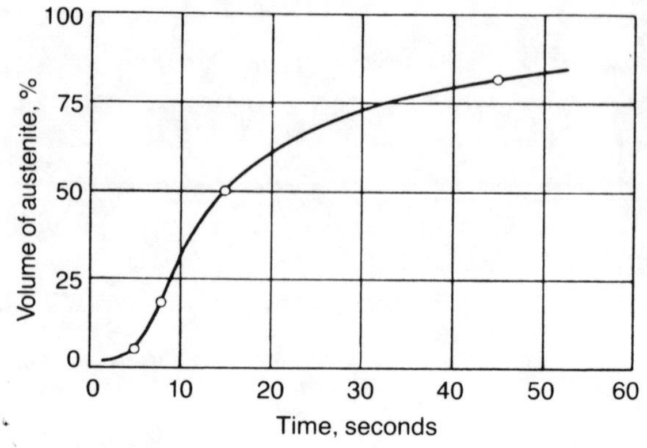

Fig. 7.6. Austenite formation from a spheroidized microstructure as a function of time. (Ref 7.10)

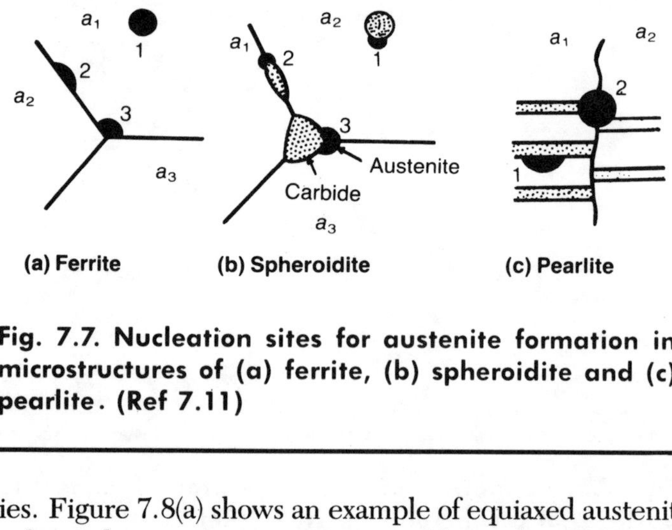

Fig. 7.7. Nucleation sites for austenite formation in microstructures of (a) ferrite, (b) spheroidite and (c) pearlite. (Ref 7.11)

boundaries. Figure 7.8(a) shows an example of equiaxed austenite grains that have formed in martensitic 4340 steel heated to 715 °C (1319 °F). The dark etching component of the microstructure in Fig. 7.8(a) is fine tempered martensite with the structure shown in Fig. 7.8(b). The other austenite morphology is acicular or lath-like and forms by nucleation between the laths of martensite. In martensitic carbon steels which readily temper even on heating to the austenitizing temperature, the mechanism of the austenite formation appears to be diffusion controlled and associated with fine carbide particles (Ref 7.14, 7.15). However, in carbon steels with high nickel content, it has been suggested (Ref 7.16) that the lath-like austenite may be formed by a shear

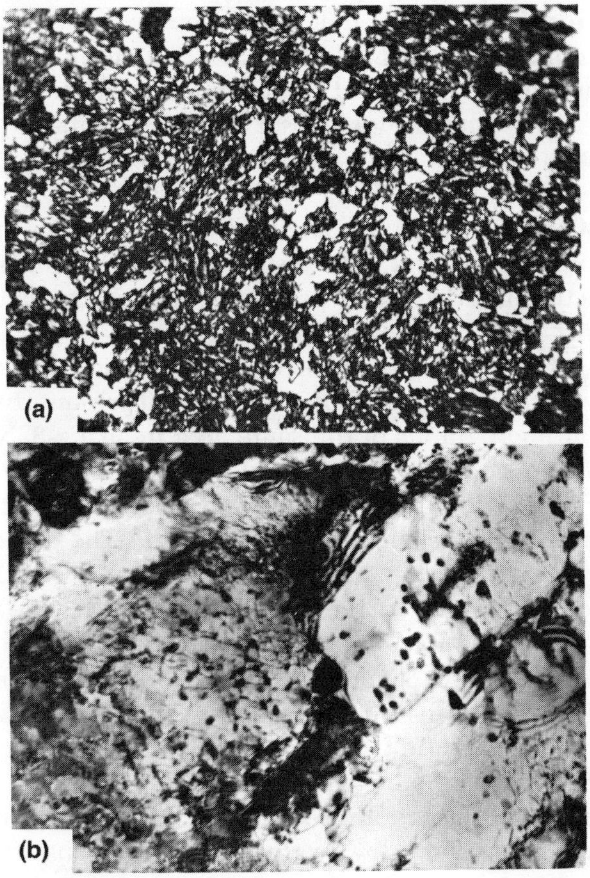

Fig. 7.8. (a) Formation of equiaxed austenite grains (white patches) in tempered martensite of a hardened 4340 steel specimen rapidly heated to 715 °C (1319 °F), held for 6 s, and oil quenched. Light micrograph. Nital etch. Magnification, 1300 ×; shown here at 75%. (b) Tempered martensite in same specimen. Very fine equiaxed grains of ferrite with spheroidized carbides formed. Transmission electron micrograph. Magnification, 20 000 ×; shown here at 75%. (T. Scoonover and G. Krauss, unpublished research)

process similar to that described below for Fe-Ni alloys. At any rate, the formation of austenite from martensite or tempered martensite is extremely rapid. In a study of AISI 4340 steel (Ref 7.14), austenite was formed in as little as 2 s at temperatures between 790 and 870 °C (1450 and 1600 °F) by rapidly heating thin specimens in a lead bath. The martensite that formed from that austenite was harder than normally obtained, an indication that the rapid austenite formation introduced imperfections that subsequently affected martensite strength. The higher than normal hardness observed after high-rate short-time austenitizing, as sometimes produced by induction heating, has been referred to as superhardness (Ref 7.17).

In Fe-30Ni alloys, the formation of austenite from martensite is clearly accomplished by a martensitic or shear mechanism. Figure 7.9(a) shows tilting surface relief produced when many plates of austenite formed in a large plate of martensite heated to 340 °C (644 °F). When the surface relief was polished away, platelets of austenite were revealed by etching (see Fig. 7.9b). Figure 7.10 presents two transmission electron micrographs of the same area in an Fe-33Ni alloy, one taken before and one after the transformation of the martensite to austenite. Figure 7.10(a) shows a portion of a martensite plate with very fine twins and the adjacent retained austenite. The circle shows the area from which a diffraction pattern was taken to confirm the presence of the martensite and the austenite. When the martensite was heated, austenite formed (see Fig. 7.10b). The new austenite is characterized by a very high density of tangled dislocations which substantially raises the strength of the shear-formed austenite relative to annealed austenite (Ref 7.18, 7.19), and provides the strain energy for recrystallization of the austenite during annealing (Ref 7.19).

The shear transformation of austenite from martensite in Fe-Ni alloys occurs even at very slow heating rates in the absence of carbon. In Fe-Ni-C alloys (22 to 32% nickel, 0.004 to 0.6% carbon), however, the shear transformation occurs only at high heating rates, and is replaced by a diffusion-controlled formation of austenite at slower heating rates (Ref 7.20). The latter type of austenite formation is associated with an equiaxed morphology, relatively low dislocation density and a lack of tilting surface relief, and is probably characteristic of equiaxed austenite formation from martensite in carbon steels.

Austenite Grain Size

Austenite grain size is usually reported in terms of ASTM grain size numbers, where n, the grain size number, is obtained from the expression $2^{(n-1)}$ which gives the number of grains per square inch in a microstructure examined at a magnification of 100×. Figure 7.11 shows prior austenite grain sizes ranging from ASTM No. 1 to ASTM No. 9 in a hardened medium-carbon steel. In this series of micrographs

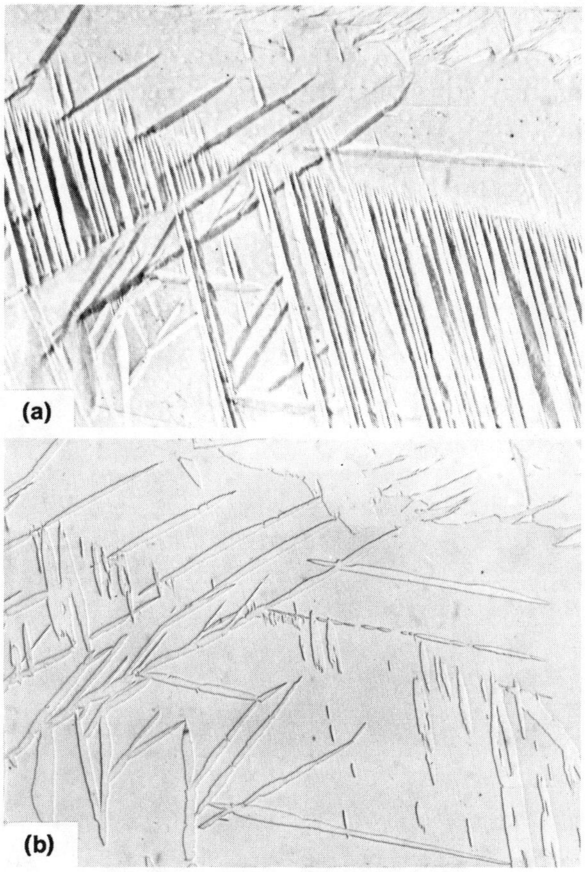

Fig. 7.9. (a) Tilting surface relief produced by formation of plates of austenite in a large plate of martensite in an Fe-32Ni alloy heated to 340 °C (644 °F). (b) Same area after polishing and etching. Both light micrographs. Magnification, 500×; shown here at 75%. (G. Krauss and W. Pitsch, unpublished research at Max-Planck Institut für Eisenforschung, Düsseldorf, Germany)

the austenite grains are revealed rather clearly by the etching differences associated with various orientations of martensite in the different austenite grains.

Frequently, however, austenite grain size is difficult to reveal, especially in quench and tempered steels, and special etching or heat

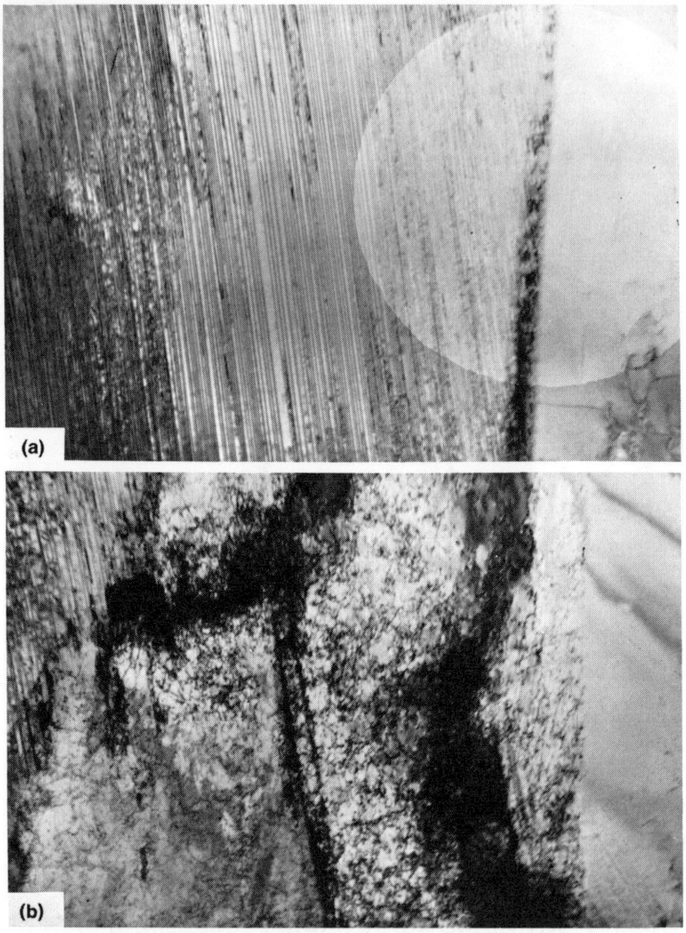

Fig. 7.10. (a) Portion of a martensite plate with fine transformation twins and adjacent retained austenite in a Fe-33Ni alloy. (b) Same area after transformation of martensite to austenite. Both transmission electron micrographs. Magnification, 40 000 ×; shown here at 60%. (G. Krauss and W. Pitsch, unpublished research at Max-Planck Institut für Eisenforschung, Düsseldorf, Germany)

treatments must be used to develop the austenitic structure. For example (Ref 7.21), Fig. 7.12 shows the microstructure of a quenched Fe-1.22C alloy etched to show the austenite grain boundaries, (a), and to show the plate martensite, (b). The austenitic grain boundaries were revealed by etching untempered specimens in a boiling solution of 25 g NaOH, 2 g picric acid, and 100 ml H_2O for 15 min, followed by etching

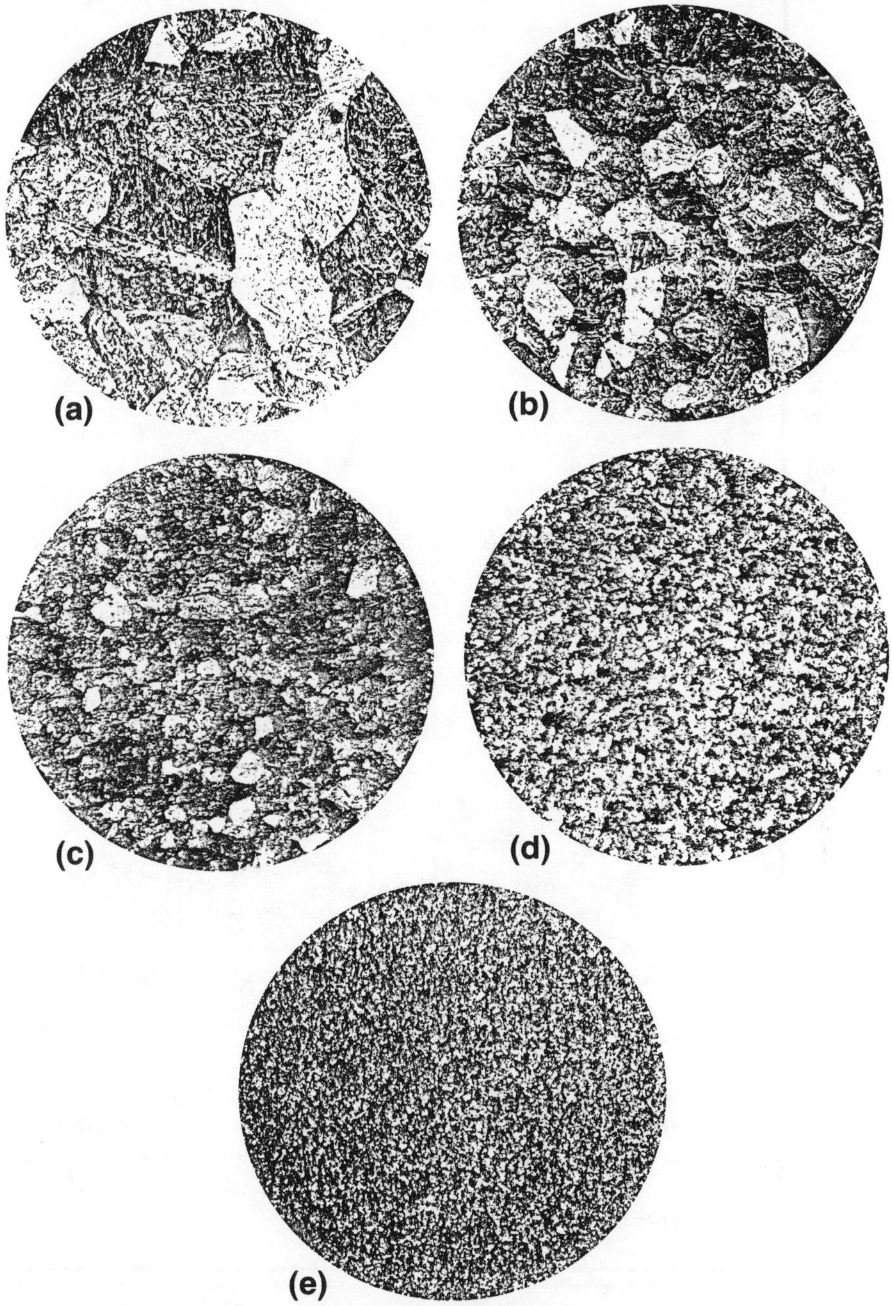

Fig. 7.11. Martensitic microstructures with prior austenite grain sizes of (a) ASTM No. 1; (b) ASTM No. 3; (c) ASTM No. 5; (d) ASTM No. 7; and (e) ASTM No. 9. These microstructures were prepared by lightly tempering and etching in a hydrochloric-picric acid solution in alcohol. Magnification, 100×; shown here at 50%. (Ref 7.10)

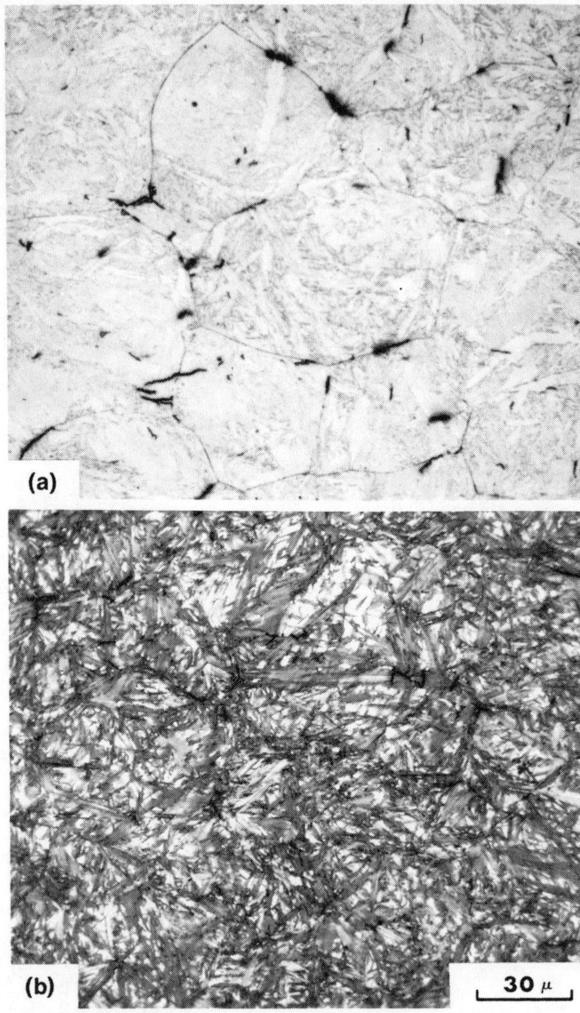

Fig. 7.12. Microstructure of Fe-1.22C alloy austenitized at 890 °C (1740 °F) for 2 min and water quenched. (a) Etched to show austenite grain boundaries (b) Etched to show martensite and retained austenite. Etchants given in text. (Ref 7.21)

lightly in nital (Ref 7.22). When nital alone was used to etch the structure, as in Fig. 7.12(b), the austenite grain boundaries were not at all obvious. A similar set of micrographs for a martensitic Fe-0.2C alloy is shown in Fig. 7.13 (Ref 7.2). In this case, the prior austenite grain boundaries were revealed after tempering by etching in a solution of 80

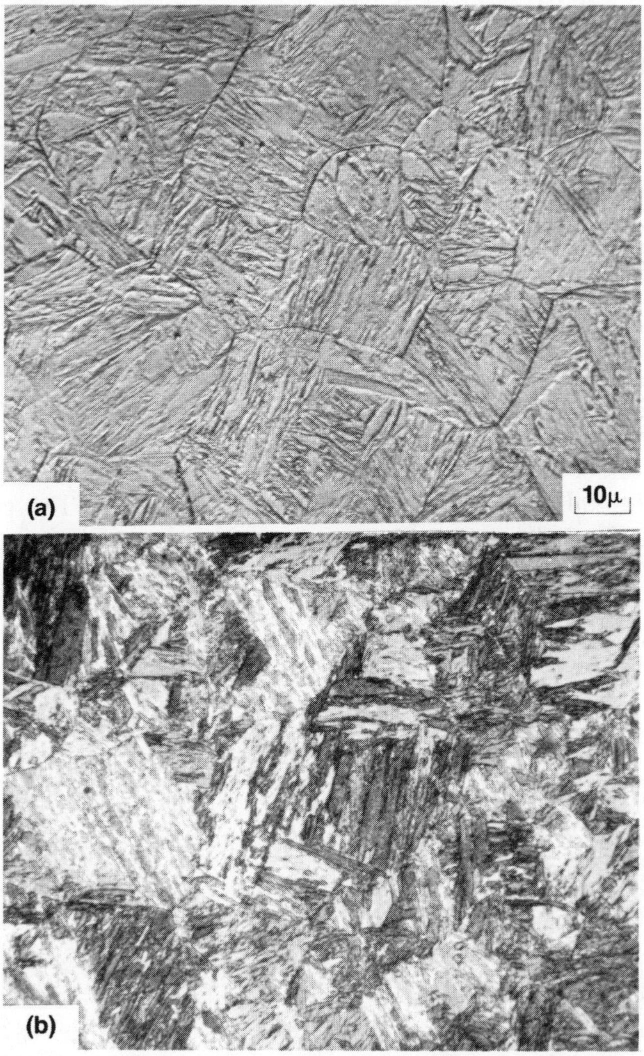

Fig. 7.13. Microstructure of an Fe-0.2C alloy quenched to form martensite. (a) Etched to show austenite grain boundaries. (b) Etched to show martensite. Etchants given in text. (Ref 7.2)

Ml H_2O, 28 ml oxalic acid (10%), and 4 ml H_2O_2, and the martensite was best developed by etching in 2% nital. The effectiveness of many of the special etchants used to reveal austenite grain boundaries may be due to impurity atom segregation at austenite grain boundaries during austenitizing or tempering.

The various techniques available to reveal austenitic grain size, including special etching techniques, have recently been reviewed by Millsop (Ref 7.23). Table 7.1, taken from his work, lists the various methods, the steels or structures most responsive to each technique, and some of the limitations. The McQuaid-Ehn test, based on carburizing for 8 h at 925 °C (1690 °F) and slow cooling to allow a proeutectoid cementite network to mark the grain boundaries, sometimes results in coarsening of the original austenite grain size. The oxidation and vacuum grooving methods are based on preferential oxidation and preferential evaporation of the austenite grain boundaries, respectively, and are useful as indicated. Slow cooling, to reveal austenite grain boundaries by the formation of proeutectoid ferrite or cementite networks, is very effective provided the carbon content of the steel is not too low. In the latter case, too much ferrite forms and the outlining effect of a thin network is lost.

A useful table for interrelationships between ASTM grain size number and other measurements of grain size is included in ASTM specification E 112−63 (Ref 7.24) and included as Table 7.2 here. Frequently grain size is measured by an intercept technique, and Table 7.2 makes possible the conversion of intercept sizes to ASTM grain size number.

Austenite Grain Size Control

Steels may be coarse grained or fine grained. Fine-grained steels are deoxidized with aluminum, and contain fine aluminum nitride particles that restrain austenite grain growth. Coarse-grained steels are generally deoxidized with silicon, a practice which does not produce particle dispersions effective in inhibiting austenite grain growth. Two types of austenite grain coarsening behavior occur in the two types of steels. Figure 7.14 shows that coarse-grained steels coarsen gradually with increasing austenitizing temperature, but fine-grained steels show no grain growth until a relatively high austenitizing temperature is reached. When coarsening finally occurs in a fine-grained steel, it occurs very rapidly. The temperature of the abrupt coarsening is referred to as the "grain coarsening temperature". Sometimes the abrupt coarsening is referred to as "discontinuous grain growth" or "secondary recrystallization". The latter term is used because the kinetics of discontinuous grain growth are very similar to recrystallization which has a slow initial or incubation stage, a more rapid second stage, and finally another slow stage.

Table 7.1. Methods to Reveal Austenitic Grain Size (Ref 7.23)

Method	Comments
Picric acid solutions	(a) Used for a wide range of steels having martensitic or bainitic structures (b) Room temperature etch (c) Unpredictable, does not always work (d) May give information on unrecrystallized grain shape
McQuaid-Ehn carburization	(a) Used for a limited range of steels (mainly hypo-eutectoid) (b) Lengthy 8 h treatment at 925 °C (1690 °F) (c) May not reflect true grain size of as-received steels
Oxidation	(a) Used for a limited range of steels (mainly hypo-eutectoid) (b) Heat treatment for 1 h at 855 °C (1575 °F) (c) May not reflect true grain size of as-received steels
Vacuum grooving	(a) Used for a wide range of steels (b) Heat treatment for 1 h or less at 900 °C (1652 °F) (c) Full potential not known (d) May not reflect true grain size of as-received steels
Delineation by ferrite or cementite	(a) Used for a range of hypo- and hyper-eutectoid steels (b) Heating above Ac_3, followed by controlled cooling (c) Used for some as-received steels (carbon composition range limited)

In many finishing heat treatments, the austenitizing temperature never exceeds 980 °C (1800 °F). Fine-grained steels, therefore, retain a fine austenite grain size, but the austenite of coarse-grained steels might coarsen significantly, especially during carburizing treatments that are frequently performed over many hours at 930 to 955 °C (1700 to 1750 °F). Figure 7.15 shows an example of a great difference in grain size that developed when steels of different coarsening behavior were carburized.

Recognition of the two different types of austenite grain coarsening behavior has led to an almost universal use of steel-making practices that produce fine-grained steels for critical heat treated parts. The mechanism of grain size control is based on the fact that aluminum that does not combine with oxygen during deoxidation combines with nitrogen and forms a dispersion of fine aluminum nitride particles. The

Table 7.2. Micrograin Size Relationships (Ref 7.24)

ASTM micro-grain size No.	Calculated "diameter" of average grain, mm	Calculated "diameter" of average grain, in. × 10^{-3}	Average intercept distance(a), mm	Average intercept distance(a), in. × 10^{-3}	Calculated area of average grain section, mm × 10^{-3}	Calculated area of average grain section, in. × 10^{-6}	Average number of grains per mm³	Nominal grains per mm² at i×	Nominal grains per in.² at 100×
00(b).......	0.508	20.0	0.451	17.8	258	400	7.63	3.88	0.250
0..........	0.359	14.1	0.319	12.6	129	200	21.6	7.75	0.50
0.5	0.302	11.9	0.268	10.6	91.2	141	36.3	11.0	0.707
1.0	0.254	10.0	0.226	8.88	64.5	100	61.0	15.5	1.0
. . . .	0.250	9.84	0.222	8.74	62.5	96.9	64.0	16.0	1.03
1.5	0.214	8.41	0.190	7.47	45.6	70.7	103	21.9	1.41
. . . .	0.200	7.87	0.178	6.99	40.0	62.0	125	25.0	1.61
. . . .	0.180	7.09	0.160	6.29	32.4	50.2	171	30.9	1.99
2.0	0.180	7.07	0.160	6.28	32.3	50.0	172.3	31.0	2.0
2.5	0.151	5.95	0.134	5.30	22.8	35.4	290	43.8	2.83
. . . .	0.150	5.91	0.133	5.24	22.5	34.9	296	44.4	2.87
3.0	0.127	5.00	0.113	4.44	16.1	25.0	488	62.0	4.0
. . . .	0.120	4.72	0.107	4.20	14.4	22.3	578.9	69.4	4.48
3.5	0.107	4.20	0.0948	3.73	11.4	17.7	821	87.7	5.66
. . . .	0.090	3.54	0.0799	3.15	8.10	12.6	1 370	123	7.97
4.0	0.0898	3.54	0.0797	3.14	8.06	12.5	1 380	124	8.0
4.5	0.076	2.97	0.0671	2.64	5.70	8.84	2 320	175	11.3
. . . .	0.070	2.76	0.0622	2.45	4.90	7.59	2 920	204	13.2
5.0	0.064	2.50	0.0564	2.22	4.03	6.25	3 910	248	16.0
. . . .	0.060	2.36	0.0533	2.10	3.60	5.58	4 630	278	17.9
5.5	0.0534	2.10	0.0474	1.87	2.85	4.42	6 570	351	22.6
. . . .	0.050	1.97	0.0444	1.75	2.50	3.88	8 000	400	25.8
6.0	0.045	1.77	0.0399	1.57	2.02	3.13	11 000	496	32.0
. . ..	0.040	1.58	0.0355	1.40	1.60	2.48	15 600	625	40.3
6.5	0.038	1.49	0.0335	1.32	1.43	2.21	18 600	701	45.3
. . . .	0.035	1.38	0.0311	1.22	1.23	1.90	23 000	816	52.7
7.0	0.032	1.25	0.0282	1.11	1.01	1.56	31 000	992	64.0
. . . .	0.030	1.18	0.0267	1.05	0.90	1.40	37 000	1 110	71.7
7.5	0.027	1.05	0.0237	0.933	0.713	1.10	52 500	1 400	90.5
. . . .	0.025	0.984	0.0222	0.874	0.625	0.969	64 000	1 600	103
8.0	0.0224	0.884	0.0199	0.785	0.504	0.781	88 400	1 980	128
. . . .	0.0200	0.787	0.0178	0.699	0.40	0.620	125 000	2 500	161
8.5	0.0189	0.743	0.0168	0.660	0.356	0.552	149 000	2 810	181
9.0	0.0159	0.625	0.0141	0.555	0.252	0.391	250 000	3 970	256
. . . .	0.0150	0.591	0.0133	0.524	0.225	0.349	296 000	4 440	287
9.5	0.0134	0.526	0.0119	0.467	0.178	0.276	420 000	5 610	362
10.0	0.0112	0.442	0.00997	0.392	0.126	0.195	707 000	7 940	512
. . . .	0.0100	0.394	0.00888	0.350	0.10	0.155	1.00×10^6	10 000	645
10.5	0.00944	0.372	0.00838	0.330	0.089	0.138	1.19×10^6	11 200	724
. . . .	0.00900	0.354	0.00799	0.315	0.081	0.126	1.37×10^6	12 300	797
. . . .	0.00800	0.315	0.00710	0.280	0.064	0.0992	1.95×10^6	15 600	1 010
11.0	0.00794	0.313	0.00705	0.278	0.063	0.0977	2.00×10^6	15 900	1 020
. . . .	0.00700	0.276	0.00622	0.245	0.049	0.0760	2.92×10^6	20 400	1 320
11.5	0.00667	0.263	0.00593	0.233	0.045	0.0691	3.36×10^6	22 400	1 450
. . . .	0.00600	0.236	0.00533	0.210	0.036	0.0558	4.63×10^6	27 800	1 790
12.0	0.00561	0.221	0.00498	0.196	0.031	0.0488	5.66×10^6	31 700	2 050
. . . .	0.00500	0.197	0.00444	0.175	0.025	0.0388	8.00×10^6	40 000	2 580
12.5	0.00472	0.186	0.00419	0.165	0.022	0.0345	9.51×10^6	44 900	2 900
. . . .	0.00400	0.158	0.00355	0.140	0.0160	0.0248	15.62×10^6	62 500	4 030
13.0	0.00397	0.156	0.00352	0.139	0.0158	0.0244	16.0×10^6	63 500	4 100
13.5	0.00334	0.131	0.00296	0.117	0.011	0.0173	26.9×10^6	89 800	5 800
. . . .	0.00300	0.118	0.00266	0.105	0.009	0.0140	37.0×10^6	111 000	7 170
14.0	0.00281	0.111	0.00249	0.0981	0.0079	0.0122	45.2×10^6	127 000	8 200
. . . .	0.00250	0.098	0.00222	0.0874	0.00625	0.00969	64.0×10^6	160 000	10 300

(a) Value of Heyn intercept for equiaxed grains. (b) The use of 00 is recommended instead of "– 1" or "minus 1" to avoid confusion.

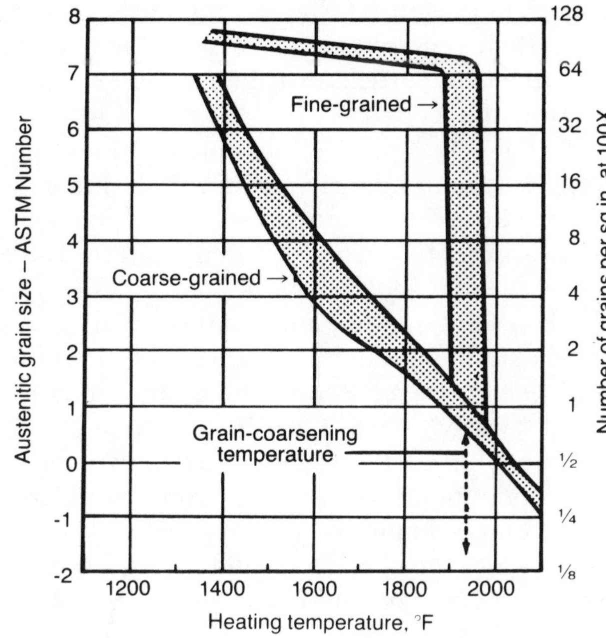

Fig. 7.14. Austenite grain size as a function of austenitizing temperature for coarse-grained and fine-grained steels. (Ref 7.25)

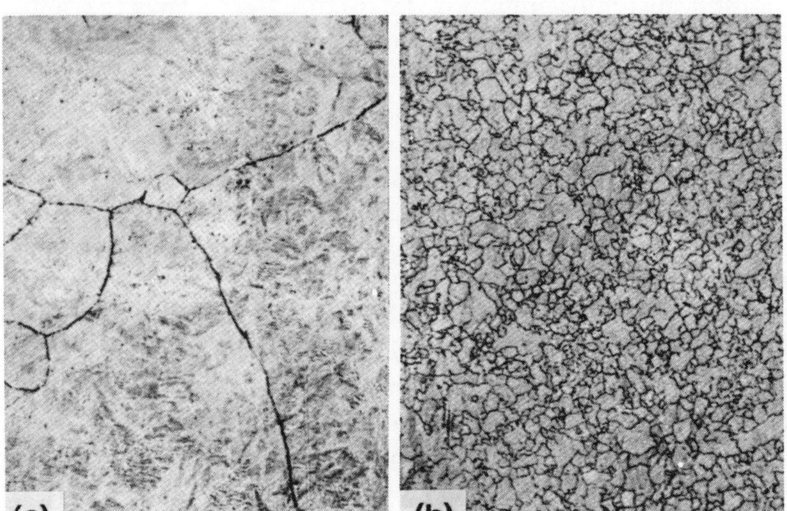

Fig. 7.15. Comparison of austenitic grain size in (a) coarse-grained SAE 1015 steel and (b) fine-grained SAE 4615 steel after carburizing. (Ref 7.10)

fine nitride particles in turn pin the austenite grain boundaries, and thereby inhibit grain growth. However, with increasing temperature the aluminum particles coarsen and/or dissolve, and very rapid grain growth occurs when the particle distribution becomes too coarse to restrain grain growth effectively.

Figure 7.16 shows the effect of aluminum content on the grain coarsening temperature in mild steel (Ref 7.26). Aluminum contents up to 0.08% increase the coarsening temperature, but higher additions cause a slight lowering of the coarsening temperature. The reaction Al + N = AlN is reversible, and at high temperatures some of the AlN particles go back in solution to permit grain coarsening, especially in steels with lower aluminum contents. Figure 7.17 shows the solubility of aluminum nitride at three different aluminum contents. The grain coarsening temperatures (arrows) are indicated on the curves (Ref 7.26). At high aluminum contents grain coarsening precedes substantial solution of aluminum nitride, and therefore the reduced coarsening temperature at high aluminum content is believed to be the result of coarse particles produced during solidification and hot working rather than coarsening and/or solution of the aluminum nitride particles during austenitizing.

Austenite grain size may be controlled by elements other than aluminum. In particular, the transition metals such as titanium, vanadium and niobium that are strong carbide and nitride formers behave similarly to aluminum with respect to producing particle dispersions that inhibit austenite grain growth. Figure 7.18 shows the solubility products for aluminum nitride and various transition metal carbides and nitrides in austenite as a function of temperature, as collected by Aronsson from a number of sources (Ref 7.27). The behavior of the titanium, niobium and vanadium compounds is similar to that of aluminum nitride: solubility in austenite decreases with decreasing temperature. Thus precipitation of alloy carbides or nitrides will occur at low austenitizing temperatures if sufficient quantities of the alloying elements are present.

Small amounts, frequently less than 0.05%, of niobium, vanadium and titanium are used in high-strength low-alloy (HSLA) steel (Ref 7.28). The very small additions of the alloying elements have led to the description of HSLA steels as microalloyed steels. HSLA steels have low carbon contents and therefore have largely ferritic microstructures. However, very fine ferrite grain sizes together with precipitation of alloy carbonitride particles raise the strengths of the HSLA steels well above those of normally processed mild steels. Mild steels with ferrite-pearlite microstructures usually have yield strengths on the

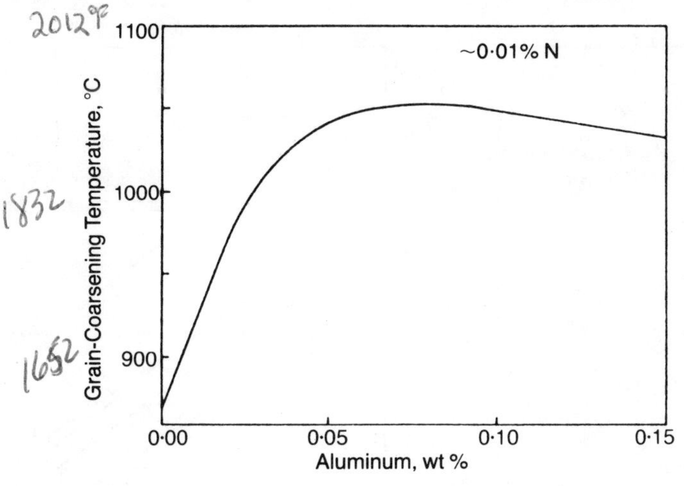

Fig. 7.16. The effect of aluminum content on the grain-coarsening temperature of austenite. (Ref 7.26)

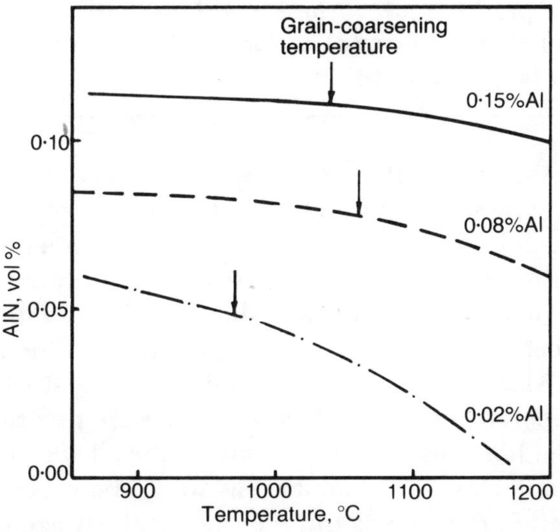

Fig. 7.17. Change in volume percent AlN as a function of temperature in mild steel containing 1.01% nitrogen and aluminum as shown. Grain-coarsening temperatures are indicated by arrows. (Ref 7.26)

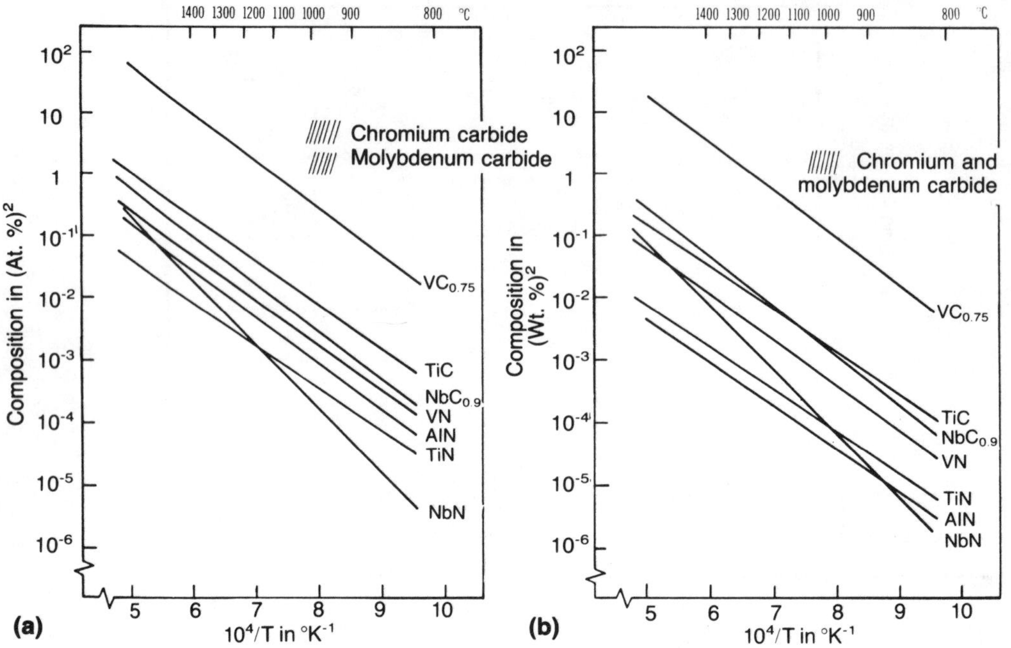

Fig. 7.18. Solubility products of various carbides and nitrides in austenite. (Ref 7.27) Ordinate in (a) is content in atomic percent; ordinate in (b) is content in weight percent.

order of 30 ksi (207 MPa) while the HSLA steels may have yield strengths ranging from 50 to 80 ksi (345 MPa to 550 MPa).

The very fine ferrite grain sizes in HSLA steels are obtained through control of austenite grain size by the interaction of alloy carbonitride precipitation and hot rolling. Figure 7.19 shows schematically the effects of processing and temperature on the austenite and ferrite grain size that develops in a microalloyed steel. At high hot working temperatures, where heavy sections are first reduced in size and maximum hot workability is required, the alloy carbides, nitrides and/or carbonitrides are in solution as would be expected from Fig. 7.18. Austenite grain size is quite coarse at this stage of processing. Without the restraining effect of fine particles, recrystallization and grain growth occur almost simultaneously with deformation at this stage of hot working. However, if finish rolling is performed just above or close to the Ar_3 temperature, fine alloy carbonitrides precipitate from the austenite because of their reduced solubility (see Fig. 7.18) at low austenitizing temperatures. The fine particles prevent austenite grain growth and even recrystallization, the amount of retardation

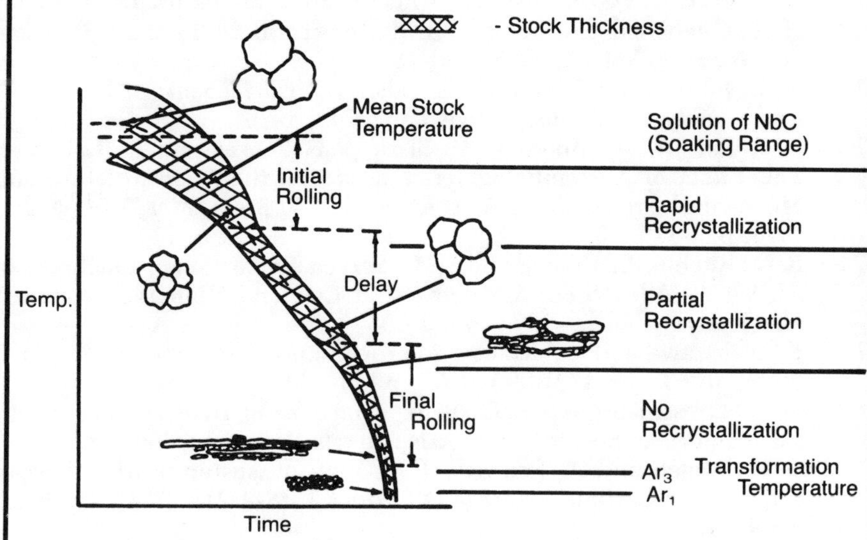

Fig. 7.19. Schematic temperature-time-processing history leading to very fine ferrite grain structure in HSLA steels. (Ref 7.29)

depending on alloy concentration, amount of deformation and the rolling temperature. In the extreme case the austenite grains become highly deformed and elongated as shown in Fig. 7.19. On cooling below Ar_3, and through Ar_1, then, ferrite grains form on the closely spaced grain boundaries of the unrecrystallized austenite. As a result, very fine ferrite grain sizes that give the HSLA steels their high strength and toughness are produced.

References

7.1 P. J. Brofman, G. S. Ansell, T. J. Nichol and G. Judd, The Effect of Fine Grain Size on the Martensitic Transformation in Fe-Ni and Fe-Ni-C alloys, *Mechanical Behavior of Metals and Alloys Associated with Displacive Phase Transformations*, Joint U.S.-Japan Conference, Rensselaer Polytechnic Institute, Troy, N.Y., June 1979

7.2 T. E. Swarr and G. Krauss, Boundaries and the Strength of Low Carbon Ferrous Martensites, in *Grain Boundaries in Engineering Materials*, J. L. Walter, J. H. Westbrook and D. A. Woodford (Eds.), Claitors Publishing Division, Baton Rouge, La., 1975, p 127-138

7.3 M. J. Roberts, Effects of Transformation Substructure on the Strength and Toughness of Fe-Mn Alloys, *Met Trans*, Vol 1, 1970, p 3287-3294

7.4 R. A. Grange, Strengthening Steel by Austenite Grain Refinement, *Trans ASM*, Vol 59, 1966, p 26-48

7.5 T. E. Swarr and G. Krauss, The Effect of Structure on the Deformation of As-Quenched and Tempered Martensite in an Fe-0.2% C Alloy, *Met Trans A*, Vol 7A, 1976, p 41-48

7.6 W. C. Johnson and J. M. Blakely (Eds.), *Interfacial Segregation*, American Society for Metals, Metals Park, Oh., 1979

7.7 G. Y. Lai, W. E. Wood, R. A. Clark, V. F. Zackay and E. R. Parker, The Effect of Austenitizing Temperature on the Microstructure and Mechanical Properties of As-Quenched 4340 Steel, *Met Trans*, Vol 5, 1974, p 1663-1670

7.8 R. O. Ritchie, B. Francis and W. L. Server, Evaluation of Toughness in AISI 4340 Alloy Steel Austenitized at Low and High Temperatures, *Met Trans A*, Vol 7A, 1976, p 831-838

7.9 K. Nakazawa and G. Krauss, Microstructure and Fracture of 52100 Steel, *Met Trans A*, Vol 9A, 1978, p 681-689

7.10 M. A. Grossmann and E. C. Bain, *Principles of Heat Treatment*, 5th Ed., American Society for Metals, Metals Park, Oh., 1964

7.11 G. R. Speich and A. Szirmae, Formation of Austenite from Ferrite and Ferrite-Carbide Aggregates, *Trans TMS-AIME*, Vol 245, 1969, p 1063-1074

7.12 G. A. Roberts and R. F. Mehl, The Mechanism and Rate of Formation of Austenite from Ferrite-Cementite Aggregates, *Trans ASM*, Vol 31, 1943, p 613

7.13 R. R. Judd and H. W. Paxton, Kinetics of Austenite Formation from a Spheroidized Ferrite-Carbide Aggregate, *Trans TMS-AIME*, Vol 242, 1968, p 206-215

7.14 T. M. Scoonover and G. Krauss, High-rate Short-time Austenitizing of 4340 Steel, *Met Eng Q*, Vol 12, 1972, p 41-48

7.15 S. Matsuda, Microstructural and Kinetic Studies of Reverse Transformation in a Low Carbon Alloy Steel, in *New Aspects of Martensitic Transformation*, Japan Institute of Metals, 1976, p 363-367

7.16 S. Watanabe, Y. Ohmori and T. Kunitaki, Formation of Austenite from Lath-like Martensite, in *New Aspects of Martensitic Transformation*, Japan Institute of Metals, 1976, p 368-374

7.17 *Induction Hardening and Tempering*, ASM Committee on Induction Hardening, American Society for Metals, Metals Park, Oh., 1964, p 133

7.18 G. Krauss, Fine Structure of Austenite Produced by the Reverse Martensitic Transformation, *Acta Met*, Vol 11, 1963, p 499-509

7.19 G. Krauss and M. Cohen, Strengthening and Annealing of Austenite Formed by Reverse Martensitic Transformation, *Trans AIME*, Vol 224, 1962, p 1212-1221

7.20 C. A. Apple and G. Krauss, The Effect of Heating Rate on the Martensite to Austenite Transformation in Fe-Ni-C Alloys, *Acta Met*, Vol 20, 1972, p 849-856

7.21 R. P. Brobst and G. Krauss, The Effect of Austenite Grain Size on Microcracking in Martensite of an Fe-1.22C Alloy, *Met Trans*, Vol 5, 1975, p 457-462

7.22 A. Benscoter of Bethlehem Steel Corporation, Bethlehem, Pa., private communication 1975

7.23 R. Millsop, A Survey of Austenite Grain Size Measurements, in *Hardenability Concepts with Applications to Steels*, TMS-AIME, Warrendale, Pa., 1978, p 316-333

7.24 Estimating the Average Grain Size of Metals, ASTM Designation: E 112-63, American Society for Testing and Materials, Philadelphia, 1963
7.25 G. F. Melloy, *Austenite Grain Size – Its Control and Effects*, Metals Engineering Institute, American Society for Metals, Metals Park, Oh., 1968
7.26 T. Gladman, The Effect of Aluminum Nitride on the Grain Coarsening Behavior of Austenite, in *Metallurgical Developments in Carbon Steels*, Special Report 81, The Iron and Steel Institute, London, 1963, p 68-70
7.27 B. Aronsson, Gefügeaufbau und Mechanische Eigenschaften einiger martensitischer Stähle unter besonderer Berücksichtigung des Einflusses von Niob und Molybdän, in *Steel Strengthening Mechanisms*, Climax Molybdenum Co., Greenwich, Conn., 1969, p 77-87
7.28 *Microalloying 75*, Union Carbide Corporation, 1977, distributed by American Society for Metals, Metals Park, Oh.
7.29 J. D. Baird and R. R. Preston, Relationships Between Processing, Structure and Properties in Low Carbon Steels, in *Processing and Properties of Low Carbon Steel*, J. M. Gray (Ed.), TMS-AIME, Warrendale, Pa., 1973, p 1-46
7.30 A. R. Marder and G. Krauss, The Effect of Morphology on the Strength of Lath Martensite, in *Proceedings of Second International Conference on the Strength of Metals and Alloys*, American Society for Metals, Metals Park, Oh., 1970, p 822-823

CHAPTER 8

Tempering of Steel

Virtually all steels that are hardened are subjected to a subcritical heat treatment referred to as tempering. Tempering improves the toughness of as-quenched martensitic microstructures but lowers strength and hardness. This chapter describes the mechanical property and microstructural changes that develop during tempering. The most important structural change is the formation of various distributions of iron and alloy carbides as the supersaturation of the as-quenched martensite is relieved and equilibrium mixtures of phases are approached with increased tempering. Finally, several types of embrittlement that may occur on tempering are described.

Mechanical Property Changes

Martensite, the object of the quenching treatments described in Chapter 6, is quite hard but it is also very brittle. The brittleness of martensitic microstructures is due to a number of factors that may include the lattice distortion caused by carbon atoms trapped in the octahedral sites of the martensite (see Fig. 6.5), impurity atom segregation at austenite grain boundaries, carbide formation during quenching and residual stresses produced during quenching. Tempering is the heat treatment of hardened steels that has reduction of brittleness or increased toughness as its major objective. Any temperature up to the lower critical may be used for tempering, thus an extremely wide variation in properties and microstructure ranging from those of as-quenched martensite to spheroidized carbides in ferrite can be produced by tempering. Ultimately it is the balance of hardness (or strength) and toughness required in service that determines the conditions of tempering for a given application.

Figure 8.1 shows impact toughness as a function of tempering temperature for hardened 0.4 and 0.5% carbon steels (Ref 8.1). There are two tempering temperature ranges that produce significant improvement in toughness from that of the as-quenched state. Tempering in the range of 150 to 200 °C (300 to 400 °F) produces a modest increase in toughness that is adequate for applications that require high strength and fatigue resistance (medium-carbon steels) or where loading is primarily compressive as in bearings and gears (high-carbon steels). The latter applications require the high hardness and associated good wear resistance that high-carbon martensite and light tempers provide. Tempering above 425 °C (800 °F) is the other important tempering temperature range. Figure 8.1 shows that toughness improves significantly after tempering in this range, but as noted below hardness and strength also decrease significantly. Therefore, tempering above 425 °C (800 °F) is used where high toughness is of major concern, and strength and hardness are important but of secondary concern.

Figure 8.1 shows that toughness may actually decrease if steels are tempered in the range of 260 to 370 °C (500 to 700 °F). This decrease in toughness is referred to as tempered martensite embrittlement, 350 °C embrittlement or 500 °F embrittlement and is discussed in more detail later in this chapter. As a result of this embrittlement, the tempering range between 260 to 370 °C (500 to 700 °F) is generally avoided in commercial practice. Another type of embrittlement, temper embrittlement, may develop in martensitic steels tempered above 425 °C (800 °F). Temper embrittlement occurs in certain alloy steels as a result of holding on slow cooling through certain tempering temperature ranges, and is also discussed in more detail later in this chapter.

Finally, Figure 8.1 also shows the substantial effect that increasing carbon content has on impact toughness by comparing the results of tempering 0.5% carbon steels to those of 0.4% carbon steels. Steels with carbon contents of 0.5% or greater have very low impact toughness and are used only where high hardness, wear resistance and/or edge retention are of prime importance. For example, hand tools, such as screw driver blades and cutting blades of all sorts, are made from quenched and low-temperature tempered high-carbon steels. Wear resistance and cutting edge retention are excellent in these applications, but any misuse of the tools such as prying or bending frequently leads to immediate fracture.

Figure 8.2, taken from a variety of sources by Grossmann and Bain (Ref 8.1), shows how hardness decreases from the maximum associated with as-quenched martensite with increasing tempering temperature.

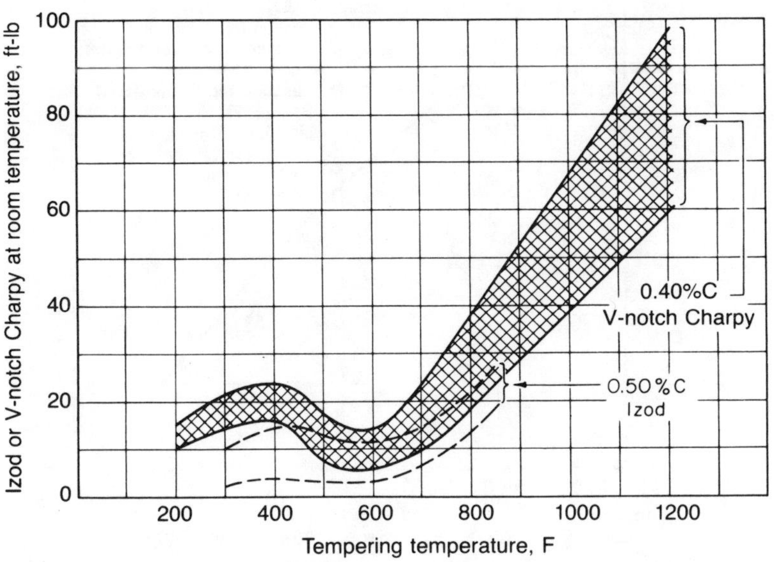

Fig. 8.1. Impact toughness as a function of tempering temperature of hardened, low-alloy, medium-carbon steels. (Ref 8.1)

The effect of carbon content is also shown. The lower hardness of low-carbon steels in the as-quenched condition and throughout tempering is emphasized in the curves. Therefore, if maximum hardness is desired, a high-carbon steel should be selected and tempering should be restricted to the 150 to 200 °C (300 to 400 °F) temperature as noted above.

Figure 8.2 indicates a slight hardness increase on low-temperature tempering of the highest carbon steels. Figure 8.3, from an investigation of the early stages of tempering of martensite in an Fe-1.22C alloy (Ref 8.2), shows tempering times and temperatures that produce an increase in hardness above that associated with the as-quenched state. This increase in hardness is a result of the precipitation of a dense distribution of very fine transition carbide particles within the martensite plates.

Generally, the interplay of hardness and toughness is of major concern in the heat treatment and application of quench and tempered steels. However, the changes in other mechanical properties with increasing tempering are also tabulated for common grades of carbon and alloy steel bars (Ref 8.3) and are quite important for the selection of

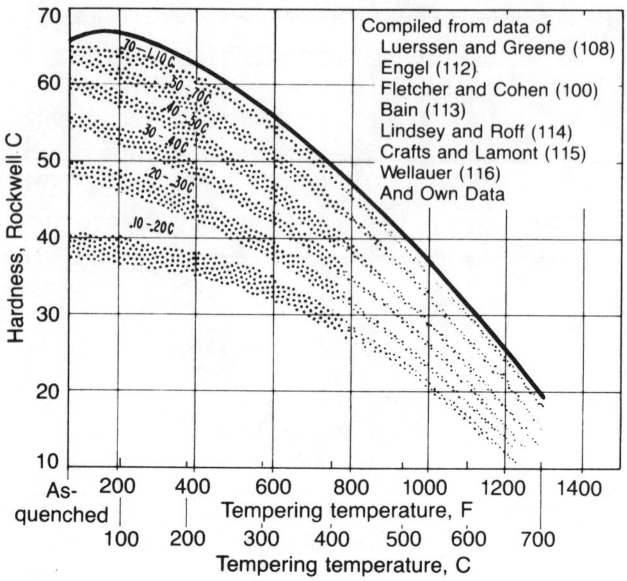

Fig. 8.2. Decrease in hardness with increasing tempering temperature for steels of various carbon contents. Reference numbers after investigators are from list in Grossmann and Bain. (Ref 8.1)

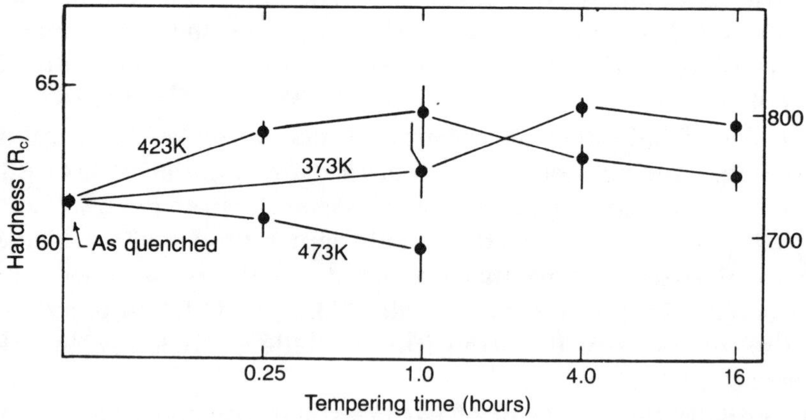

Fig. 8.3. Hardness as a function of time at three tempering temperatures for martensite in an Fe-1.22C alloy. (Ref 8.2)

steels and design of heat treatments for some applications. Figure 8.4 shows the changes in mechanical properties that occur when an oil-quenched AISI 4340 steel is tempered at temperatures above 200 °C (400 °F). Both yield strength and tensile strength decrease continuously and elongation and reduction of area increase with increasing tempering temperature. The as-quenched hardness and the mechanical properties of 4340 steel for selected tempering treatments as a function of bar size are listed in Table 8.1. The strength properties for a given treatment decrease with increasing bar diameter (see Chapter 6).

Figure 8.4 shows two other aspects of the mechanical behavior of tempered carbon steels. One is the fact that there is no decrease in ductility produced by tempering in the temperature range that produces tempered martensite embrittlement. Specimen design and testing account for this observation. The toughness data shown in Fig. 8.1 are based on impact toughness testing accomplished by loading notched specimens at a high strain rate. Figure 8.4, on the other hand, is based on tensile testing of smooth round specimens at relatively slow strain rates. Thus at slow strain rates, without the stress concentrating effect of a notch, the microstructure of a steel tempered even in the range 260-370 °C (500 to 700 °F) can accommodate loading without undue embrittlement. On impact loading, however, the reverse is true and disregard of strain rate and notch effects may lead to unexpected failure in certain applications.

Figure 8.4 also shows that the yield and tensile strengths of the tempered 4340, at first well separated after tempering at low temperatures, tend to approach each other after tempering at high temperatures. This effect is a common characteristic of hardened carbon and low-alloy steels, and is related to differences in work-hardening behavior that develop on tempering. Figure 8.5 shows stress-strain curves that illustrate the changes in work hardening that develop with tempering of lath martensite in an Fe-0.2C alloy (Ref 8.4). In this case, the as-quenched martensite was obtained by quenching in a NaOH-NaCl solution and tempering was performed by heating in lead at 400 °C (750 °F) for 1 min. The work-hardening rate in the as-quenched specimen was quite high, as shown by the rapid increase in stress with increasing strain, while the stress-strain curve for the tempered specimen was almost flat, indicating a very low rate of work hardening. This difference in work hardening behavior is attributed to interaction of dislocations with relatively coarse particles of cementite that form on tempering. In as-quenched specimens, dislocations tangle and form a tight substructure of fine cells with increasing deformation, but with

Table 8.1. Mechanical Properties of Various Size Rounds of a Single Heat(a) of 4340 Steel After Various Heat Treatments Illustrating Mass Effects (Ref 8.3)

Size round, mm	Size round, in.	Tensile strength, MPa	Tensile strength, ksi	Yield strength, MPa	Yield strength, ksi	Elongation in 50 mm (2 in.), %	Reduction in area, %	Hardness, HB
Annealed: heated to 810 °C (1490 °F), furnace cooled 12 °C/h (20 °F/h) to 354 °C (670 °F), air cooled								
25.4	1	745	108	472	68.5	22.0	49.9	217
Normalized: heated to 871 °C (1600 °F), air cooled								
12.7	½	1448	210	972	141	12.1	35.3	388
25.4	1	1282	186	862	125	12.2	36.3	363
50.8	2	1220	177	793	115	13.5	37.3	341
101.6	4	1110	161	710	103	13.2	36.0	321
Oil quenched from 800 °C (1475 °F), tempered at 538 °C (1000 °F)								
12.7	½	1255	182	1165	169	13.7	45.0	363
25.4	1	1207	175	1145	166	14.2	45.9	352
50.8	2	1172	170	1103	160	16.0	54.8	341
101.6	4	1138	165	1000	145	15.5	53.4	331
Oil quenched from 800 °C (1475 °F), tempered at 593 °C (1110 °F)								
12.7	½	1145	166	1117	162	17.1	57.0	331
25.4	1	1138	165	1096	159	16.5	54.1	331
50.8	2	1014	147	958	139	19.0	60.4	293
101.6	4	924	134	793	115	19.7	60.7	269
Oil quenched from 800 °C (1475 °F), tempered at 650 °C (1200 °F)								
12.7	½	1000	145	938	136	20.0	59.3	285
25.4	1	958	139	883	128	20.0	59.7	277
50.8	2	931	135	834	121	20.5	62.5	269
101.6	4	855	124	730	106	21.7	63.0	255

As-quenched hardness (oil), HRC

Size round, mm	Size round, in.	Surface	½ radius	Center
12.7	½	58	58	56
25.4	1	57	57	56
50.8	2	56	55	54
101.6	4	53	49	47

(a) Ladle composition: 0.40 C; 0.68 Mn; 0.020 P; 0.013 S; 0.28 Si; 1.87 Ni; 0.74 Cr; 0.25 Mo; grain size 7-8.

large cementite particles present, the dislocations remain uniformly distributed and a well-defined dislocation cell structure never develops. Figure 8.6 shows a uniform distribution of dislocations in the tempered Fe-0.2C martensite. This distribution did not change on deformation (Ref 8.4).

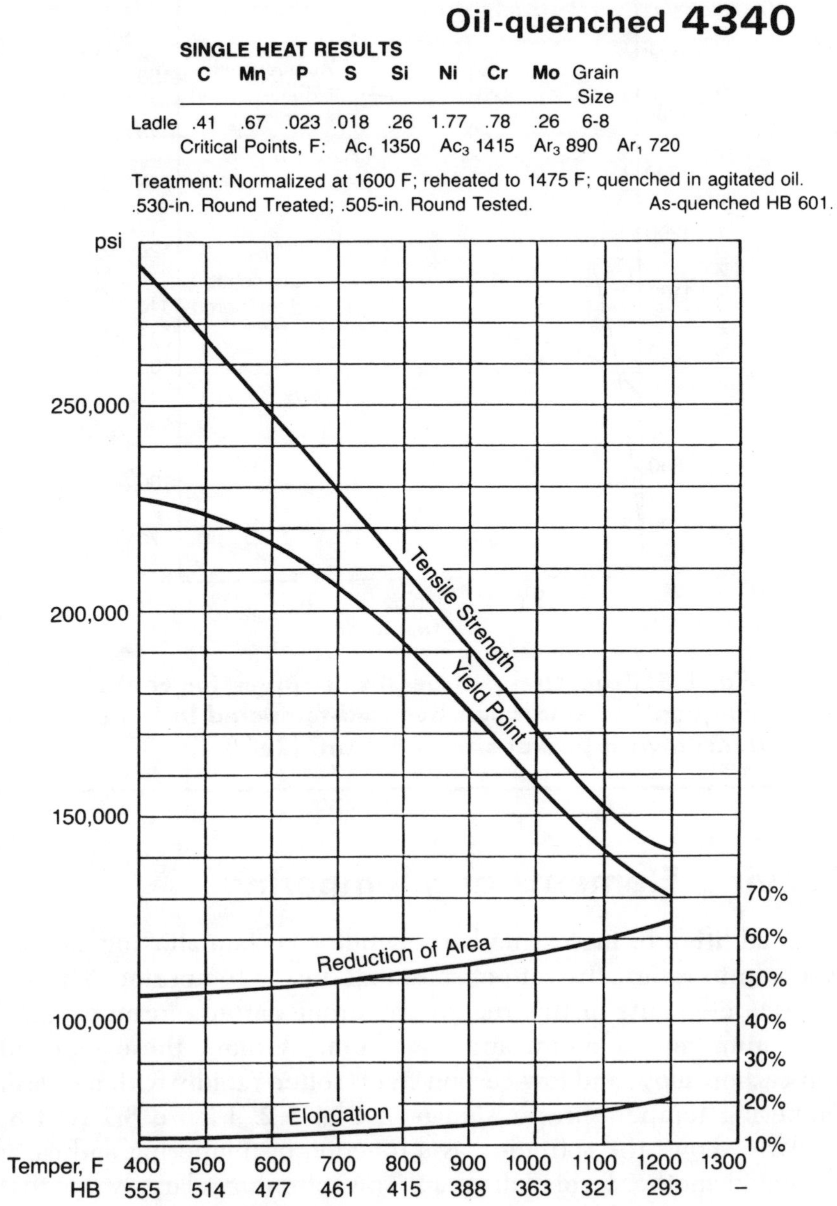

Fig. 8.4. Change in mechanical properties with tempering temperature for oil-quenched 4340 steel. (Ref 8.3)

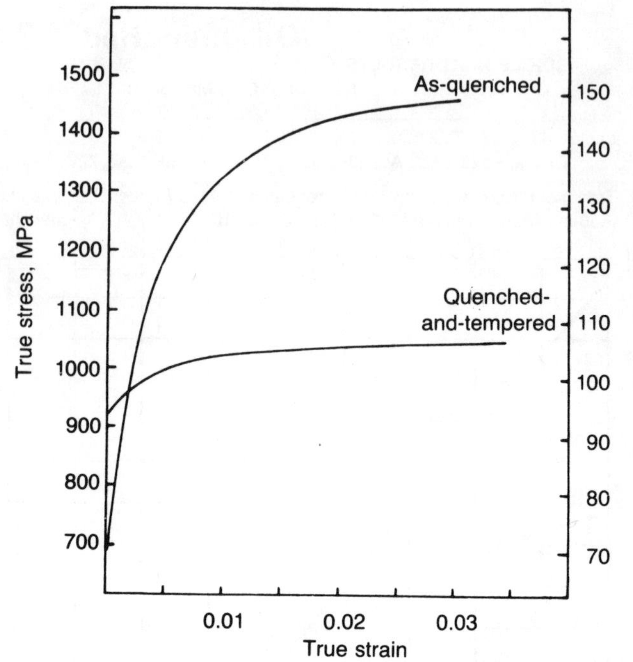

Fig. 8.5. True stress – true strain curves for Fe-0.2C as-quenched and quenched-and-tempered lath martensite with packet size of 8.2μm. (Ref 8.4)

Alloying Elements and Tempering

In addition to increasing hardenability, certain alloying elements also help to retard the rate of softening during tempering. The most effective elements in this regard are strong carbide formers such as chromium, molybdenum and vanadium. Without these elements, iron-carbon alloys and low-carbon steels soften rapidly with increasing tempering temperature as shown in Fig. 8.2. Figure 8.7 (Ref 8.5) similarly shows the softening as a function of tempering and carbon content in another form of diagram. This softening is largely due to the rapid coarsening of cementite with increasing tempering temperature, a process dependent on the diffusion of carbon and iron. If present in a steel in sufficient quantity, however, the carbide-forming elements not only retard softening but also form fine alloy carbides that produce a hardness increase at higher tempering temperatures. This hardness increase is frequently referred to as secondary hardening.

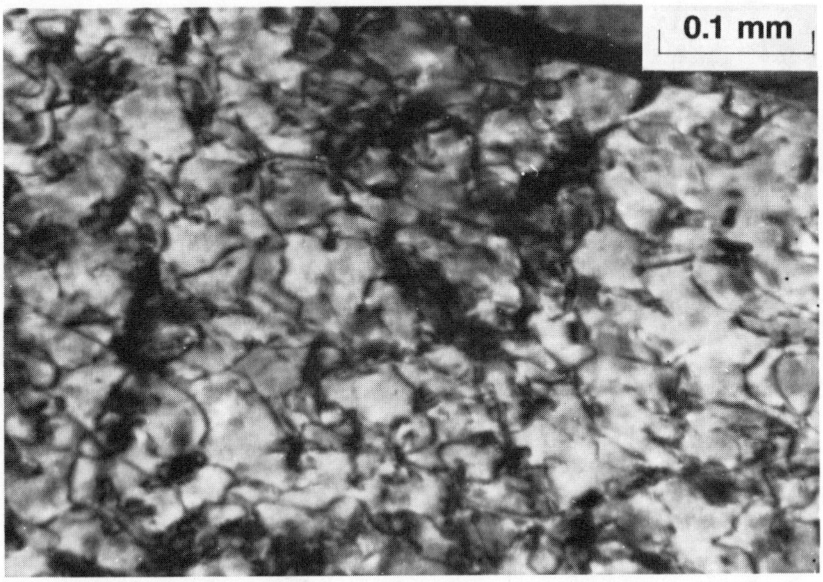

Fig. 8.6. Uniform distribution of dislocations and cementite particles in tempered Fe-0.2C martensite. Transmission electron micrograph. (Ref 8.4)

Figure 8.8 shows secondary hardening in a series of steels containing molybdenum (Ref 8.6, 8.7). The higher the molybdenum content, the higher the hardness associated with the secondary hardening peak, and even at 0.47% molybdenum when no hardness peak is observed, a significant retardation of softening is apparent. The secondary hardening peaks develop only at high tempering temperatures because alloy carbide formation depends on the diffusion of the carbide-forming elements, a more sluggish process than that of carbon and iron diffusion. As a result, not only is a finer dispersion of particles produced, but also once formed, the alloy carbides are quite resistant to coarsening. The latter characteristic of the fine alloy carbides is used to advantage in tool steels that must not soften even though high temperatures are generated by their use in hot working dies or high-speed machining. Also, ferritic low-carbon steels containing chromium and molybdenum are used in pressure vessels and reactors operated at temperatures around 540 °C (1000 °F) because the alloy carbides resist coarsening at those temperatures and therefore provide good creep resistance.

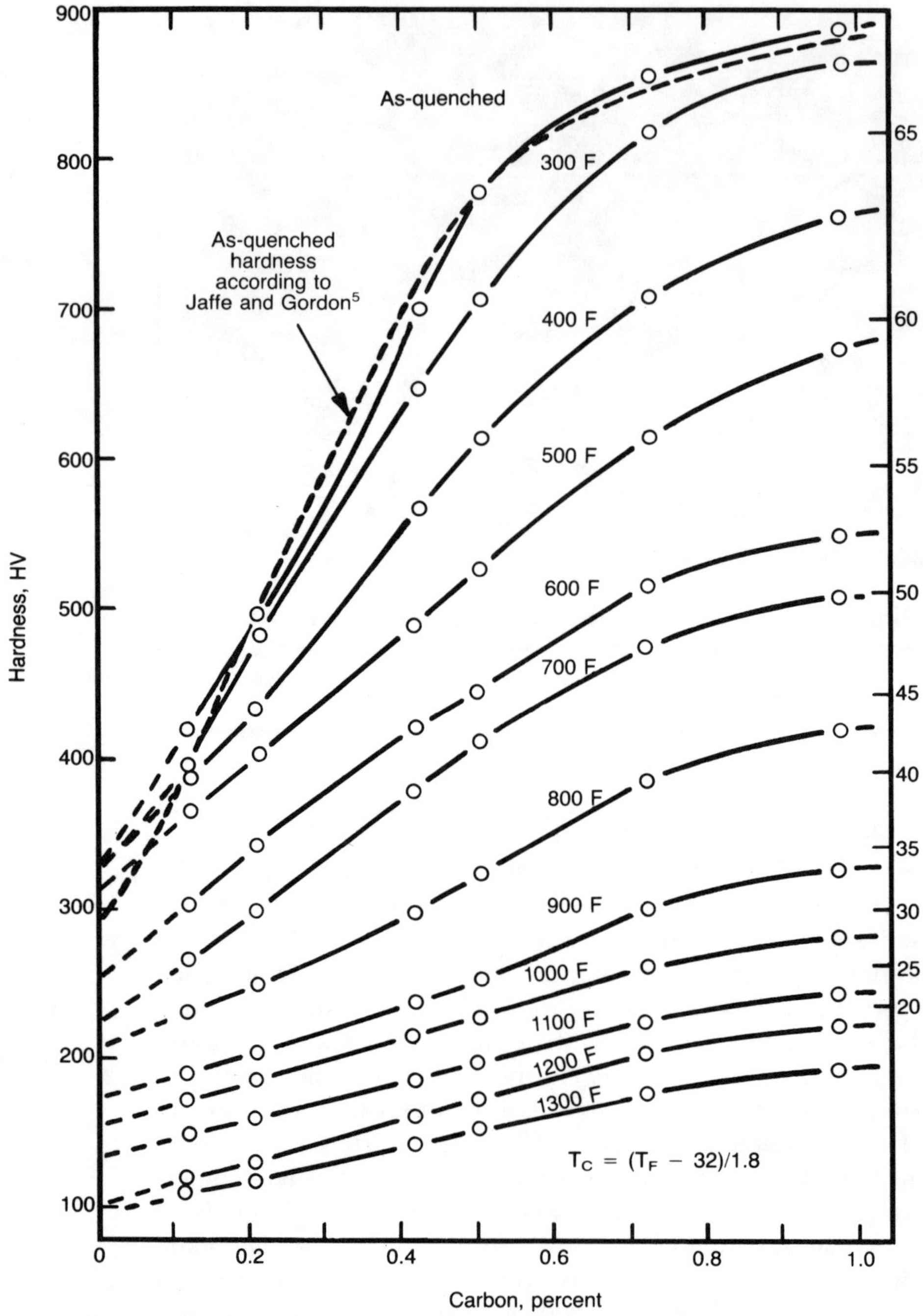

Fig. 8.7. Hardness as a function of carbon content of martensite in Fe-C alloys tempered at various temperatures. (Ref 8.5)

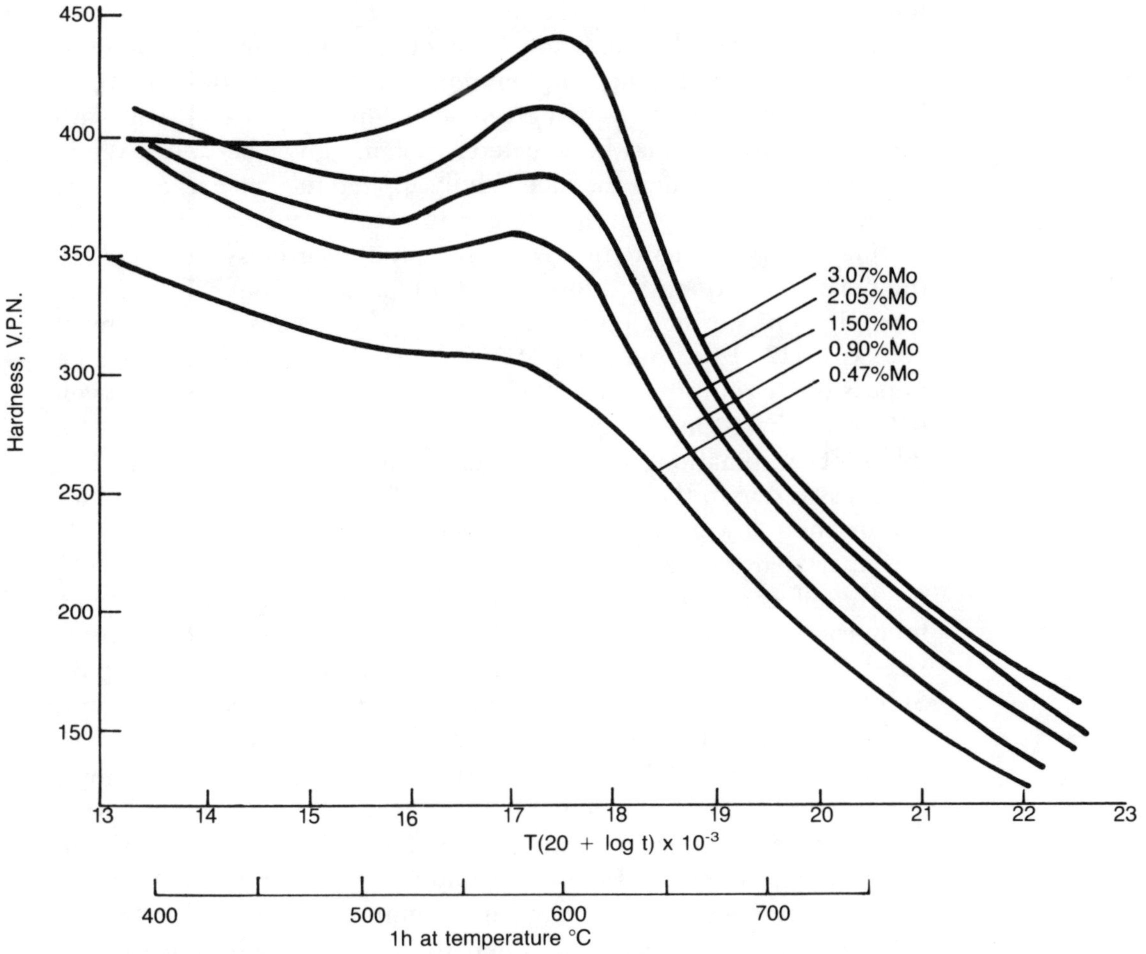

Fig. 8.8. Retardation of softening and secondary hardening during tempering of steels with various molybdenum content. (Ref 8.6)

Up to this point, tempering has been discussed with temperature as the major variable. The structural changes responsible for the property changes, however, are thermally activated and therefore dependent on both temperature and time. For example, if a single mechanism of structural change is operating during a stage of tempering, say the coarsening of cementite, a given hardness may be obtained by tempering at a high temperature for a short time or by tempering at a lower temperature for a longer time. Generally if time is not mentioned, as is the case for most of the preceding figures, a constant tempering time of 1 h is assumed.

The interchangeability of time and temperature is accomplished by use of a tempering parameter, $T(20 + \log t) \times 10^{-3}$, where T is temperature in Kelvin, and t is time in hours. Figure 8.8 shows hardness changes plotted as a function of the tempering parameter as well as a function of tempering temperature where the time has been held constant at 1 h. Thus in a given alloy, tempering treatments with times other than 1 h may be selected to obtain a given hardness. While the tempering parameter is successfully applied to plain carbon steels, caution must be used in applying it to secondary hardening steels. During secondary hardening, the maximum hardness obtained on tempering is frequently a function of temperature (Ref 8.7, 8.8). For example, a higher maximum hardness may be obtained by holding at 600 °C (1112 °F) rather than at 700 °C (1292 °F), and it would be impossible to reproduce the 600 °C (1112 °F) hardness maximum even with very short-time tempering at 700 °C (1292 °F). This inability of different combinations of time and temperature to reproduce the same hardness is due to the somewhat coarser distribution of alloy carbides and/or their lower degree of coherency with the matrix at higher secondary hardening temperatures.

The effect of alloying elements on hardness changes produced by tempering martensitic carbon and low-alloy steels has been summarized in the investigation by Grange, Hribal and Porter (Ref 8.5). Steels with silicon, manganese, phosphorus, nickel, chromium, molybdenum and vanadium additions up to 1.5% were examined. Graphs of hardness differences (ΔH_v) relative to tempered Fe-C alloys, as a function of alloying element content, were obtained for tempering temperatures between 200 and 700°C (400 and 1300°F) for constant tempering times of 1 h. Figures 8.9 and 8.10 show plots of ΔH_v versus alloying element content for martensite tempered at 260°C (500°F) and 540°C (1000°F), respectively. When the ΔH_v for each element at a given tempering temperature from plots similar to those of Fig. 8.9 and 8.10 is added to the base tempered hardness as determined by carbon content and tempering temperature (see Fig. 8.7), the final tempered hardness of a plain-carbon or low-alloy steel may be readily estimated.

Figures 8.9 and 8.10 reflect interesting differences due to the various alloying elements. The strong carbide formers, as discussed relative to secondary hardening, do not have a large effect until high temperatures are reached. Nickel has a very small and constant effect on tempered hardness at all temperatures, and since it is not a carbide-forming element, its influence is considered to be due to a weak solid solution strengthening effect. Silicon has a substantial retarding effect on softening around 316 °C (600 °F), an effect attributed (Ref 8.9) to its inhibition of the transformation of the low-temperature transition carbide to the more stable cementite. Manganese at low tempering

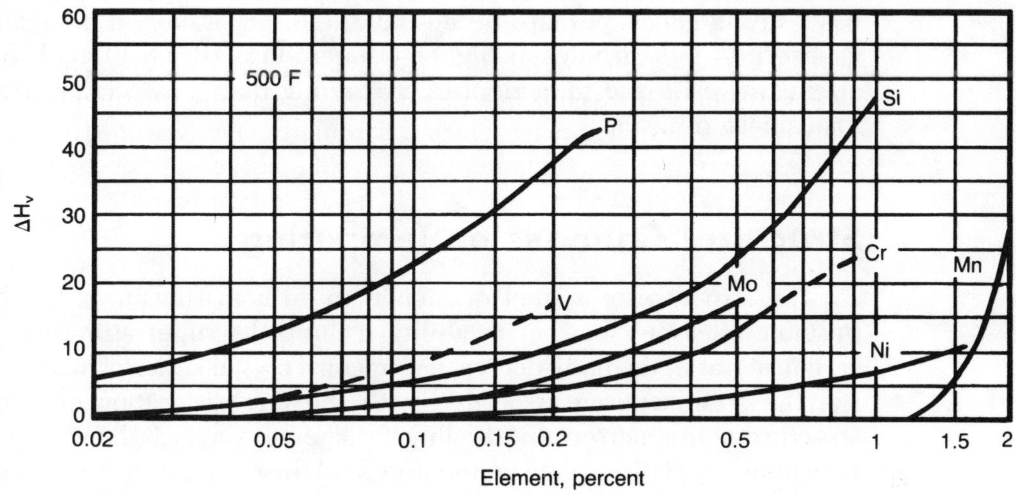

Fig. 8.9. Effect of alloying elements on the retardation of softening during tempering at 260 °C (500 °F) relative to Fe-C alloys. (Ref 8.5)

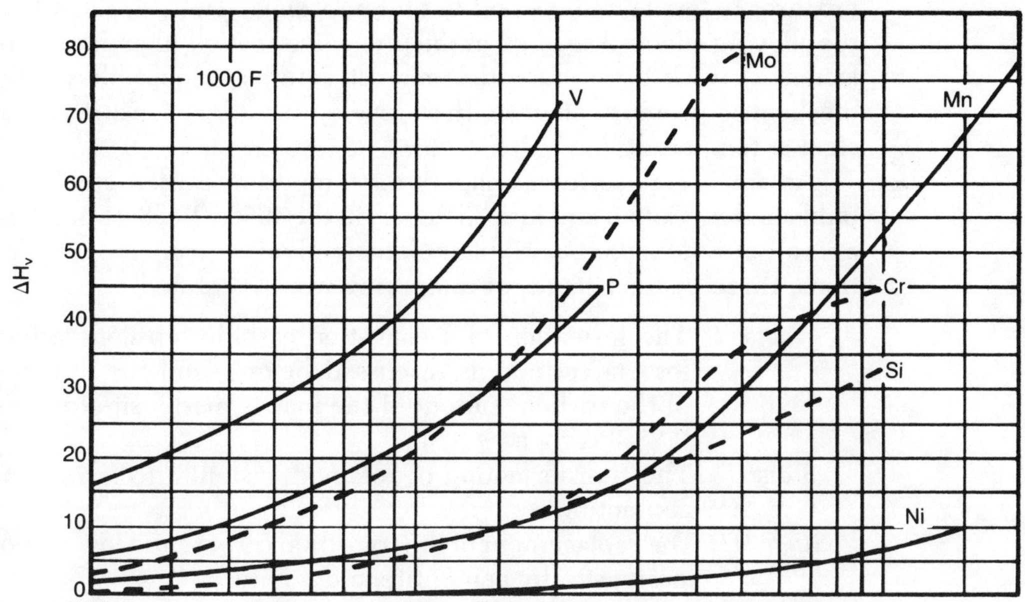

Fig. 8.10. Effect of alloying elements on the retardation of softening during tempering at 540 °C (1000 °F) relative to Fe-C alloys. (Ref 8.5)

temperatures has little effect on softening, but at higher temperatures has a strong effect, perhaps because of the incorporation of the manganese into the carbides at higher temperatures (Ref 8.10) and the attendant resistance to cementite coarsening that is associated with manganese diffusion.

Structural Changes on Tempering

The structure of a steel quenched to form martensite is highly unstable. Reasons for the instability include the supersaturation of carbon atoms in the body-centered tetragonal crystal lattice of martensite, the strain energy associated with the fine dislocation or twin structure of the martensite, the interfacial energy associated with the high density of lath or plate boundaries, and the retained austenite that is invariably present even in low-carbon steels. The supersaturation of carbon atoms provides the driving force for carbide formation; the high strain energy the driving force for recovery; the high interfacial energy the driving force for grain growth or coarsening of the ferrite matrix; and the unstable austenite the driving force for transformation to mixtures of ferrite and cementite on tempering. Thus, even without the alloying effects discussed in the preceding section, there are many factors at work to produce the microstructures responsible for the mechanical property changes that develop when martensitic carbon steel is tempered.

An important series of papers on tempering carbon steels was published by Cohen and his colleagues in the 1950s (Ref 8.11-8.14). As a result of systematic x-ray, dilatometric and microstructural observations, three distinct stages of tempering were identified:

Stage I: The formation of a transition carbide, epsilon carbide (or eta carbide as discussed below), and the lowering of the carbon content of the matrix martensite to about 0.25% carbon.

Stage II: The transformation of retained austenite to ferrite and cementite.

Stage III: The replacement of the transition carbide and low-carbon martensite by cementite and ferrite.

The temperature ranges for the three stages overlap, depending on the tempering times used, but the temperature ranges of 100 to 250 °C (212 to 482 °F), 200 to 300 °C (392 to 572 °F) and 250 to 350 °C (482 to 662 °F) are generally accepted for the first, second and beginning third stages, respectively (Ref 8.15). The formation of the alloy carbides responsible for secondary hardening is sometimes referred to as the fourth stage of tempering. Also it is now recognized that carbon atom segregation to dislocations and various boundaries may occur during quenching and/or holding at room temperature (Ref 8.15-8.17) and carbon atom clustering in as-quenched martensite may precede carbide formation (Ref 8.18, 8.19) that occurs in the first stage of tempering. Thus, tempering involves much more than three stages of tempering, but because of their central importance to understanding the behavior of tempered steels the three stages listed above will be discussed in more detail.

The transition carbide that forms in the first stage of tempering was first identified as having a hexagonal structure and designated epsilon (ϵ) carbide by Jack (Ref 8.20). More recently, Hirotsu and Nagakura (Ref 8.21, 8.22) have shown that the transition carbide has an orthorhombic structure isomorphous with transition metal carbides of the M_2C type. The transition carbide with the latter structure was designated as eta (η) carbide. The structures of the epsilon and eta carbides are very similar and are differentiated primarily by electron diffraction spots that come from a regular array of carbon atoms (or sublattice of carbon atoms) in the eta carbide (Ref 8.21, 8.22). Both the epsilon carbide, $Fe_{2.4}C$ (Ref 8.11), and the eta carbide, Fe_2C (Ref 8.21), have carbon contents substantially higher than that of the cementite, Fe_3C, that forms at higher temperatures. Kinetic studies show that the first stage of tempering is dependent on the diffusion of carbon through the martensite with an activation energy of 16 000 cal/mol (Ref 8.11).

Figures 8.11 through 8.13 are transmission electron micrographs that show various aspects of the transition carbide formation in the martensite of an Fe-1.22C alloy tempered at 150 °C (302 °F) for 16 h. Figure 8.11 shows a typical plate martensitic microstructure with plates of a variety of sizes and patches of retained austenite (black areas) between the plates. Each of the plates contains a highly uniform distribution of fine carbide particles. Figure 8.12 shows a typical array of transition carbides, identified as eta carbides (Ref 8.2), in a single plate of martensite. The carbides appear to be in the form of fine platelets, but Fig. 8.13, a dark field micrograph taken with illumination from a carbide diffraction spot, shows that the eta carbide is actually present as rows of fine spherical particles about 2 nm in diameter (Ref 8.2, 8.21). The dark contrast in the plate-like morphology associated with the carbides in Fig. 8.12 is apparently due to strain effects between the martensitic matrix and the rows of particles.

The transformation of retained austenite during tempering occurs only after the transition carbide is well established. Figure 8.14 shows the rate of transformation of the retained austenite in an Fe-1.22C alloy

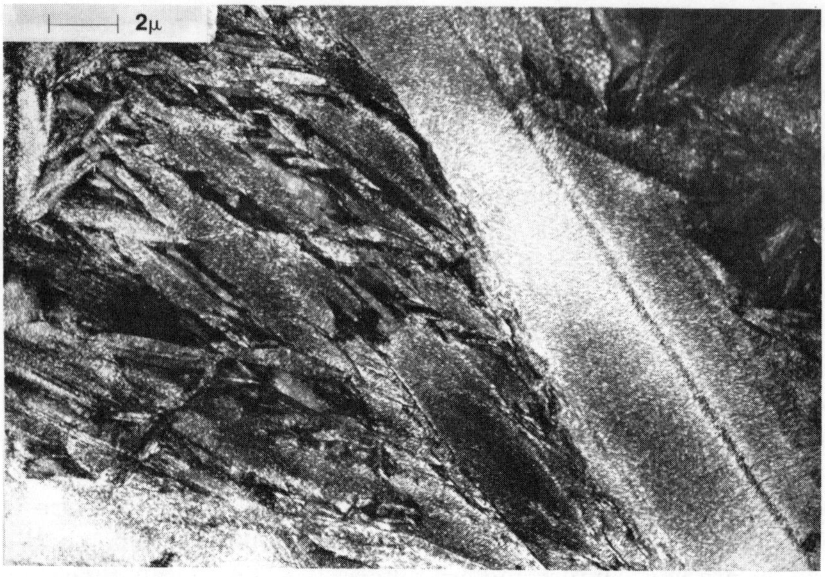

Fig. 8.11. Martensitic microstructure in an Fe-1.2C alloy tempered at 150 °C (302 °F). The microstructure consists of plates of various sizes containing uniform arrays of very fine carbides and retained austenite (black patches). Transmission electron micrograph. (Ref 8.23)

at three different tempering temperatures. About 19% retained austenite, distributed as shown in Fig. 8.11, was initially present in the as-quenched structure. Even at 180 °C (356 °F) the retained austenite transformed completely to mixtures of ferrite and cementite if held for sufficiently long times. Analysis (Ref 8.23) of the austenite transformation kinetics in Fig. 8.14 yielded an activation energy of 1.15×10^5 J/mol (27 kcal/mol) in good agreement with the activation energies for the diffusion of carbon in austenite (Ref 8.24) and the activation energy for the second stage of tempering reported by Roberts, Averbach and Cohen (Ref 8.11). Figure 8.15 shows that retained austenite is present in small amounts, about 2 and 4%, in as-quenched specimens of 4130 and 4340 steels, respectively, and that for tempering times of 1 h the transformation of retained austenite in these low-alloy medium-carbon steels begins only above 200 °C (392 °F). Transformation is complete at about 300 °C (572 °F) (Ref 8.25), and cementite becomes an important part of the microstructure, after tempering at 300 °C (572 °F) and higher temperatures.

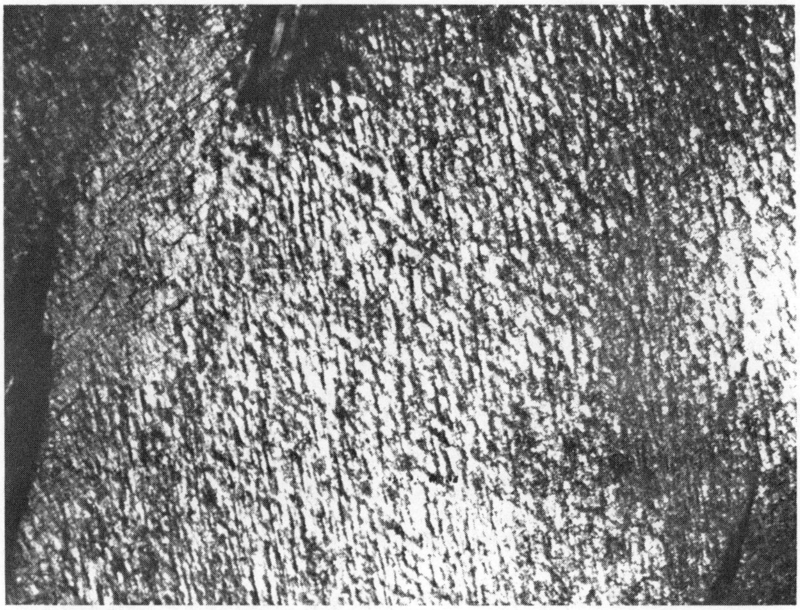

Fig. 8.12. Distribution of eta carbide in martensite plate of an Fe-1.22C alloy tempered at 150 °C (302 °F) for 16 h. Transmission electron micrograph. Magnification, 80 000 ×; shown here at 75%. (Ref 8.2)

The third stage of tempering consists of the formation of ferrite and cementite as required by the Fe-C diagram. However, there is some evidence, especially in high-carbon steels, that Hägg or chi (X)-carbide formation precedes cementite or theta (Θ)-carbide formation (Ref 8.26, 8.27). The chi carbide has a monoclinic structure, and the composition Fe_5C_2. However, despite the differences between cementite and chi carbide, the relatively complex structures of the two carbide phases are similar and difficult to separate by x-ray or electron diffraction techniques. Therefore, in view of the experimental difficulty in separating the presence of chi carbide from that of cementite, the temperature and compositions of the steels in which chi carbide forms are not yet completely defined.

Figure 8.16 shows the dense carbide distribution that has formed in the martensite of an Fe-1.22C alloy tempered at 350 °C (662 °F). In this case, the carbides were best identified as chi carbide (Ref 8.28). Two carbide morphologies are present: those that have nucleated and grown within the martensite plates, and very long planar carbides that

Fig. 8.13. Rows of fine spherical eta carbide particles in a martensite plate of an Fe-1.22C alloy tempered at 150 °C (302 °F) for 16 h. Dark-field transmission electron micrograph. Magnification, 80 000×; shown here at 75%. (Ref 8.2)

have formed along the plate interfaces, perhaps as a result of the transformation of retained austenite in the second stage of tempering. A third morphology of chi carbide and/or cementite in tempered high-carbon steels consists of parallel arrays of carbides formed on transformation twins sometimes present in high-carbon martensite, especially in the midrib portions of the plates (Ref 8.27). The carbides that have formed within the plates are coarser than the transition carbides and will eventually spheroidize if tempering is performed at higher temperatures.

The carbide structures and distributions that form in alloy steels and retard softening and/or produce secondary hardening during tempering are quite varied. Many of the alloy carbides and their formation on tempering have been characterized by Honeycombe and his colleagues. Much of this work, including descriptions of the carbide structures produced by tempering vanadium, molybdenum, tungsten, chromium and titanium steels, is reviewed in Ref 8.29. The alloy carbide distributions formed in the secondary hardening range, 500 to 650 °C (932 to 1202 °F), depend on the nature of the cementite

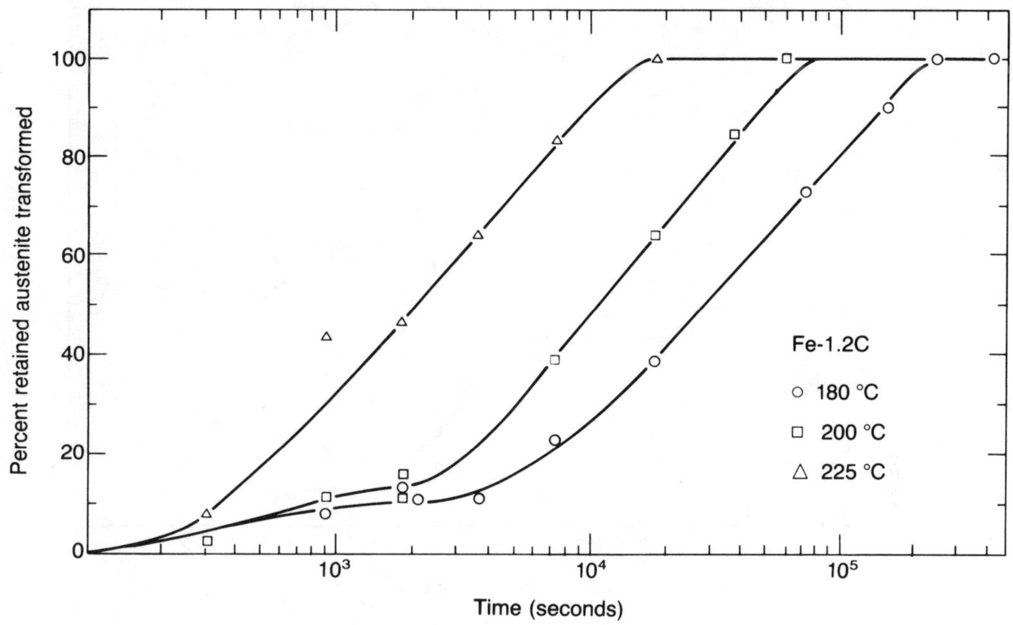

Fig. 8.14. Transformation of retained austenite in an Fe-1.22C alloy as a function of time at three tempering temperatures. (Ref 8.23)

distribution formed at lower tempering temperatures, and the nature of the transformation of cementite to the alloy carbide. Honeycombe (Ref 8.29) presents evidence for two basic modes of alloy carbide formation on tempering. The carbides may form directly from the cementite, a mode referred to as *in situ* transformation, or the carbides may form by separate nucleation, after the cementite particles dissolve in the ferrite matrix. The independently nucleated alloy carbide particles are often nucleated on the dislocations residual from the as-quenched martensite, and tend to be much finer than the alloy carbides nucleated on the cementite particles.

Most of the structural changes discussed above have involved the formation of various types of carbides during tempering. There are also important changes in the martensitic matrix that accomplish the formation of fully tempered structures consisting of spheroidized carbides in a matrix of equiaxed ferrite grains. Figures 8.17 through 8.20 show changes in the matrix structure that developed during the tempering of lath martensite in an Fe-0.2C alloy (Ref 8.30). Figure 8.17 shows that tempering at 400 °C (752 °F) for 15 min produces little change from the appearance of as-quenched lath martensite (see Chapter 3) on the scale

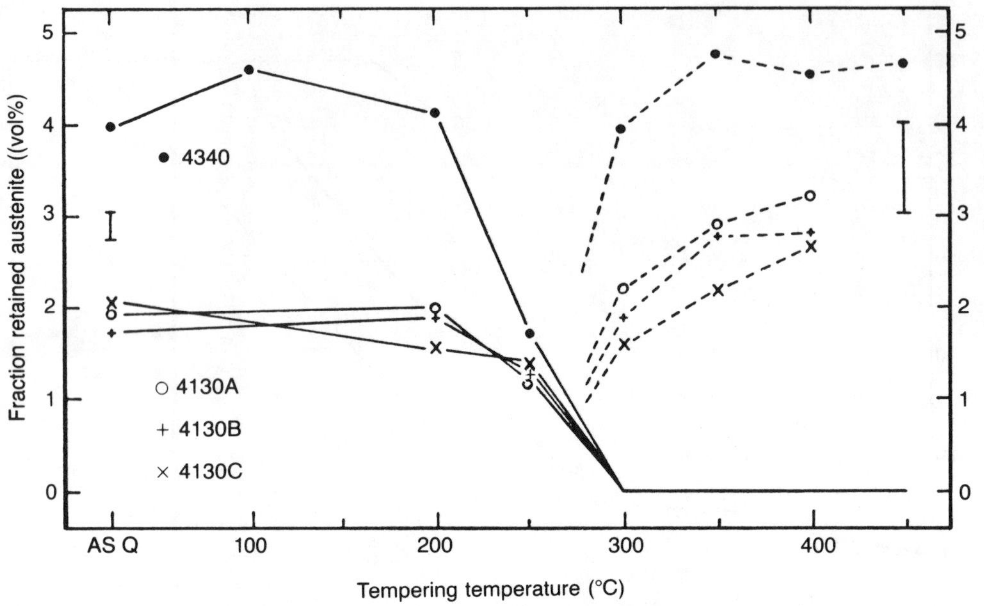

Fig. 8.15. Retained austenite and cementite as a function of tempering temperature in 4340 and 4140 type steels. The amounts of the phases were determined by Mössbauer spectroscopy. (Ref 8.25)

resolvable with the light microscope. More pronounced changes are visible in a specimen tempered at 700 °C (1292 °F) (see Fig. 8.18) but even after this rather severe temper, the packet morphology with its parallel sub-units is still clearly visible. The major effects of tempering have been to eliminate many of the smaller laths and to produce coarse, spherical cementite particles at the prior austenite grain boundaries and within the packets. More severe tempering, 700 °C (1292 °F) for 12 h, begins to break up the remaining parallel blocks of crystals within the packets and more equiaxed ferrite grains begin to form (see Fig. 8.19). The equiaxed grains contain subboundaries made up of regular dislocation arrays as shown in the electron micrographs of Fig. 8.20.

Systematic measurement of the change in lath boundary per unit volume as a function of tempering of the Fe-0.2C martensite shows that the very high lath boundary area per unit volume of the fine laths in as-quenched martensite decreased very rapidly on tempering (Ref 8.30). This initial rapid decrease is primarily due to the elimination of the low-angle boundaries between laths of similar orientation.

Fig. 8.16. Cementite and/or chi-carbide formation in martensitic structure of an Fe-1.22C alloy tempered at 350 °C (662 °F) for 1 h. Transmission electron micrograph. Magnification, 30 000 ×; shown here at 75%. (Ma, Ando and Krauss, unpublished research, Colorado School of Mines, Golden, Colo.)

Simultaneously, fine carbides precipitate and help to stabilize the surviving lath boundaries to maintain their parallel orientation within the packets. All of these initial matrix changes occur as a result of recovery mechanisms. The dislocation density is effectively lowered not only by the reduction of dislocations within the laths but also by the elimination of the low-angle lath boundaries. Eventually, with coarsening of the carbide particles, the remaining large-angle boundaries rearrange themselves to produce more equilibrium junctions between grains as typical of the mechanisms associated with grain growth (Ref 8.31). Any residual dislocations within the laths then rearrange themselves into low-angle boundaries within the equiaxed grains as shown in Fig. 8.20. Such subdivision of large grains by dislocation boundaries is referred to as polygonization. Thus, the formation of the equiaxed ferrite matrix that develops after long-time high-temperature temper-

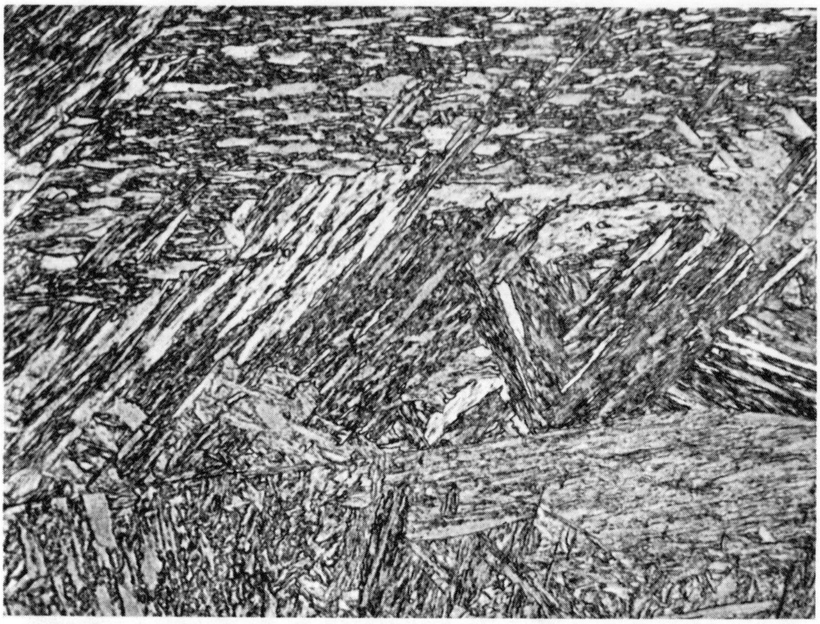

Fig. 8.17. Microstructure of lath martensite in an Fe-0.2C alloy after tempering at 400 °C (752 °F) for 15 min. Light micrograph. Nital etch. Magnification, 500×. (Ref 8.30)

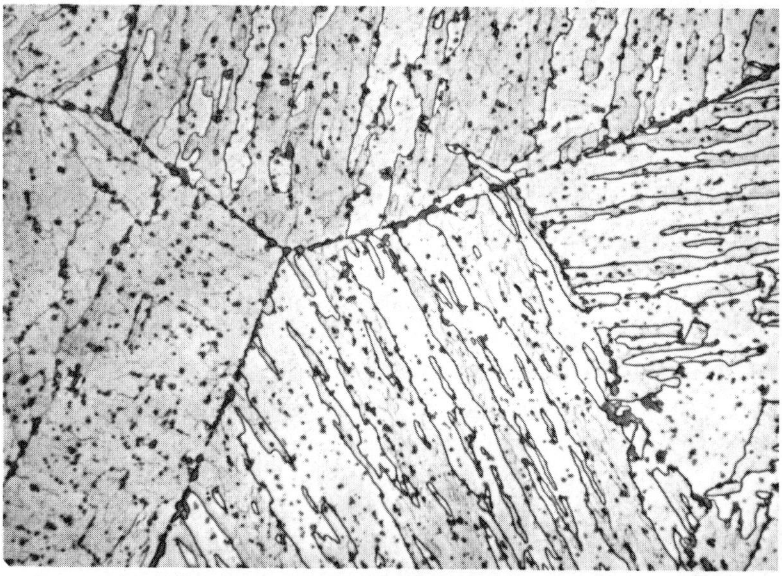

Fig. 8.18. Microstructure of lath martensite in an Fe-0.2C alloy after tempering at 700 °C (1292 °F) for 2 h. Light micrograph. Nital etch. Magnification, 500×. (Ref 8.30)

Fig. 8.19. Microstructures of lath martensite in an Fe-0.2C alloy after tempering at 700 °C (1292 °F) for 12 h. Light micrograph. Nital etch. Magnification, 500×. (Ref 8.30)

ing of a low-carbon lath martensite is accomplished by recovery and grain growth mechanisms (Ref 8.30, 8.32). Apparently, the recovery mechanisms that operate early in tempering lower the strain energy of the as-quenched martensite to the point where there is no longer sufficient driving force for recrystallization.

Embrittlement Phenomena and Tempering

High-strength quench and tempered steels are susceptible to a number of different types of embrittlement. Some of the embrittlement mechanisms are due to structural changes introduced during processing and tempering, and some are due to the interaction of the environment with the quench and tempered microstructures. Examples of the first type of embrittlement are tempered martensite embrittlement, temper embrittlement, and embrittlement due to aluminum nitride formation. Examples of the second type are hydrogen embrittlement and liquid metal embrittlement. Often there is overlap be-

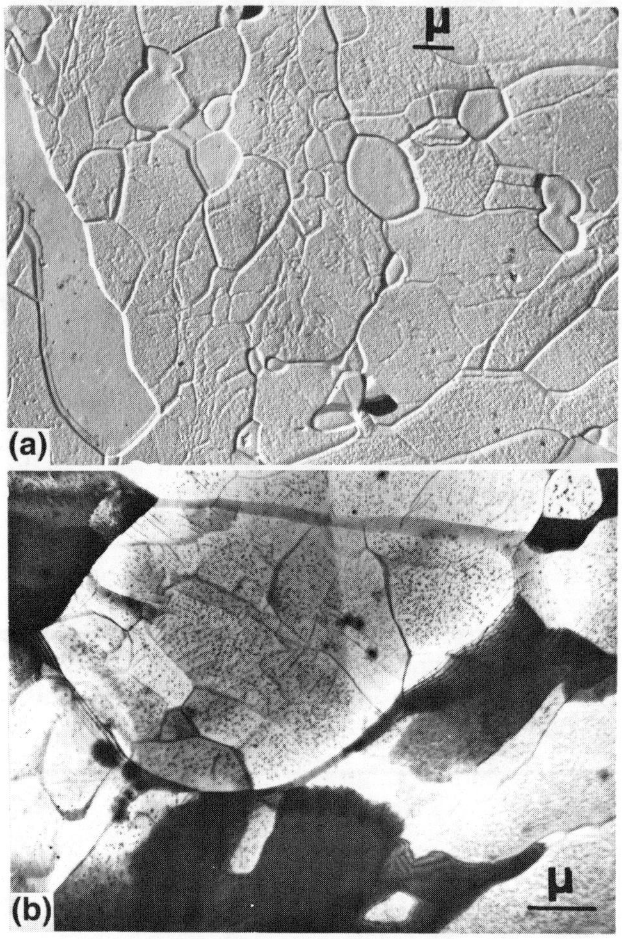

Fig. 8.20. Structure of lath martensite in an Fe-0.2C alloy after tempering at 700 °C (1292 °F) for 12 h. (a) Single stage replica of polished and etched (nital) surface. (b) Transmission electron micrograph.

tween two types of embrittlement. For example, tempered martensite embrittlement may be severely aggravated by exposure to hydrogen. This section compares and describes general features of the two classic embrittlement phenomena, tempered martensite embrittlement and temper embrittlement. Each of the embrittlement phenomena is then described in more detail in separate sections.

The characteristics of tempered martensite embrittlement and temper embrittlement are well known, but the causes continue to receive intensive study. Tempered martensite embrittlement (TME) occurs after tempering between 260 to 370 °C (500 to 700 °F) and is also referred to as 350 °C embrittlement or 500 °F embrittlement. Temper embrittlement (TE) occurs after tempering in or cooling through the temperature range of 375 to 575 °C (707 to 1070 °F). In an effort to introduce a terminology more related to the steps necessary to induce the two types of embrittlement, McMahon has suggested the term "one-step embrittlement" for TME and the term "two-step embrittlement" for TE, the latter term because two tempering treatments or a heating step and a cooling step are sometimes required to induce temper embrittlement (Ref 8.33).

Both TME and TE are associated with a shift in the impact transition temperature, as shown schematically for two hypothetical steels in Fig. 8.21. Steel A represents an inherently tougher steel than Steel B in that its transition temperature in the unembrittled state is lower than that of Steel B. Thus, when an embrittlement phenomenon shifts the transition temperature to higher temperatures by an equal amount for both steels, Fig. 8.21 shows that Steel A does not show as sharp a drop in room temperature toughness as does Steel B. Figure 8.21, therefore, emphasizes the importance of the testing temperature relative to the transition temperature, and shows that the results of room temperature testing may be misleading. For example, if Steel A is subjected to impact loading in service below room temperature, it will behave in a much more brittle fashion than indicated by the room temperature test.

Although there are similarities in the effects of the two types of embrittlement, from a practical standpoint TME and TE are separable into two different phenomena because they occur in two different temperature ranges and because TME is a much more rapid process than TE. TME develops within the normal 1-h time period for tempering, while TE takes many hours to accomplish. TE is therefore of major concern in heavy sections that are tempered at high temperatures to insure a good balance of strength and toughness. In particular, large shafts and rotors for power generating equipment are susceptible to TE because even after tempering above the critical temperature range, the heavy sections cool very slowly over a period of many hours through the critical range for embrittlement. TME, on the other hand, develops during tempering for short times in the critical range and, therefore, is independent of section size and/or cooling rate after tempering.

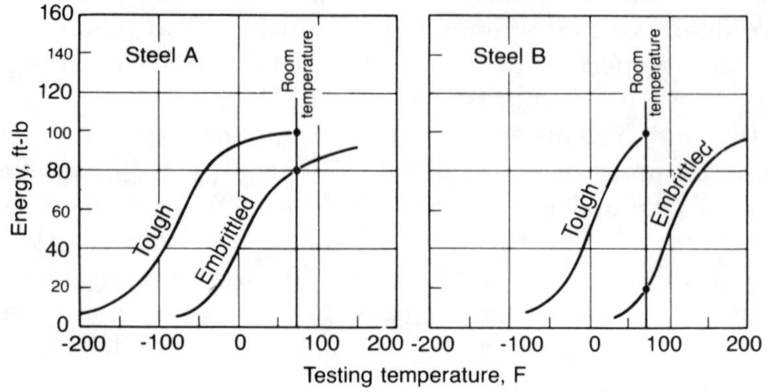

Fig. 8.21. Transition in fracture behavior for two hypothetical steels in the tough and embrittled conditions. (Ref 8.1)

Tempered Martensite Embrittlement

Tempered martensite embrittlement may or may not be associated with impurity atom segregation to prior austenitic grain boundaries and may be associated with three different modes of fracture through the tempered martensite of specimens tempered between 260 to 370 °C (500 and 700 °F). The common factor for all the manifestations of TME, at least in medium-carbon hardenable steels, appears to be the formation of cementite during tempering. In steels with low impurity content and/or the effect of impurities minimized by the gettering action of an alloying element, for example molybdenum interaction with phosphorus (Ref 8.34), the source of the cementite that leads to brittle transgranular fracture is the decomposition of retained austenite in the second stage of tempering. This explanation for the cause of TME was first proposed by Thomas (Ref 8.35) after transmission electron microscopy revealed the presence of thin films of retained austenite between laths of martensite in as-quenched medium-carbon steels and the transformation of this retained austenite to thin plates of cementite on tempering. Figure 8.15 shows graphically that retained austenite is replaced by cementite in the critical temperature range for TME.

Figure 8.22 shows several characteristics of TME in two 4340 steels of the same composition except for phosphorus content (Ref 8.36). The impact toughness of the steel with the higher phosphorus content (0.03%) is inferior to that of the steel with the lower phosphorus

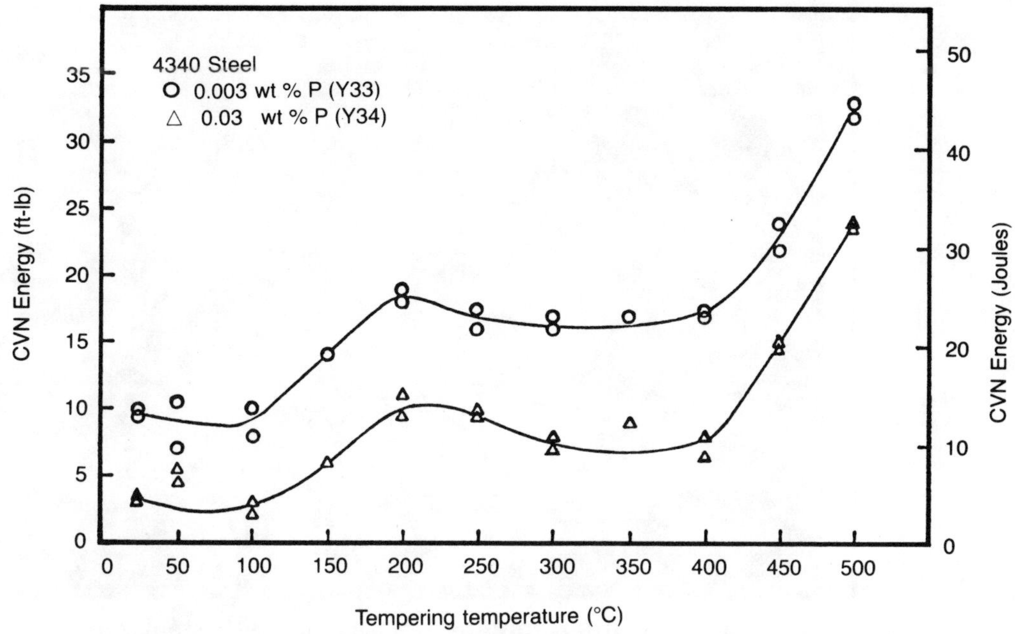

Fig. 8.22. Room temperature Charpy impact energy versus tempering temperature for 4340 steel containing different amounts of phosphorus as shown. Specimens were austenitized at 870 °C (1598 °F), oil quenched, and tempered 1 h. at temperatures shown. (Ref 8.36)

content (0.003%) after tempering over the entire range of temperatures up to 500 °C (932 °F). Also, both steels show a trough or plateau in energy absorbed between 250 and 400 °C (482 and 752 °F). The lower toughness of the higher phosphorus steel was related to a substantial amount of intergranular fracture, about 20% after tempering up to 200 °C (392 °F), and over 80% after tempering between 300 and 400 °C (572 and 752 °F). Figure 8.23 shows the intergranular fracture along prior austenite grain boundaries of the high phosphorus 4340 steel broken at room temperature after tempering at 400 °C (752 °F). The intergranular mode of fracture associated with TME is quite common and has been related to phosphorus segregation to the austenite grain boundaries during austenitizing (Ref 8.37-8.39). The phosphorus, therefore, is present at the grain boundaries in the as-quenched martensite and after tempering up to 200 °C (392 °F), but only after tempering in the range where cementite forms in tempered martensite does TME fully develop. The latter observation indicates that an interaction between

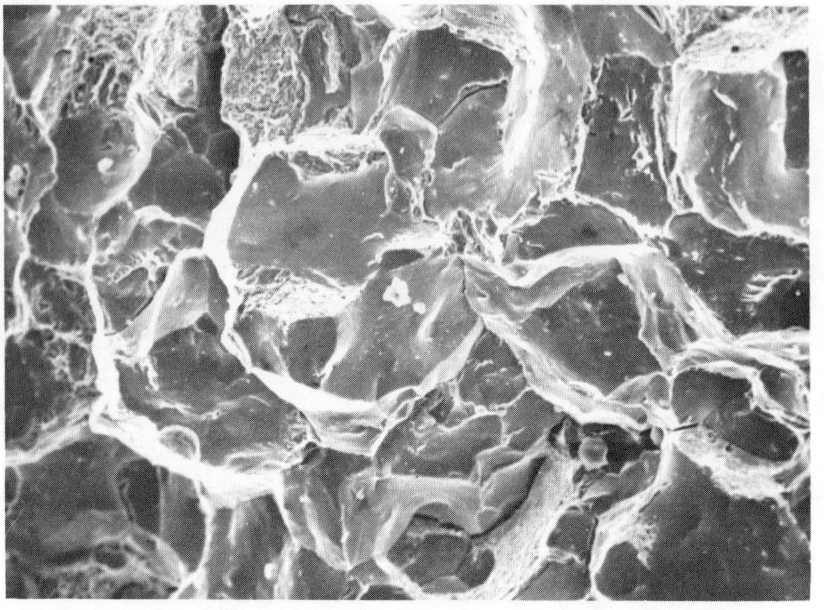

Fig. 8.23. Intergranular fracture of 4340 steel containing 0.03% phosphorus after tempering at 400 °C (752 °F). Specimen was broken by impact loading at room temperature. (Ref 8.36)

phosphorus and cementite is necessary for the intergranular mode of TME.

Figure 8.24 shows the fracture surface of the low-phosphorus 4340 steel broken by impact testing at room temperature after tempering at 350 °C (662 °F). No intergranular fracture is apparent, but flat cleavage facets are interspersed among regions of ductile fracture. These cleavage facets were found to be oriented across the laths of a packet (Ref 8.36). The initiation of the translath cleavage was attributed to cracking of the relatively thick cementite particles as suggested by King, Smith and Knott (Ref 8.40). Figure 8.25 shows interlath carbides in a specimen of low phosphorus content 4340 tempered at 350 °C (662 °F). The carbides formed between laths as a result of the transformation of the 4% austenite retained in the as-quenched condition (see Fig. 8.15).

Another type of transgranular fracture mode associated with TME is interlath cleavage induced by cracking parallel to the cementite formed by the transformation of the retained austenite. This mode has been observed in 4130-type steels (Ref 8.41), and originally proposed by Thomas (Ref 8.35) as the cause of TME. In the 4130 steels, flat, thin

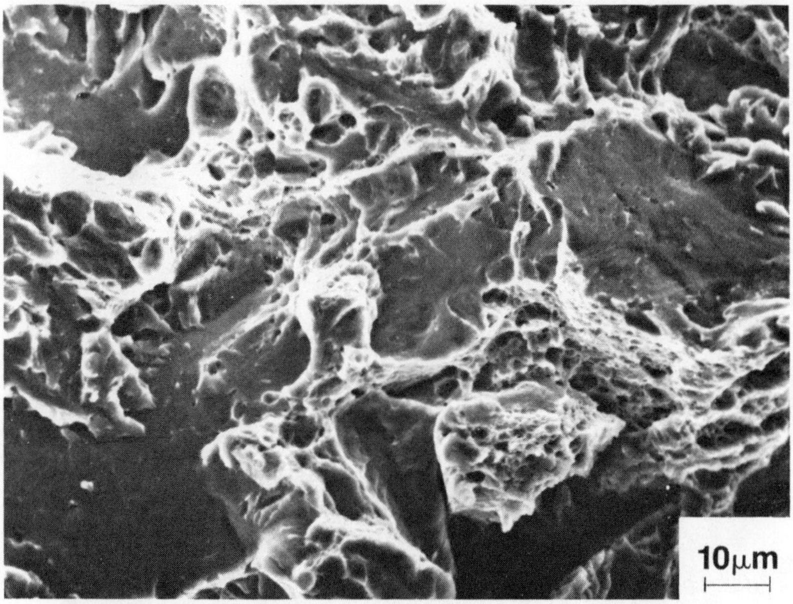

Fig. 8.24. Flat cleavage facets in 4340 steel containing 0.003% phosphorus after tempering at 350 °C (662 °F). Specimen was broken by impact loading at room temperature. (Ref 8.36)

carbides were extracted from the flat fracture facets in embrittled specimens, as shown in Fig. 8.26. The cause of the two types of transgranular fracture observed to accompany TME in certain steels may be related to interlath carbide thickness, the thinner carbides promoting interlath fracture and the thicker carbides promoting translath cleavage. This explanation is supported by Fig. 8.15, which shows a greater amount of cementite in the 4340 steel as a result of its higher retained austenite content as compared to the 4130 steel. Assuming similar lath morphologies and size distributions in the 4130 and 4340 steel, the greater volume fraction of carbides in the 4340 steel should translate into thicker carbides which, as observed, should initiate translath cleavage.

Temper Embrittlement

Temper embrittlement is a long-standing metallurgical problem the causes of which are just now being understood because of new theoretical approaches and the recent availability of surface analysis

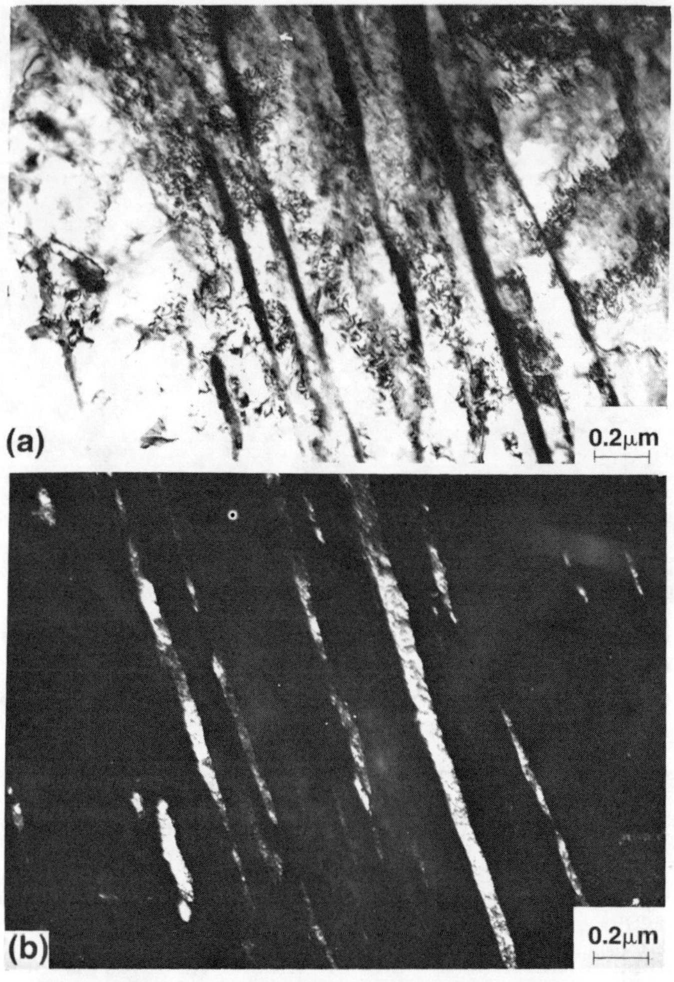

Fig. 8.25. Interlath carbides formed during tempering of 4340 steel containing 0.003% phosphorus at 350 °C (662 °F). (a) Bright field image and (b) dark field image using a cementite diffracted beam for illumination. Transmission electron micrographs. (Ref 8.36)

equipment capable of detecting grain boundary segregation on an atomic scale (Ref 8.42). However, the many review articles (Ref 8.33, 8.42-8.46) on TE show that the tempering conditions and compositional factors that induce TE are well known even if the exact mechanisms still remain to be discovered.

Heat treatment factors include, as stated earlier, the fact that susceptible steels must be heated in or cooled through the critical

Fig. 8.26. Carbides extracted from cleavage fracture facets of 4130-type steel tempered at 350 °C (662 °F). Magnification, 11 000×; shown here at 75%. (Ref 8.41)

temperature range of 375 to 575 °C (706 to 1070 °F) in order to develop TE. The embrittlement that results is detectable primarily by an increase in impact transition temperature as shown in Fig. 8.27 for a 3140 steel (containing nominally 1.15% nickel and 0.65% chromium) embrittled by both isothermal tempering and slow cooling through the critical temperature range. The embrittling kinetics follow C-curve behavior with tempering time and temperature, with a nose or minimum time for embrittlement at about 550 °C (1022 °F). One study (Ref 8.47) shows that it takes about an hour at 550 ° C (1022 °F) for the first increase in transition temperature to be noticeable, and several hundred hours for the first signs of embrittlement at around 375 °C (706 °F), the lower temperature limit for TE. TE is reversible, and de-embrittlement may occur on heating to about 575 °C (1070 °F) after holding only a few minutes at temperature.

Compositional factors affecting TE include the requirement that specific impurities must be present for a steel to be susceptible. The impurities most detrimental are antimony, phosphorus, tin and arsenic. Relatively small amounts of these elements, on the order of 100 ppm (0.01%) or less have been shown to cause TE. Silicon and manganese in large amounts also appear to be detrimental. Plain carbon

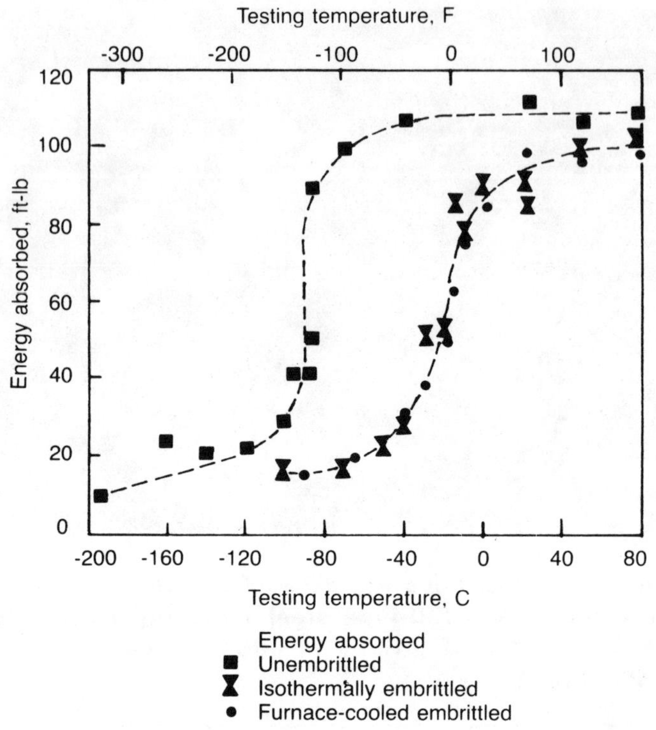

Fig. 8.27. Shift in impact transition curve to higher temperature as a result of TE produced in SAE 3140 steel by isothermal holding and furnace cooling through the critical range. (Ref 8.1)

steels are not considered to be susceptible to TE, provided the manganese content is held below 0.5%. Alloy steels are most susceptible, especially the chromium-nickel steels that are frequently used for heavy rotors. Molybdenum, however, reduces susceptibility to TE and, in amounts of 0.5% or less, is an important alloying element added to steels to minimize TE.

The cause of TE has been difficult to identify because there is no readily resolvable microstructural feature identifiable with the characteristic intergranular fracture of embrittled specimens. The only metallographic evidence of embrittlement has been the ability of certain etchants to reveal prior austenite grain boundaries containing segregated phosphorus (Ref 8.42). Currently available surface analysis techniques, however, such as Auger electron spectroscopy (AES), which is capable of analyzing composition of atomic layers adjacent to the intergranular fracture surfaces, have contributed greatly to the understanding of TE. AES shows not only high concentrations of impurity atoms segregated to the fracture surfaces, but also gradients of the alloying elements such as nickel that may in fact stimulate the impurity element segregation to the prior austenite grain boundaries. For example, grain boundary carbides may reject the nickel as they grow and, therefore, produce nickel concentration gradients that in turn cause the concentration of impurity atoms (Ref 8.48). Further, the interactions of impurities and alloying elements that produce segregation have been treated in a thermodynamic model developed by Guttmann (Ref 8.49) with good success. Thus, present research indicates that not just impurity elements but the interaction of impurities and alloying elements may be responsible for the segregation that leads to grain boundary decohesion in temper embrittled steels. Further research in these directions should continue to deepen the understanding of the causes of TE.

Aluminum Nitride Embrittlement

Another type of intergranular embrittlement sometimes encountered in hardened steels is that associated with aluminum nitride precipitation on prior austenite grain boundaries. Whereas a fine dispersion of aluminum nitride particles is desirable to control grain growth in steels (see Chapter 7), aluminum nitrides precipitated in the form of sheet-like particles on cooling from solidification, or reprecipitated on cooling after solution at high austenitizing temperatures, may significantly reduce toughness. The intergranular fracture due to aluminum nitride particles is most frequently encountered in carbon steel castings (Ref 8.50), and is sometimes referred to as "rock-candy" fracture because the coarse intergranular facets of the castings produce a macroscopically crystalline appearance. Figure 8.28 shows aluminum nitride and carbide particles extracted from the intergranular fracture surface of a medium-carbon steel casting. The aluminum nitride particles are characteristically very thin, with a plate or sheet-like morphology, and are readily identified by electron diffraction (Ref 8.50). Although the intergranular precipitation of aluminum nitrides is most closely tied to intergranular fracture in cast steels, plate-like aluminum nitride particles have also been observed in wrought medium-carbon steels (Ref 8.51) and, if present in high concentration because of high nitrogen and aluminum content, may contribute to the intergranular fracture associated with TME (Ref 8.37, 8.52).

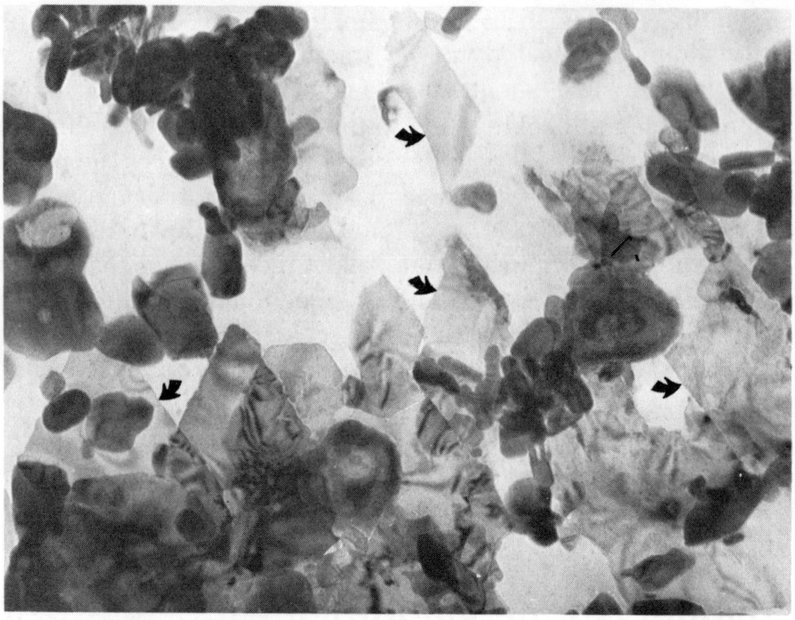

Fig. 8.28. Thin aluminum nitride particles (arrows) extracted from the intergranular fracture surface of a medium-carbon cast steel. Thick or dark particles are carbides. Extraction replica electron micrograph. Magnification, 82 500 × ; shown here at 75%. (G. Krauss, unpublished research, Colorado School of Mines, Golden, Colo.)

Liquid Metal Embrittlement

The exposure of steels to liquid metals may also cause brittle fracture by intergranular cracking (Ref 8.53). Plain carbon and low-alloy steels may be embrittled by exposure to liquid lead, cadmium, brass, aluminum bronze, copper, zinc, lead-tin solders and lithium (Ref 8.54). The initiation of fracture by liquid metal embrittlement is not time dependent but begins immediately on wetting of the steels. Often very low stress is sufficient to cause fracture by this mechanism.

Of special interest with respect to heat treatment of carbon steels is the work of Breyer and his colleagues concerning embrittlement of steels by lead (Ref 8.55-8.57). Lead may cause embrittlement if externally applied or if present internally in a steel. The latter cause of embrittlement is magnified by the widespread use of leaded steels for

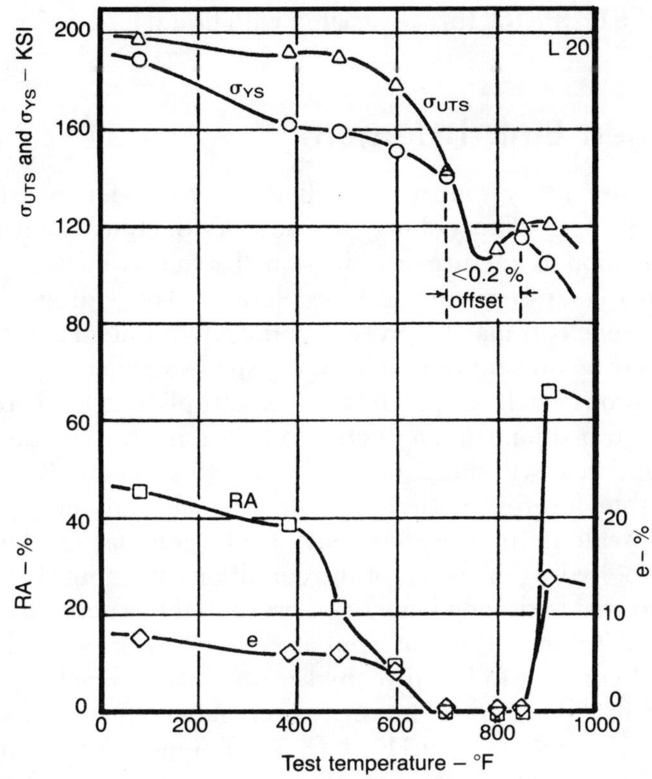

Fig. 8.29. Tensile properties of leaded 4145 steel quenched and tempered to strengths of 200 ksi (1380 MPa) as a function of tensile test temperature. (Ref 8.55)

improved machinability. Figure 8.29 shows an extreme example of the effects of lead embrittlement in a leaded 4145 steel heat treated to strengths close to 200 ksi (1380 MPa). At testing temperatures between 200 and 480 °C (400 and 900 °F) the ductility is significantly reduced with the most severe reduction to zero ductility occurring at and above the melting point of lead, 327 °C (621 °F). Generally the embrittlement is more severe the higher the strength level of the steel and therefore quench and tempered steels, if leaded, are especially susceptible. The fractures associated with the embrittlement are generally intergranular (Ref 8.57).

In summary, three conditions are necessary for lead embrittlement: (1) the presence of lead either externally or internally in the steel, (2) tensile loading, and (3) temperatures between 200 and 480 °C

(400 and 900 °F). Avoiding any one of these three conditions will prevent the brittle fracture associated with liquid lead embrittlement.

Hydrogen Embrittlement

There are many embrittling effects of hydrogen on steels: the ultimate strength of a steel may be reduced; ductility as measured by total elongation to fracture or reduction of area may be decreased; and crack growth may be significantly accelerated. The hydrogen responsible for these effects may be present in the environment external to the steel or may be present internally as a result of steelmaking or processing operations such as pickling or electroplating. Hydrogen may promote a transition from a ductile to brittle fracture mode or it may reduce ductility without a change in fracture mode (Ref 8.58). Of considerable interest with regard to quench and tempered high-strength steels is the development of intergranular cracking in the presence of hydrogen. Some of the conditions affecting such cracking due to internal hydrogen have been presented in a paper by Johnson, Morlet and Troiano (Ref 8.59) and are described briefly below. More recently, hydrogen-induced or hydrogen-assisted cracking has been related to the same type of impurity segregation effects responsible in part for TME (Ref 8.37) and TE (Ref 8.53). For more information on the manifold effects of hydrogen on materials the reader is referred to the book *Effect of Hydrogen on Behavior of Materials* as identified in Ref 8.58. A collection of important papers published in the literature between 1875 and 1977 concerning hydrogen effects is also available (Ref 8.60).

An important characteristic of hydrogen embrittlement in high-strength steel is the delay time required for failure. The higher the hydrogen content of a steel the lower the stress and the shorter the time for failure. Johnson, Morlet and Troiano (Ref 8.59) studied this problem by cathodically charging with hydrogen a 4340 steel oil quenched and tempered to 230 ksi (1585 MPa) tensile strength. After charging, the specimens were immediately plated with cadmium, a procedure that delayed outgassing of the hydrogen and insured a uniform distribution of hydrogen throughout the specimen section after baking. Figure 8.30 shows the effects of baking at 150 °C (300 °F) on the static fatigue (sustained loading) of the hydrogen-charged specimens. Increasing baking time effectively lowers hydrogen content even in the plated specimens, and sufficient baking eventually restores the strength of charged specimens to that of uncharged specimens. The horizontal portions of the curves in

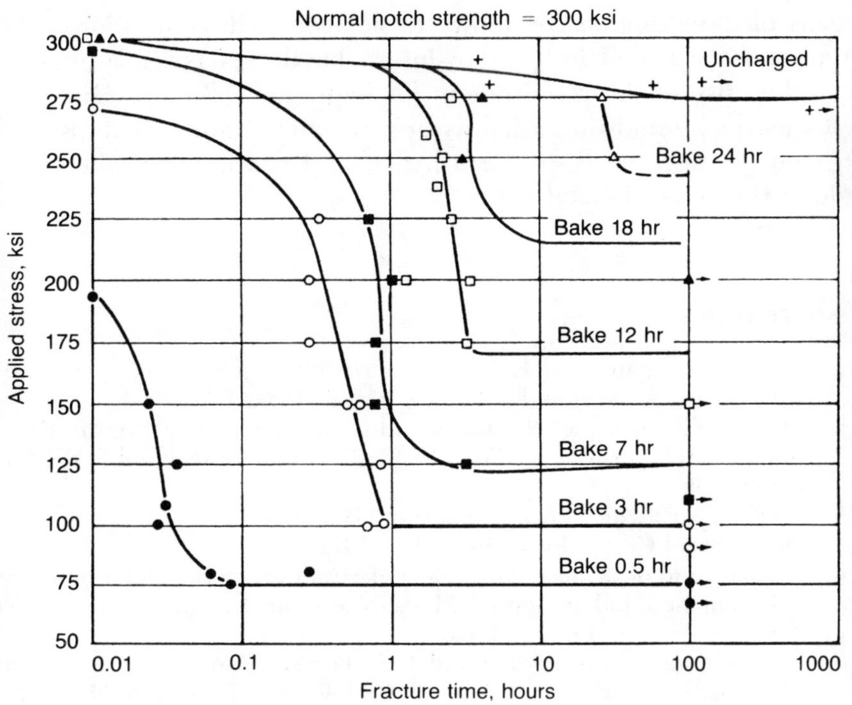

Fig. 8.30. Static fatigue curves for quenched and tempered 4340 notched specimens charged with hydrogen and baked at 150 °C (300 °F) for the times shown. (Ref 8.59)

Fig. 8.30 are designated as static fatigue or endurance limits, i.e., the stress level below which failure would not occur no matter what the duration of stress application. As hydrogen content is decreased by baking, the static fatigue limit increases.

The specimens used to obtain Fig. 8.30 were notched and therefore the static fatigue limits hold for that particular notch geometry (Ref 8.59). In general the sharper the notch, the lower the static fatigue limits, an indication that a critical combination of hydrogen concentration and triaxial stress state is required for crack initiation. Another very important aspect of hydrogen cracking revealed by the Johnson study (Ref 8.59) was the observation of an incubation period for cracking. This incubation period is dependent on hydrogen concentration (see Fig. 8.30) and is related to the time for diffusion of hydrogen to the point of crack initiation within the triaxial stress field. The need for an incubation time for crack initiation means that high-strain rate tests might not detect hydrogen embrittlement, a situation just the reverse of TME and TE. Sustained loading is, therefore, the most sensitive means of detecting susceptibility to delayed hydrogen failure. The propagation of a crack initiated by internal hydrogen is discontinuous, again because of the requirement for hydrogen diffusion. Once the crack leaves behind the initial hydrogen concentration, it stops, and does not advance until sufficient hydrogen has diffused to the stress field at the crack tip (Ref 8.61).

References

8.1 M. A. Grossmann and E. C. Bain, *Principles of Heat Treatment*, 5th Ed., American Society for Metals, Metals Park, Oh., 1964

8.2 D. L. Williamson, K. Nakazawa and G. Krauss, A Study of the Early Stages of Tempering in an Fe-1.2% C Alloy, *Met Trans A*, Vol 10A, 1979, p 1351-1363

8.3 *Modern Steels and Their Properties*, Handbook 2757, 7th Ed., Bethlehem Steel Corp., Bethlehem, Pa., 1972

8.4 T. Swarr and G. Krauss, The Effect of Structure on the Deformation of As-Quenched and Tempered Martensite in an Fe-0.2% C Alloy, *Met Trans A*, Vol 7A, 1976, p 41-48

8.5 R. A. Grange, C. R. Hribal and L. F. Porter, Hardness of Tempered Martensite in Carbon and Low-Alloy Steels, *Met Trans A*, Vol 8A, 1977, p 1775-1785

8.6 E. C. Rollason, Fundamental Aspects of Molybdenum in Transformation of Steel, Climax Molybdenum Co., London

8.7 K. J. Irvine and F. B. Pickering, The tempering characteristics of low-carbon low-alloy steels, *JISI*, Vol 194, 1960, p 137-153

8.8 E. Smith and J. Nutting, The Tempering of Low-Alloy Creep-resistant Steels containing Chromium, Molybdenum, and Vanadium, *JISI*, Vol 187, 1957, p 314-329

8.9 A. G. Allen and P. Payson, The Effect of Silicon on the Tempering of Martensite, *Trans ASM*, Vol 45, 1953, p 498-532

8.10 W. Crafts and J. L. Lamont, Effects of Alloys in Steel as Resistance to Tempering, *Trans AIME*, Vol 172, 1947, p 222-243

8.11 C. S. Roberts, B. L. Averbach and M. Cohen, The Mechanism and Kinetics of the First Stage of Tempering, *Trans ASM*, Vol 45, 1953, p 576-604

8.12 B. S. Lement, B. L. Averbach and M. Cohen, Microstructural Changes on Tempering Iron-Carbon Alloys, *Trans ASM*, Vol 46, p 851-881

8.13 F. E. Werner, B. L. Averbach and M. Cohen, The Tempering of Iron-Carbon Martensite Crystals, *Trans ASM*, Vol 49, 1957, p 823-84

8.14 B. S. Lement, B. L. Averbach and M. Cohen, Further Study of Microstructural Changes on Tempering Iron-Carbon Alloys, *Trans ASM*, Vol 47, 1955, p 291-319

8.15 G. R. Speich, Tempered Ferrous Martensitic Structures, in *Metals Handbook*, Vol 8, 8th Ed., American Society for Metals, Metals Park, Oh., p 202-204

8.16 G. R. Speich and W. C. Leslie, Tempering of Steel, *Met Trans*, Vol 3, 1972, p 1043-1954

8.17 G. R. Speich, Tempering of Low-Carbon Martensite, *Trans TMS-AIME*, Vol 245, 1969, p 2553-2564

8.18 J. Gemin and P. A. Flinn, Mössbauer Effect Study of the Clustering of Carbon Atoms during the Room-Temperature Aging of Iron-Carbon Martensite, *Trans TMS-AIME*, Vol 242, 1968, p 1419-1430

8.19 G. V. Kurdjumov and A. G. Khachaturyan, Phenomena of Carbon Atom Redistribution in Martensite, *Met Trans*, Vol 3, 1972, p 1069-1076

8.20 K. H. Jack, Structural Transformations in the Tempering of High Carbon Martensitic Steel, *JISI*, Vol 169, 1951, p 26-36

8.21 Y. Hirotsu and S. Nagakura, Crystal Structure and Morphology of the Carbide Precipitated from Martensite High Carbon Steel during the First Stage of Tempering, *Acta Met*, Vol 20, 1972, p 645-655

8.22 Y. Hirotsu and S. Nagakura, Electron Microscopy and Diffraction Study of the Carbide Precipitated at the First Stage of Tempering of Martensitic Medium Carbon Steel, *Trans Jpn Inst Met*, Vol 15, 1974, p 129-134

8.23 T. A. Balliett and G. Krauss, The Effect of the First and Second Stages of Tempering on Microcracking in Martensite of an Fe-1.22% C Alloy, *Met Trans A*, Vol 7A, 1976, p 81-86

8.24 C. Wells, W. Batz and R. F. Mehl, Diffusion Coefficient of Carbon in Austenite, *Trans AIME*, Vol 188, 1950, p 553-560

8.25 D. L. Williamson, R. G. Shupmann, J. P. Materkowski and G. Krauss, Determination of Small Amounts of Austenite and Carbide in a Hardened Medium Carbon Steel by Mössbauer Spectroscopy, *Met Trans A*, Vol 10A, 1976, p 379-382

8.26 Y. Imai, Phases in Quenched and Tempered Steels, *Trans Jpn Inst Met*, Vol 16, 1975, p 721-734

8.27 Y. Ohmori, Hägg Carbide Formation and Its Transformation into Cementite During the Tempering of Martensite, *Trans Jpn Inst Met*, Vol 13, 1972, p 119-127

8.28 C. Ma, T. Ando, D. L. Williamson and G. Krauss, Research in progress, Colorado School of Mines, Golden

8.29 R. W. K. Honeycombe, *Structure and Strength of Alloy Steels*, Climax Molybdenum Co., London

8.30 R. W. Caron and G. Krauss, The Tempering of Fe-C Lath Martensite, *Met Trans*, Vol 3, 1972, p 2381-2389

8.31 P. G. Shewmon, *Transformations in Metals*, McGraw-Hill, New York, 1969, p 220

8.32 R. M. Hobbs, G. W. Lorimer and N. Ridley, Effect of Silicon on the Microstructure of Quenched and Tempered Medium-Carbon Steels, *JISI*, Vol 210, 1972, p 757-764

8.33 C. J. McMahon, Jr., Strength of Grain Boundaries in Iron-base Alloys, in *Grain Boundaries in Engineering Materials*, Claitors Publishing Division, Baton Rouge, La., 1975, p 525-552

8.34 C. J. McMahon, Jr., A. K. Cianelli and H. C. Feng, The Influence of Mo on P-induced Temper Embrittlement in Ni-Cr Steel, *Met Trans A*, Vol 8A, 1977, p 1055-1057

8.35 G. Thomas, Retained Austenite and Tempered Martensite Embrittlement, *Met Trans A*, Vol 9A, 1978, p 439-450

8.36 J. P. Materkowski and G. Krauss, Tempered Martensite Embrittlement in SAE 4340 Steel, *Met Trans A*, Vol 10A, 1979, p 1643-1651
8.37 S. K. Banerji, C. J. McMahon, Jr. and H. C. Feng, Intergranular Fracture in 4340-type Steels: Effects of Impurities and Hydrogen, *Met Trans A*, Vol 9A, 1978, p 237-247
8.38 B. J. Schultz and C. J. McMahon, Jr., STP-499, ASTM, 1972, p 104
8.39 H. Ohtani and C. J. McMahon, Jr., Modes of Fracture in Temper Embrittled Steels, *Acta Metallurgica*, Vol 23, p 377-386
8.40 J. E. King, R. F. Smith and J. F. Knott, Toughness Variations During Tempering of a Plain Carbon Martensitic Steel, *Fracture 1977*, Vol 2, ICF4, Waterloo, Canada
8.41 R. G. Schupmann, "A Study of Tempered Martensite Embrittlement in 4230 Type Steels", M. S. Thesis, Colorado School of Mines, Golden, 1978
8.42 I. Olefjord, Temper Embrittlement, Review 231, *Inter Met Rev*, Vol 23, 1978, p 149-163
8.43 J. H. Hollomon, Temper Brittleness, *Trans ASM*, Vol 36, 1946, p 473-540
8.44 B. C. Woodfine, Temper Brittleness: A Critical Review of the Literature, *JISI*, Vol 173, 1953, p 229-240
8.45 J. M. Capus, The Mechanism of Temper Brittleness, in *Temper Embrittlement in Steel*, STP 407, ASTM, Philadelphia, Pa., 1968, p 3-19
8.46 C. J. McMahon, Jr., Temper Brittleness – An Interpretive Review, in *Temper Embrittlement in Steel*, STP 407, ASTM, 1968, p 127-167
8.47 F. L. Carr, M. Goldman, L. D. Jaffee and D. C. Buffum, Isothermal Temper Embrittlement of SAE 3140 Steel, *Trans TMS-AIME*, Vol 197, 1953, p 998
8.48 C. J. McMahon, Jr., E. Furubayashi, H. Ohtani and H. C. Feng, A Study of Grain Boundaries During Temper Embrittlement of a Low Carbon Ni-Cr Steel Doped with Antimony, *Acta Met*, Vol 24, 1976, p 695-704
8.49 M. Guttmann, The Link between Equilibrium Segregation and Precipitation in Ternary Solutions Exhibiting Temper Embrittlement, *Met Science*, Vol 10, 1976, p 337-341
8.50 J. A. Wright and A. G. Quarrell, Effect of Chemical Composition on the Occurrence of Intergranular Fracture in Plain Carbon Steel Castings Containing Aluminum and Nitrogen, *JISI*, Vol 197, 1962, p 299-307
8.51 J. P. Materkowski, "Tempered Martensite Embrittlement in 4340 Steel as Related to Phosphorus Content and Carbide Morphology", M. S. Thesis, Colorado School of Mines, 1978, Golden
8.52 J. M. Capus and G. Mayer, The Influence of Trace Elements on Embrittlement Phenomena in Low-Alloy Steels, *Metallurgia*, Vol 62, 1960, p 133-138
8.53 C. L. Briant and S. K. Banerji, Intergranular Failure in Steel: the Role of Grain Boundary Composition, Review 232, *Inter Met Rev*, Vol 23, 1978, p 164-199
8.54 Liquid Metal Embrittlement, in *Metals Handbook*, Vol 10, 8th Ed., American Society for Metals, Metals Park, Oh., p 228-229
8.55 S. Mostovoy and N. N. Breyer, The Effect of Lead on the Mechanical Properties of 4145 Steel, *Trans ASM*, Vol 61, 1968, p 219-232
8.56 W. R. Warke and N. N. Breyer, Effect of Steel Composition on Lead Embrittlement, *JISI*, Vol 209, 1971, p 779-784

8.57 R. D. Zipp, W. R. Warke and N. N. Breyer, A Comparison on Elevated Temperature Tensile Fractures in Nonleaded and Leaded 4145 Steel, STP 453, ASTM, 1969, p 111-133

8.58 I. M. Bernstein, R. Garber and G. M. Pressouyre, Effect of Dissolved Hydrogen on Mechanical Behavior of Metals in *Effect of Hydrogen on Behavior of Materials*, A. W. Thompson and I. M. Bernstein (Eds.), *TSM-AIME*, Warrendale, Pa., 1976, p 37-58

8.59 H. H. Johnson, J. G. Morlet and A. R. Troiano, Hydrogen, Crack Initiation and Delayed Failure in Steel, *Trans TMS-AIME*, Vol 212, 1958, p 528-536

8.60 *Hydrogen Damage*, American Society for Metals, Metals Park, Oh., 1977

8.61 A. R. Troiano, The Role of Hydrogen and Other Interstitials in the Mechanical Behavior of Metals, *Trans ASM*, Vol 52, 1960, p 54-80

CHAPTER 9

Special Heat Treatments

An important consideration in many of the heat treatments described up to this point is the variation in cooling rate between the surface and center of heat treated parts. The differential cooling produces stresses that may cause distortion and even cracking. A number of precautions may be taken to reduce the residual stress, but two special heat treatments, martempering and austempering, have been developed specifically to minimize residual stress formation and/or to increase the toughness of heat treated steel parts. This chapter describes the origin of residual stress and various approaches to minimizing it. Martempering and austempering are then discussed. Also described briefly are thermomechanical treatments that use both mechanical and thermal processing to produce desirable combinations of properties, and a new approach to heat treating low-carbon steels based on intercritical annealing.

Residual Stress, Distortion and Quench Cracking

Two processes that occur during the cooling of heat treated steel parts are responsible for the dimensional changes that ultimately result in residual stresses and distortion. One is the volume expansion that occurs when the close-packed, fcc structure of austenite transforms to the more open crystal structures of ferrite, cementite and martensite. The other is the normal thermal contraction that occurs during cooling of any single phase or combination of phases in the absence of a phase transformation. The volume expansion due to austenite transformation is the dominant factor in any heat treatment that involves cooling from the austenite phase field, while the thermal contraction is the dominant factor in subcritical heat treatments. Residual stresses and distortion arise because cooling rate is a function of section size or position in a part (as discussed in Chapter 6) and therefore the volume changes occur at different times in different locations during the cooling process. Heat treatments that involve austenitizing produce residual stress patterns quite different from those produced by subcritical annealing.

When a steel part is cooled from the austenite phase field, the surface cools most rapidly and austenite transformation with the attendant volume expansion occurs there first. Generally, the still hot, ductile austenite of the untransformed interior accommodates readily to the changes in surface dimensions. However, when the interior transforms at some later time in the cooling process, its expansion is restrained by the hardened, transformed surface layer. This restraint places the interior of the part in compression while at the same time the interior expansion places the surface in tension. Just the reverse is true for subcritically heat treated parts. The surface cools and contracts first, and again the still hot and ductile interior accommodates readily. However, when the interior eventually cools, its contraction is opposed by the higher strength surface. The restraint on the contraction places the interior in tension while the surface is, in turn, placed in compression by the contracting interior.

The stresses produced by the different rates of cooling throughout a part produce distortion if the stresses generated are high enough to produce nonuniform yielding or plastic deformation on cooling. Even without yielding, stresses up to the yield point may be present on reaching room temperature, and such residual stresses will be superimposed on the applied stresses in service unless tempering or stress-relief treatments are performed. Surface compressive stresses are desirable, and oppose tensile stresses applied in service. Residual tensile stresses, however, add to the applied tensile stresses and effectively lower strength and fatigue resistance in service.

The most critical situation with respect to cooling stresses is in steels austenitized and quenched to form martensite. There the surface tensile stresses, especially on severe quenching, may become high enough to cause quench cracking rather than yielding. High-carbon and tool steels that are inherently hard and brittle are especially susceptible to quench cracking. The quench cracks are invariably intergranular, and their formation may be related to some of the same factors that cause intergranular fracture as a result of tempered martensite embrittlement (discussed in Chapter 8).

Practically, the tendency to distortion, quench cracking, and/or high residual stress formation during heat treatment may be reduced by any change in processing that reduces the differences in the rates of cooling between the surface and interior of a part. Quite effective in this regard is a change to more moderate quenching for hardening, even to the point of air cooling for certain tool steels, if problems are encountered with more severe quenching. The effectiveness of a less severe quench in hardening may require the use of a more hardenable steel, and the matching of steel composition, section size, and cooling rates is an important application of hardenability. Reducing the temperature difference between austenitizing and the quenchant is also sometimes effective in reducing quench cracking or distortion. For example, steels carburized at 950 °C (1750 °F) are often cooled to 840 °C (1550 °F) prior to quenching and may be quenched in oil heated to 70 °C (170 °F) instead of oil held at room temperature. The austempering and martempering treatments described below are also used to minimize differential cooling effects throughout a heat treated part.

The discussion to this point has described the effects of variations in cooling rate through the cross section of a heat treated steel part. Exactly the same considerations apply during heating with the direction of the thermal and transformation volume changes reversed. Thus distortion or cracking may develop in the heating portion of a heat treatment cycle if a steel part cannot accommodate the stresses generated by uneven heating through its cross section. Reheat cracking during stress relief heat treatments of welded low-alloy heat resisting steels is one example of difficulties caused by heating for a subcritical heat treatment (Ref 9.1). Apparently the coarse grained structures of the heat-affected zone have such poor ductility that the deformation required for stress relief cannot be accommodated.

Another situation where difficulty on heating may be encountered is in the austenitizing of high-carbon and tool steels (Ref 9.2). In order to minimize distortion or cracking problems on heating, these steels are preheated prior to heating to the final austenitizing temperature. The work is heated slowly to the preheat temperature and held long enough to ensure that the temperature is uniform throughout the part. Hot work die steels and high-speed steels, especially those containing large amounts of tungsten and molybdenum that lower thermal conductivity, are susceptible to distortion and/or cracking on heating, and preheating of these steels prior to high-temperature austenitizing is essential. Preheating temperatures range between 600 to 800 °C (1110 to 1470 °F) and are tabulated together with complete hardening cycles for the various grades of tool steels by Wilson (Ref 9.2).

Martempering

Martempering or interrupted quenching is a hardening treatment that consists of quenching to a temperature above the M_s, usually by quenching into a salt bath, holding for a time sufficient for the temperature to become uniform, and then air cooling through the M_s to room temperature (Ref 9.3). Tempering is then performed as required. Figure 9.1 shows schematically the temperature-time sequence for martempering superimposed on a transformation diagram. As noted, the surface and the center of a part cool at different rates to the hold temperature above M_s, and then cool uniformly as martensite is formed on air cooling. The equalization of temperature throughout a part prior to martensite formation ensures that the transformation stresses across the part will be minimal and therefore that cracking and distortion will also be minimized.

An important aspect of martempering is that no transformation product other than martensite should form. Therefore, steels that are suitable for martempering must have sufficient hardenability not only with respect to higher temperature transformation products such as ferrite and pearlite, but also with respect to bainite that might form just above the M_s. Another consideration in this regard is that hot salt has a quenching severity slightly lower than oil quenching. A steel for martempering, therefore, must also have sufficient hardenability to compensate for the reduced rate of cooling. The air cooling through the martensite transformation range is also very important. Water quenching, even though the temperature is uniform throughout the part just above M_s, will almost invariably lead to cracking (Ref 9.3).

Austempering

Austempering is another hardening treatment designed to reduce distortion and cracking in higher carbon steels. The object of austempering, however, is to form bainite rather than martensite. Figure

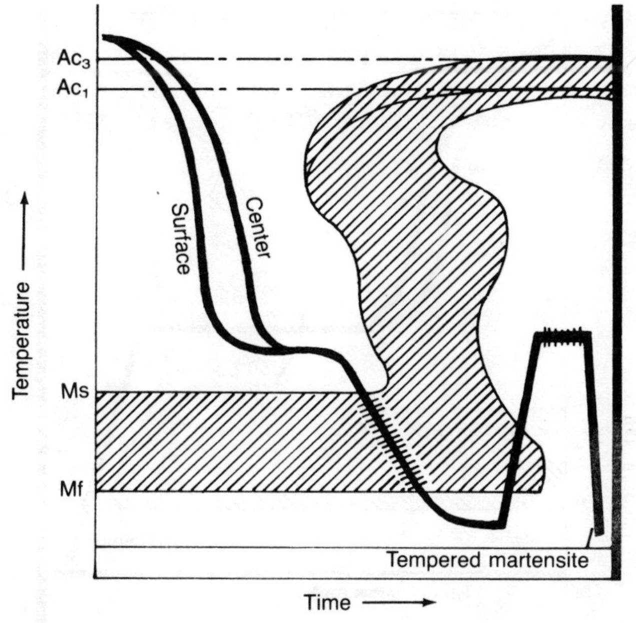

Fig. 9.1. Schematic diagram of martempering heat treatment cycle superimposed on an IT diagram for a medium-carbon steel. (Ref 9.4)

9.2 shows schematically the heat treatment cycle for austempering. The steel is austenitized, quenched in molten salt held at a temperature above M_s, and then allowed to transform to bainite at that temperature. No tempering is required. As shown, the temperatures of the center and surface of an austempered part come together at the holding temperature, and the absence of thermal gradients in the subsequent transformation to bainite minimizes the stresses generated during austempering.

As in martempering, a steel suitable for austempering must have sufficient hardenability to avoid the higher temperature austenite transformation products when quenched into heated molten salt with a relatively low quench severity. Carbon steel parts for austempering are therefore limited in size to obtain sufficiently high cooling rates to avoid pearlite. If an alloy steel is selected to compensate for the reduced quenching efficiency of the hot molten salt, the bainite hardenability may also be increased to the point where very long times are required for complete transformation. Thus a major processing advantage of

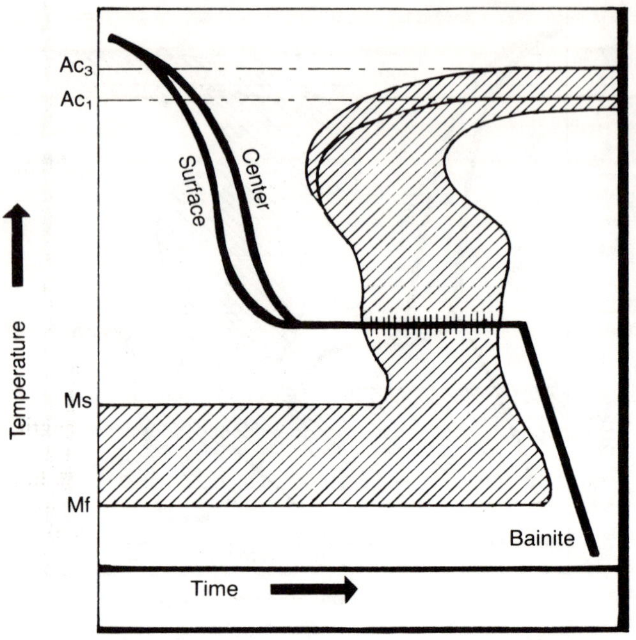

Fig. 9.2. Schematic diagram of austempering heat treatment cycle superimposed on an IT diagram for a medium-carbon steel. (Ref 9.4)

austempering, the fact that no tempering is required, may be offset by the increased holding time for bainite formation.

A major advantage of austempering, apart from significantly reducing distortion or cracking of high-carbon steel parts, is that the toughness is greatly improved relative to tempered martensite at the same hardness level. Table 9.1 shows that at a hardness of Rockwell C 50, austempered 0.74% carbon steel has much better ductility and impact toughness than the same steel in the quench and tempered condition. It should be noted that the quenched steel has a high phosphorus content and was tempered at 315 °C (600 °F), conditions that indicate the low impact toughness was most probably due to tempered martensite embrittlement as discussed in Chapter 8. Figure 9.3 is another comparison of the impact toughness produced by austempering and conventional quench and tempering of the 0.74% carbon steel (Ref 9.5). The comparison is made at different hardness levels. Austempering is clearly most beneficial in the hardness range between Rockwell C 50 and 55.

Table 9.1. Comparison of Mechanical Properties Produced by Austempering and Quench and Tempering (Ref 9.3)

Steel composition: 0.74C, 0.37 Mn, 0.145Si, 0.039S, 0.044P

New Method, Direct from Austenite		Quench-and-Temper Method	
Heat 5 min at 790 °C (1450 °F) Quench into lead alloy bath at 305 °C (580 °F) Let specimens remain in bath for 15 min Quench into water		Heat 5 min at 790 °C (1450 °F) Quench into oil at 21 °C (70 °F) Temper immediately in lead alloy bath, 30 min at 315 °C (600 °F) Quench into water	
Mechanical Properties (Average of 6 tests)		**Mechanical Properties** (Average of 6 tests)	
Rockwell C hardness	50.4	Rockwell C hardness	50.2
Ultimate strength, ksi	282.7	Ultimate strength, ksi	246.7
Yield point, ksi	151.3	Yield point, ksi	121.7
Elongation, % in 6 inches	1.9	Elongation, % in 6 inches	0.3
Reduction of area, %	34.5	Reduction in area, %	0.7
Impact, ft-lb(a)	35.3	Impact, ft-lb(a)	2.9

(a) Foot-pounds absorbed in breaking 0.180-inch round, unnotched specimens.

Thermomechanical Treatments

Thermomechanical treatments are processing treatments that combine plastic deformation with thermal processing or heat treatment in order to produce microstructures and improved properties not obtained by independently applied conventional heat treatment or working operations. Generally, increased strength with improvement in ductility and/or toughness are the objectives of thermomechanical treatments. Intensive research in the 1950s and 1960s showed that these objectives could be achieved, and resulted in a U.S. classification system that recognized three types of thermomechanical treatments (Ref 9.6, 9.7). Table 9.2, after Azrin (Ref 9.8), describes the three classes of thermomechanical treatment and Fig. 9.4 shows the treatments schematically superimposed on an idealized transformation diagram. The Soviets have also actively pursued the development of thermomechanical treatments and their classification system is shown in Table 9.3 and Fig. 9.5, also after Azrin (Ref 9.8). As Fig. 9.4 shows, the thermomechanical treatments, recognized as such, have the modification of martensite (either by the deformation of the austenite preceding transformation or by the deformation of the martensite after transformation) as the major approach to improved properties. Very

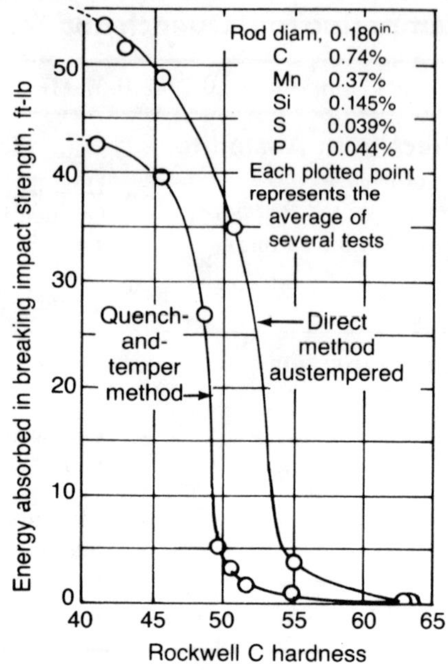

Fig. 9.3. Comparison of impact toughness of austempered and quenched-and-tempered carbon steel as a function of hardness. (Ref 9.3)

Table 9.2. U.S. Classification of Thermomechanical Treatments (Ref 9.8)

CLASS I: Deformation occurs before the austenite transformation. Austenite is deformed in the stable austenite range above the critical temperature (A_1) or in the unstable region above the pearlite nose or in the bay region between the pearlite and bainite noses.

CLASS II: Deformation during the austenite transformation. Depending on the deformation temperature, as well as the M_S and M_D temperatures, the transformation products can be either pearlite, bainite or martensite. The martensite transformation can be due to a strain-induced or stress-assisted transformation.

CLASS III: Deformation after austenite transformation to martensite or other transformation products.

(a) M_D is temperature above which martensite cannot be formed by plastic deformation.

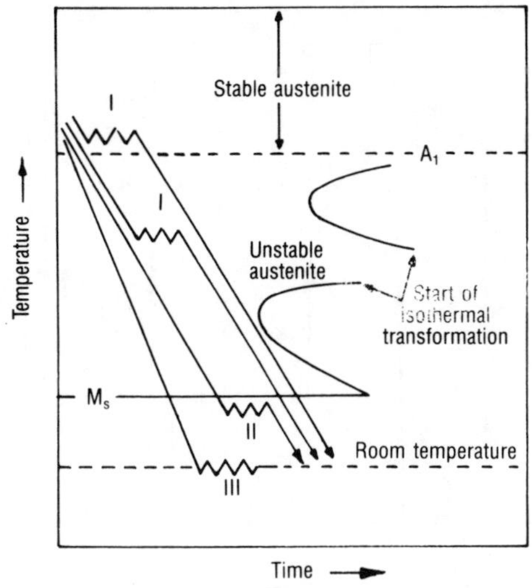

Fig. 9.4. Schematic diagram of the U.S. classification of thermomechanical treatments superimposed on a time-temperature-transformation diagram. (Ref 9.8)

Table 9.3. Soviet Classification of Thermomechanical Treatments (Ref 9.8)

SHT:	Standard Heat Treatment — Conventional heat treatment without deformation.
TMT:	Thermomechanical Treatment — A combined thermal and mechanical treatment generally involving a phase transformation.
HTTMT:	High Temperature Thermomechanical Treatment — Deformation above the recrystallization temperature.
LTTMT:	Low Temperature Thermomechanical Treatment — Deformation below the recrystallization temperature.
CTMT:	Combined Thermomechanical Treatment — HTTMT followed by LTTMT.
PTMT:	Preliminary Thermomechanical Treatment — Deformation by HTTMT or LTTMT or cold working followed by rapid reaustenitizing and quenching.
MTT:	Mechanico-thermal Treatment — Deformation at room or elevated temperature with or without subsequent annealing or aging applied to a material which does not undergo a phase transformation. As with TMT, deformation can be below (LTMTT) or above (HTMTT) the recrystallization temperature.

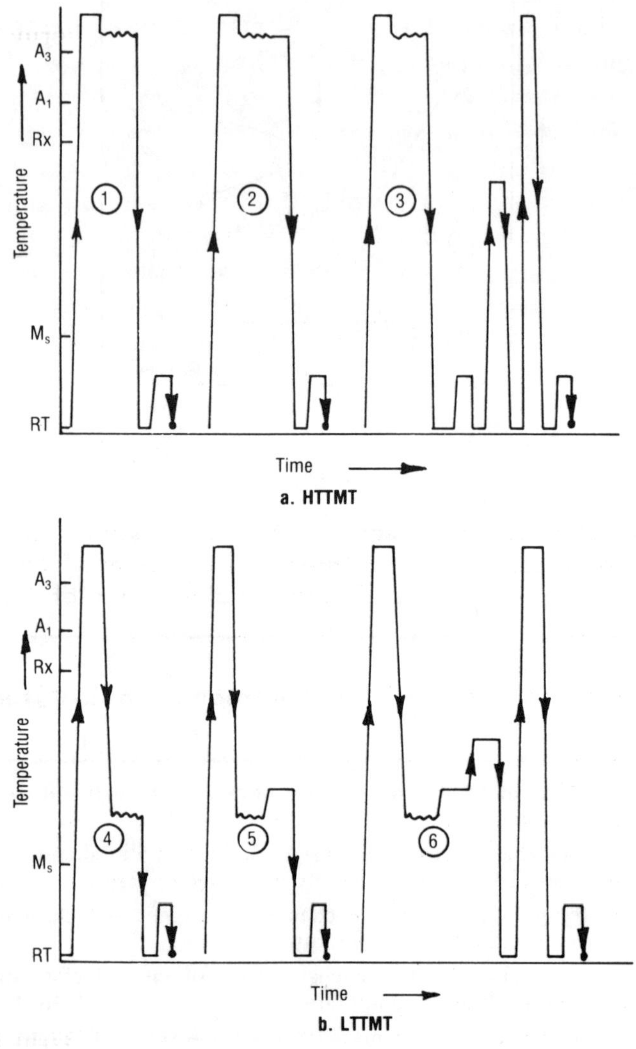

Fig. 9.5. Schematic time-temperature-deformation cycles for Soviet thermomechanical treatments. Treatments 2 and 5 include polygonization; 3 and 6 include hereditary treatments. Rx is the minimum recrystallization temperature, and HTTMT and LTTMT are defined in Table 9.3. (Ref 9.8)

high strengths therefore result from this type of thermomechanical process.

Ausforming is a special term for the class I type of thermomechanical treatment in which austenite is deformed in the bay of the time-temperature-transformation diagram prior to martensite formation. A special variant of this treatment involves the use of highly alloyed steels with an initial structure containing metastable austenite (Ref 9.9). These steels are referred to as transformation induced plasticity (TRIP) steels because they transform to martensite on straining. The strain induced martensite in turn resists plastic instability or necking and thereby produces good combinations of very high strength and ductility. Figure 9.6, after Zackay et al. (Ref 9.9), compares strengths and ductilities produced by various thermomechanical techniques. The properties of the low-alloy high-strength steels indicated are those developed by conventional quench and tempering of carbon steels. The maraging steels are highly alloyed steels, containing about 18% nickel, 8% cobalt, and small amounts of aluminum and titanium. High strength and good toughness are obtained by precipitation of compounds such as Ni_3Mo and Ni_3Ti in a matrix of carbon-free lath martensite (Ref 9.10).

While the physical metallurgy of thermomechanical treatments such as those shown in Fig. 9.4 and 9.5 is well established, very few industrial applications of thermomechanical treatments have been developed (Ref 9.8, 9.11). Reasons for this situation include technical factors such as the difficulty in machining the very hard thermomechanically treated products; the deterioration in mechanical properties when thermomechanically processed parts are joined by welding; the high alloy content of steels that respond best to thermomechanical treatment; and the difficulty in achieving the uniform deformation required for improved properties in parts with complex shapes. The status of thermomechanical treatments in the U.S. and the U.S.S.R. is reviewed in the reports by Henning (Ref 9.11) and Azrin (Ref 9.8), respectively, and the reader is referred to these publications for the practical aspects and limitations of thermomechanical treatments as well as extensive bibliographies of the literature on thermomechanical treatments.

Perhaps the most important application of thermomechanical processing actually in use is the controlled rolling of microalloyed HSLA steels (see Chapter 7) in regard to grain size control. The processing of HSLA low-carbon steels, however, depends on very fine ferrite grain sizes and precipitation for strengthening and not

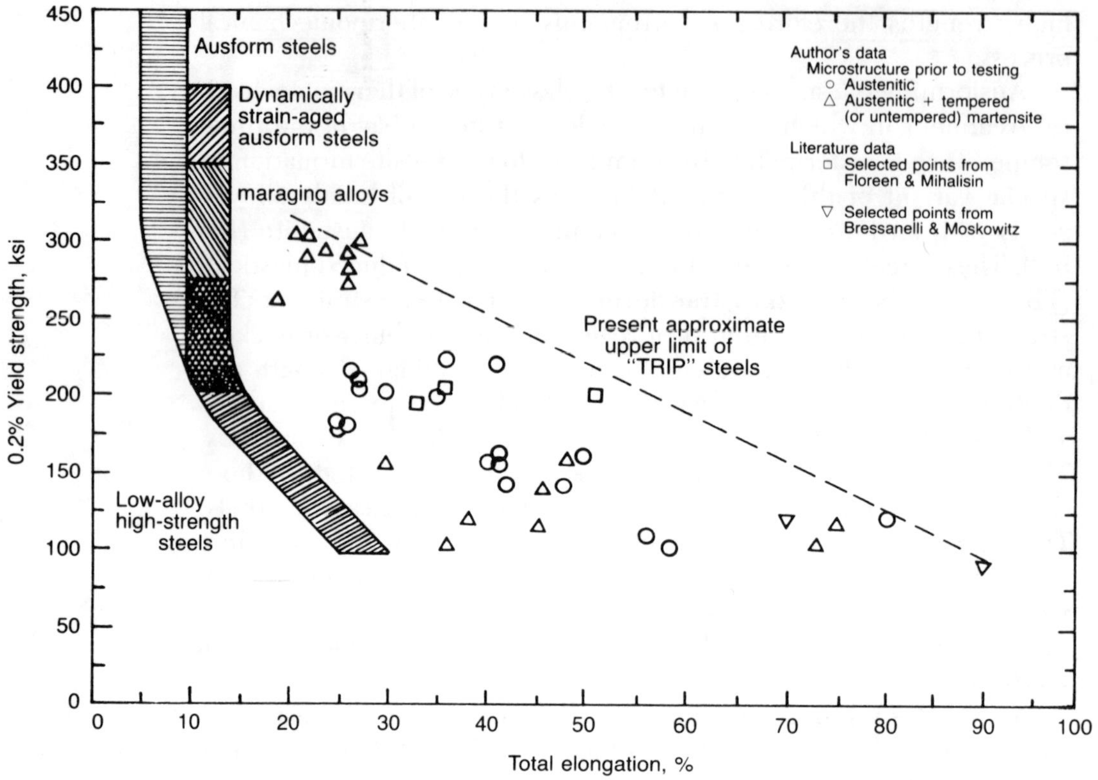

Fig. 9.6. Yield strength versus total elongation for various classes of high-strength steels and thermomechanical treatments. (Ref 9.9)

martensite formation, and therefore falls outside of the classification systems shown in Table 9.2.

Intercritical Heat Treatment

A new approach to developing good combinations of strength and ductility in low-carbon steels consists of intercritical annealing of hot or cold rolled sheet steels with microstructures consisting initially of ferrite and pearlite. Figure 9.7 shows the Fe-C diagram with four temperatures indicated between the A_1 and A_3 temperatures: the higher the intercritical annealing temperature, the more austenite formed during annealing. As discussed below, on cooling the austenite may transform into various microstructures whose formation is dependent on the cooling rate and the hardenability of the austenite.

Intercritical annealing for producing high strength, formable sheet steels was first applied by Rashid (Ref 9.12) and Hayami and Furukawa

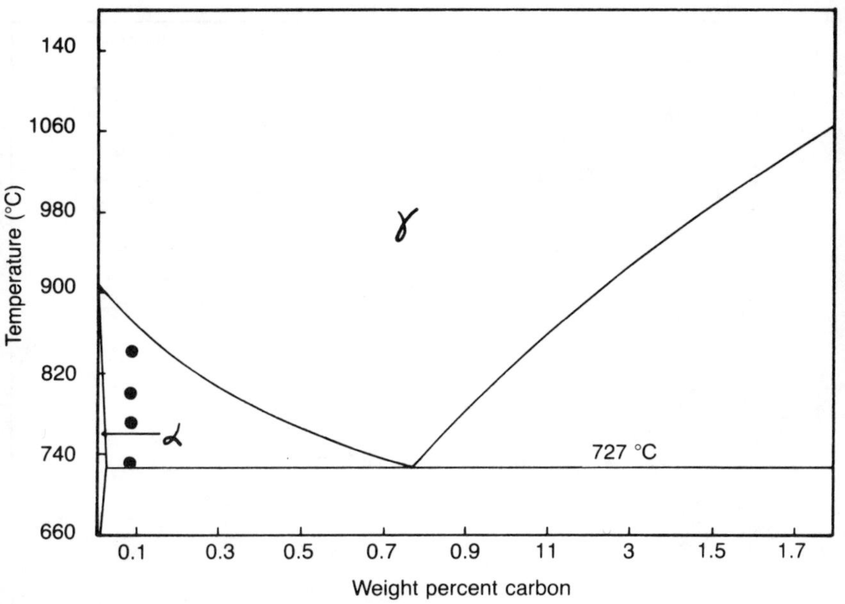

Fig. 9.7. Fe-C diagram with four intercritical annealing temperatures indicated. (Courtesy of E. Indacochea, Colorado School of Mines, Golden, Colo.)

(Ref 9.13). Figure 9.8 shows the basis for three stages in the development of ferritic low-carbon steels. The lower stress-strain curve represents the deformation behavior of mild steel with ferrite-pearlite microstructures. The yielding is discontinuous and yield strengths are typically 30 ksi (207 MPa). SAE 950X and SAE 980X are HSLA steels with yield strengths of 50 ksi (345 MPa) and 80 ksi (562 MPa), respectively. The microstructures still consist of ferrite and pearlite but the ferrite grain size is highly refined because of controlled rolling and microalloying with vanadium. GM 980X is similar to SAE 980X but has been intercritically annealed to convert the pearlite to martensite. Rashid termed the resulting microstructure "dual-phase" to distinguish the ferrite-martensite microstructure from the ferrite-pearlite microstructure of conventionally treated mild steels or HSLA steels.

Figure 9.8 shows that the intercritical annealing treatment has eliminated discontinuous yielding and significantly increased ductility relative to SAE 980X. These changes in deformation behavior are not yet fully explained, but the formation of martensite during cooling from the intercritical annealing temperature is important in that it introduces a high density of dislocations into the ferrite. These dislocations

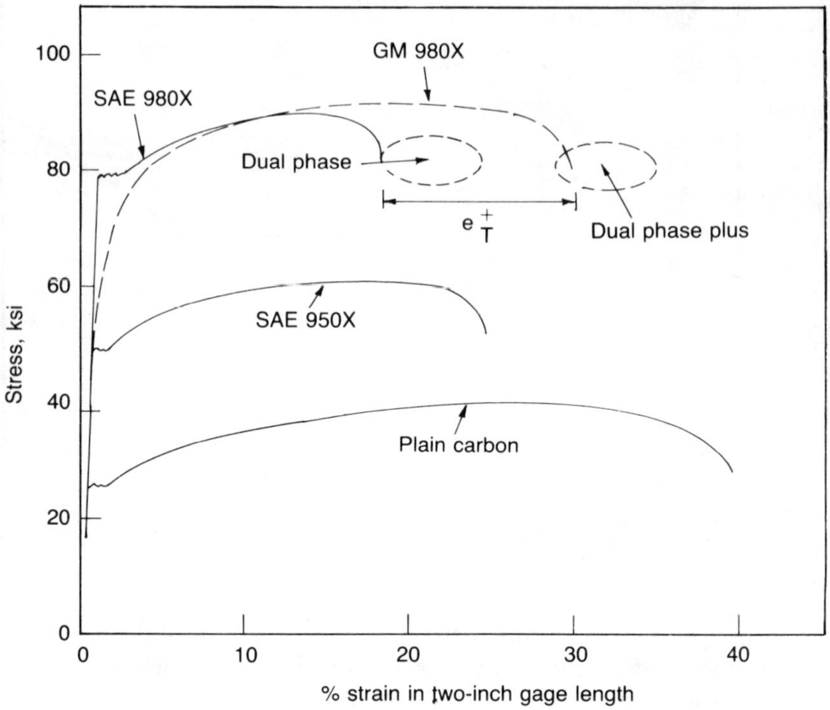

Fig. 9.8. Stress-strain curves for plain carbon, SAE 950×, SAE 980×, and GM 980×. The GM 980× has been intercritically annealed and dual-phase microstructures produced. The two dashed ellipses indicate reported ranges of elongation for dual-phase steels. (Ref 9.12)

move immediately on application of stress, and thereby produce the continuous yielding associated with intercritically annealed steels.

The microstructures produced by cooling the austenite formed during intercritical annealing are very much dependent on alloy composition and cooling rate. Severe quenching is required to form martensite in dilute alloys but air cooling is frequently sufficient to form considerable martensite in steels with higher hardenability. Even though the steels are low carbon, the intercritical annealing concentrates the carbon in the austenite, another helpful factor for improved hardenability.

Figure 9.9 shows the variety of microstructures that may form in an intercritically annealed steel when cooling rate is varied (Ref 9.14). In this case the steel contained 0.08% carbon, 1.47% manganese and

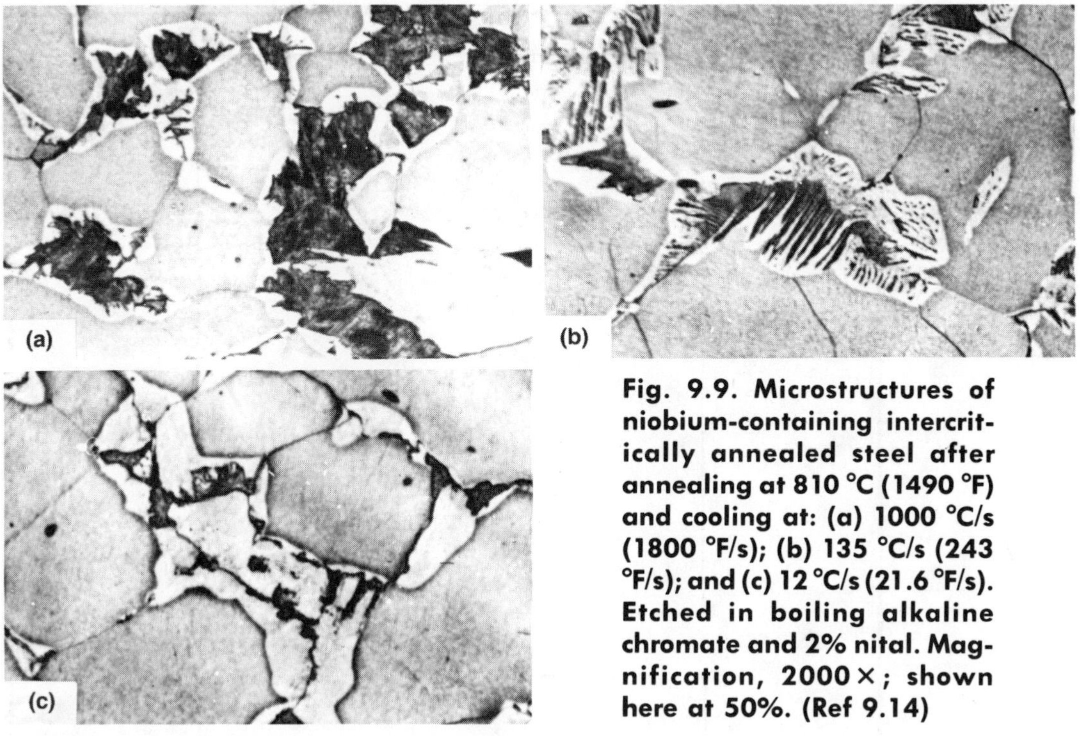

Fig. 9.9. Microstructures of niobium-containing intercritically annealed steel after annealing at 810 °C (1490 °F) and cooling at: (a) 1000 °C/s (1800 °F/s); (b) 135 °C/s (243 °F/s); and (c) 12 °C/s (21.6 °F/s). Etched in boiling alkaline chromate and 2% nital. Magnification, 2000×; shown here at 50%. (Ref 9.14)

0.053% niobium. The austenite of the specimen cooled at the highest rate has transformed primarily to martensite, the uniformly dark structure (see Fig. 9.9a). Surrounding the martensite is a white layer where some of the austenite has transformed during cooling to ferrite ("new" ferrite) prior to the martensite formation. This new ferrite is considered to grow epitaxially (with the same crystal orientation) from the ferrite retained at the intercritical annealing temperature (Ref 9.15). Figure 9.9(b) and (c) show that the amount of new ferrite increases with decreasing cooling rate. Also, less martensite and more ferrite-carbide mixtures form with decreasing cooling rate. These microstructural changes are summarized schematically in the microstructure map shown in Fig. 9.10 (Ref 9.14). Also shown in the top part of Fig. 9.10 is a schematic plot of the mechanical property changes with microstructure or cooling rate. The highest strengths and lowest ductilities are associated with microstructures that are most rapidly cooled and contain the most martensite. Good combinations of strength and ductility result from intermediate cooling rates that produce more complex microstructures. The slowest cooling rates produce structures typical of mild steels and have substantially lower strengths and higher ductilities. Discontinuous yielding also returns for the latter microstructures.

The old ferrite or ferrite retained during intercritical annealing is

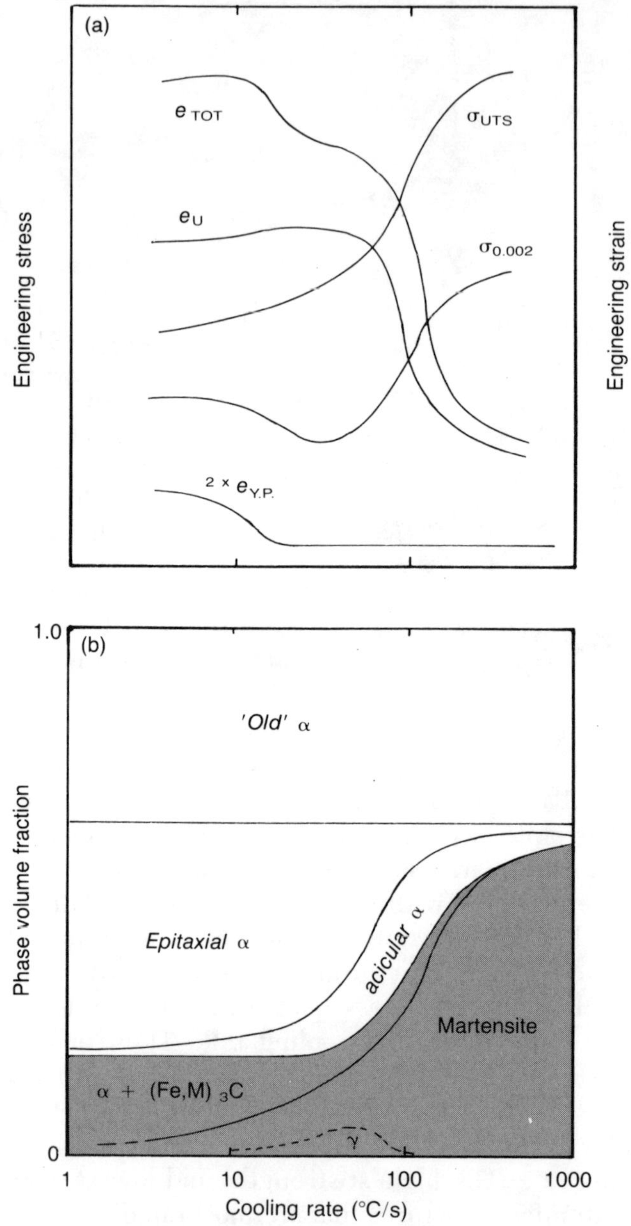

Fig. 9.10. (a) Schematic of change in mechanical properties with cooling rate from intercritical annealing. (b) Schematic map of microstructures formed by different cooling rates. (Ref 9.14)

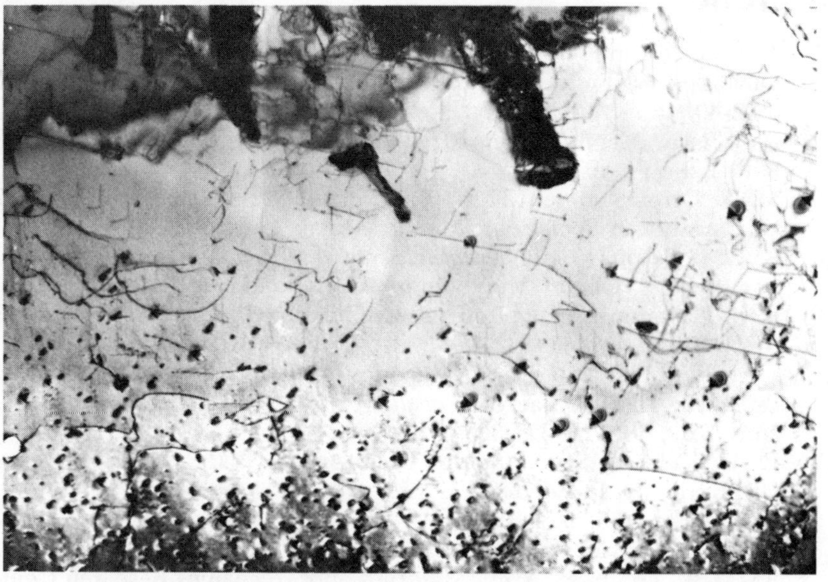

Fig. 9.11. Area adjacent to lamellar carbides formed during cooling a microalloyed steel from 810 °C (1490 °F). The ferrite immediately adjacent to the carbides is free of the precipitates that have developed in the retained ferrite matrix. Transmission electron micrograph. Magnification, 16 500 ×; shown here at 70%. (Courtesy of M. D. Geib, Colorado School of Mines, Golden, Colo.)

stained gray by the etchant (Ref 9.15) used to reveal the microstructures in Fig. 9.9. Figure 9.11 is a transmission electron micrograph taken from an area adjacent to the austenite transformation product in a specimen treated similarly to that shown in Fig. 9.9(b). The new ferrite is free of precipitate particles that have formed in the old ferrite matrix and no interfaces or boundaries separate the two types of ferrite. The dislocation density in the ferrite is also rather low because of the reduced amount of martensite that has formed during cooling at the intermediate rate.

The above discussion shows that intercritically annealed steels offer good combinations of strength and formability, a characteristic that makes this type of heat treatment attractive for vehicle applications where weight saving through increased strength is viable. The reader is referred to the proceedings of the symposium identified in Ref 9.14 for a number of papers on all aspects of intercritical annealing and dual-phase steel processing.

References

9.1 A. Vinckier and A. Dhooge, Reheat Cracking in Welded Structures During Stress Relief Heat Treatments, *J of Heat Treating*, Vol 1, 1979, p 72-80

9.2 R. Wilson, *Metallurgy and Heat Treatment of Tool Steels*, McGraw-Hill, London, 1975

9.3 M. A. Grossmann and E. C. Bain, *Principles of Heat Treatment*, 5th Ed., American Society for Metals, Metals Part, Oh., 1964, p 189-196

9.4 E. C. Rollason, *Fundamental Aspects of Molybdenum on Transformation of Steel*, Climax Molybdenum Co., London

9.5 E. S. Davenport, E. L. Roff and E. C. Bain, *Trans ASM*, Vol 22, 1934, p 289-310

9.6 S. V. Radcliffe and E. B. Kula, in *Fundamentals of Deformation Processing*, W. A. Backofen *et al*. (Eds.), Syracuse University Press, New York, 1964, p 321-363

9.7 E. B. Kula, in *Strengthening Mechanisms – Metals and Ceramics*, J. J. Burke, N. L. Reed and V. Weiss (Eds.), Syracuse University Press, New York, 1966, p 83-121

9.8 M. Azrin, "Soviet Progress in Thermomechanical Treatment of Metals" AMMRC TR 76-36, Army Materials and Mechanics Research Center, Watertown, Mass., Nov, 1976

9.9 V. F. Zackay, E. R. Park, D. Fahr and R. Busch, The Enhancement of Ductility in High-Strength Steels, *Trans ASM*, Vol 60, 1967, p 252-259

9.10 A. Magnee, J. M. Drapier, J. Dumont, D. Coutsouradis and L. Habraken, *Cobalt-containing High-Strength Steels*, Centre d'information du Cobalt, Brussels, 1974

9.11 H. J. Henning, *Applications and Potential of Thermomechanical Treatment*, DMIC Memorandum 251, Defense Metals Information Center, Battelle Memorial Institute, Columbus, Oh., Nov, 1970

9.12 M. S. Rashid, "GM980x – Potential Applications and Review", Research Publication GMR-232, General Motors Corp, Warren, Mich, 1977

9.13 S. Hayami and T. Furukawa, A Family of High Strength Cold Rolled Steels, *Microalloying* 75, Union Carbide Corp, New York, p 311-321

9.14 D. K. Matlock, G. Krauss, L. F. Ramos and G. S. Huppi, A Correlation of Processing Variables with Deformation Behavior of Dual-Phase Steels, in Proceedings of Symposium *Structure and Properties of Highly Formable Dual-Phase HSLA Steels*, AIME, Warrendale, Pa., 1980, p 62-90

9.15 R. D. Lawson, D. K. Matlock and G. Krauss, An Etching Technique for Microalloy Dual-Phase Steels, *Metallography*, Vol 13, 1980, p 71-87

CHAPTER 10

Surface Hardening

Surface hardening is used to extend the versatility of certain steels by producing combinations of properties not readily attainable in other ways. For many applications, wear and the most severe stresses act only on the surface of a part. Therefore, the part may be fabricated from a formable low- or medium-carbon steel, and is surface hardened by a final heat treatment after all other processing has been accomplished. Surface hardening also reduces distortion and eliminates cracking that might accompany through hardening, especially in large sections. Localized hardening of selected areas is also possible by means of certain surface hardening techniques. This chapter describes two major approaches to surface hardening. One approach does not change composition and consists of hardening the surface by flame or induction heating. The other approach changes surface composition and includes the applications of such techniques as carburizing, nitriding and carbonitriding.

Flame Hardening

Flame hardening consists of austenitizing the surface of a steel by heating with an oxyacetylene or oxyhydrogen torch and immediately quenching with water. A hard surface layer of martensite over a softer interior core with a ferrite-pearlite structure results. There is no change in composition, and therefore the flame-hardened steel must have adequate carbon content for the desired surface hardness. The rate of heating and the conduction of heat into the interior appear to be more important in establishing case depth than having a steel of high hardenability (Ref 10.1). Figure 10.1 shows hardness gradients pro-

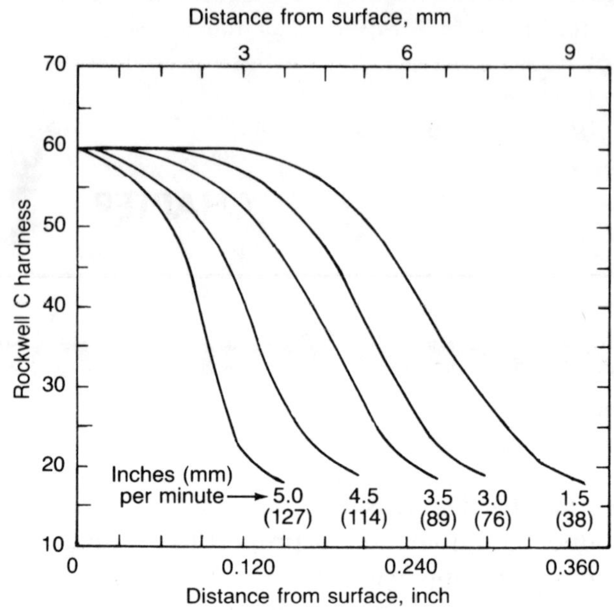

Fig. 10.1. The effect of flame speed on depth of hardening of a 1050 steel forging. (Ref 10.2)

duced by various rates of flame travel across a 1050 steel forging. The slower the rate of travel the greater the heat penetration and the depth of hardening.

A number of different methods of flame hardening have been developed (Ref 10.2). Localized or spot hardening may be performed by directing a stationary flame head to an area of a stationary work piece. Progressive methods where the torch travels over the work piece or the work piece travels under a stationary torch and quenching fixture are used for long bars. Spinning methods in which the work piece is rotated within an array of torches are often used for small rounds. In this method heating is performed first, then the flames are extinguished, and quenching is finally accomplished by water sprays or dropping the part into a quench tank. In all cases, the quenched parts are tempered to improve toughness and relieve stresses induced by the surface hardening.

Induction Heating

Induction heating is an extremely versatile method for hardening steel. Uniform surface hardening, localized surface hardening, through hardening, and tempering of hardened pieces may all be performed by induction heating. Heating is accomplished by placing a steel part in the magnetic field generated by high-frequency alternating current passing through an inductor, usually a water-cooled copper coil. The rapidly alternating magnetic field established within the coil induces current (I) within the steel. The induced currents then generate heat (H) according to the relationship $H = I^2R$ where R is the electrical resistance. Steel, consisting primarily of ferrite or bcc iron, is ferromagnetic up to its Curie temperature (768 °C or 1400 °F), and the rapid change in direction of the internal magnetization of domains in a steel within the field of the coil also generates considerable heat. When a steel transforms to austenite, which is nonmagnetic, this contribution to induction heating becomes negligible. A wide variety of heating patterns may be established by induction heating depending on the shape of the coil, the number of turns of the coil, the operating frequency and the alternating current power input (Ref 10.3). Figure 10.2 shows examples of the heating patterns produced by various types of coils.

The depth of heating produced by induction is related to the frequency of the alternating current. The higher the frequency, the thinner or more shallow the heating. Therefore, deeper case depths and even through hardening are produced by using lower frequencies. The various types of commercial equipment and the selection of operating conditions for a given application are fully described in Ref 10.3.

As in flame hardening, induction heating does not change the composition of a steel, and therefore a steel selected for induction hardening must have sufficient carbon content and alloying for the desired surface hardness distribution. Generally, medium- and high-carbon steels are selected because the high surface strengths and hardness attainable in these steels significantly improve fatigue and wear resistance. Induction hardening introduces residual compressive stresses into the surface of hardened parts. Therefore, the fatigue strengths of induction surface hardened parts may be higher than those of through hardened parts in which quenching develops residual surface tensile stresses that may only be partly relieved during tempering. The greater the depth of hardening by induction heating, however, the more the surface stress state approaches that of through hardening. Too deep a hardened case may in fact cause surface tensile stresses and even cracking of susceptible steels (Ref 10.3).

The duration of high-frequency induction heating cycles for surface hardening is extremely short, often only a few seconds. As a result the time for formation of austenite is limited, and compensation is made by

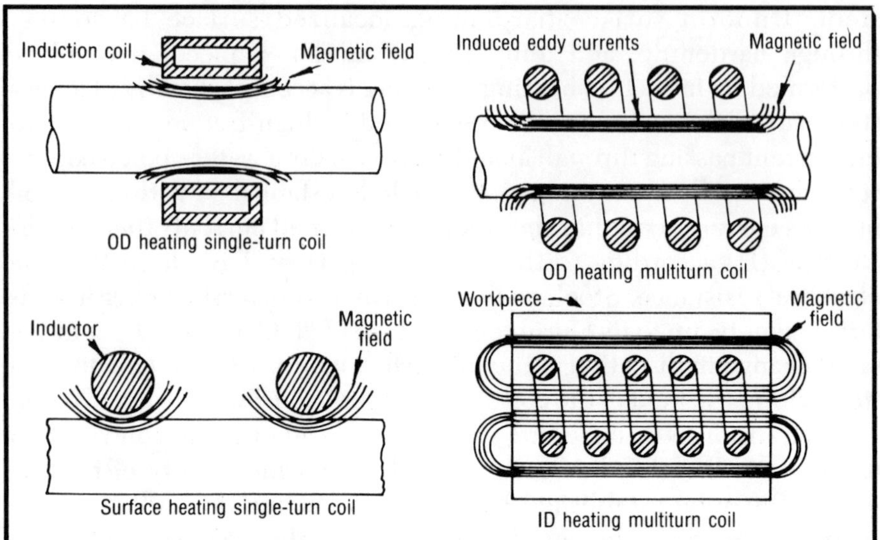

Fig. 10.2. Schematic diagram of the magnetic fields and induced currents produced by several types of induction coils. (Ref 10.3)

increasing the temperature of austenitizing. Figure 10.3 shows how the Ac_3 temperature in a 1042 steel is affected by heating rate and microstructure. The high heating rates of induction heating substantially raise Ac_3. Microstructures with coarse carbides, such as the 1042 steel in the annealed condition as shown, or steels with coarse spheroidized microstructures or alloy carbides, require higher austenitizing temperatures for carbide solution than do steels with finer microstructures. Too high an austenitizing temperature, however, may result in austenite grain coarsening (see Chapter 7). An interesting consequence of the very short austenitizing time for induction surface hardening is the development of a hardness above that normally expected for through-hardened martensite. This higher hardness is sometimes referred to as superhardness (see Fig. 10.4) and, as discussed in earlier chapters, may be a result of martensite formed in very fine-grained, imperfect austenite produced by the short-time austenitizing treatments used in induction surface hardening.

Carburizing: Processing Principles

Carburizing is a heat treatment in which the carbon content of the surface of a low-carbon steel is increased by exposure to an appropriate

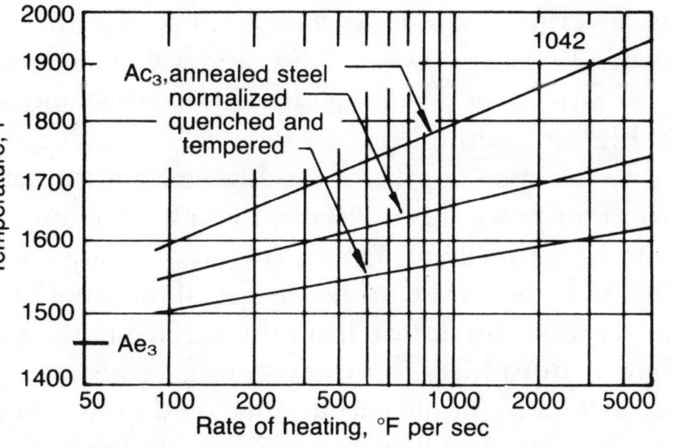

Fig. 10.3. Change in Ac_3 temperature of 1042 steel as a function of microstructure and heating rate. (Ref 10.3)

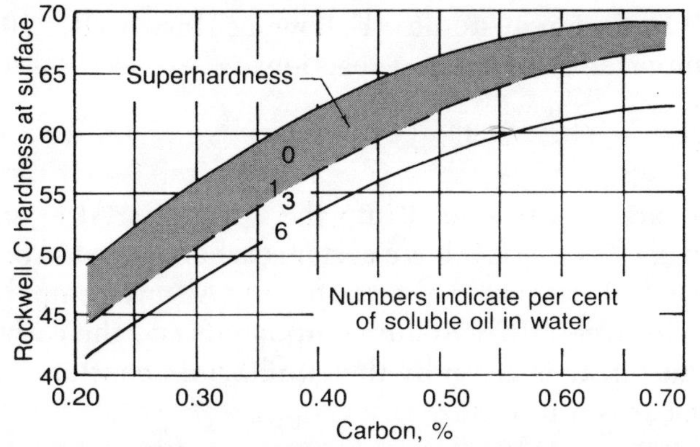

Fig. 10.4. Superhardness produced by induction hardening compared to that produced by conventional furnace hardening (lower solid curve). (Ref 10.3)

atmosphere at a temperature in the austenite phase field. Hardening is accomplished when the high-carbon surface layer is quenched to form martensite. The Fe-C diagram, (see Chapter 1) shows that the maximum solubility of carbon in austenite ranges from 0.8% at the eutectoid temperature to about 2% at the eutectic temperature. Although alloying elements reduce carbon solubility, more than enough carbon can be introduced into the austenite of plain carbon or alloy steels by carburizing to produce the maximum martensitic hardness after quenching. Complications of carbide formation, brittle martensite and retained austenite develop if carbon content is too high, and for these reasons the maximum carbon content in a carburized steel is generally controlled to between 0.8% and 1%. Carburizing is most frequently performed between 850 to 950 °C (1550 and 1750 °F), but sometimes higher temperatures are used to reduce cycle times and/or produce deeper depths of the high-carbon surface layer.

Two important processes influence the introduction of carbon into austenite during carburizing. One is the environmental reaction that causes carbon to be absorbed at the surface of the steel. Another is the rate at which carbon can diffuse from the surface to the interior of the steel. Carbon is introduced by the use of gaseous atmospheres (gas carburizing), salt baths (liquid carburizing), and solid compounds (pack carburizing) (Ref 10.4). All of these methods have limitations and advantages, but gas carburizing is used most often for large-scale production because it can be accurately controlled and involves a minimum of special handling.

Carbon is introduced into the surface of steel by gas-metal reactions between the various components of an atmosphere gas mixture and the solid solution austenite. Following Harvey (Ref 10.5), one of the most important carburizing reactions is

$$CO_2(g) + C \rightleftharpoons 2CO(g) \qquad \text{(Eq 10.1)}$$

where C is carbon introduced into the austenite. At equilibrium, a carbon ratio of CO_2 and CO has a certain carbon potential or maintains a certain level of carbon in the austenite. At any temperature, the relationship between the gaseous components and the carbon in solution of the austenite is given by the equilibrium constant K which for reaction 10.1 is written as:

$$K = \frac{P_{CO}^2}{a_c P_{CO_2}} \qquad \text{(Eq 10.2)}$$

where P_{CO} and P_{CO_2} are the partial pressures of CO and CO_2, respectively, and a_c is activity of carbon. The activity of carbon is related to the weight percent carbon in the austenite by the activity coefficient of carbon (f_c) by the following equation:

$$a_c = f_c \text{ wt. \% C} \qquad \text{(Eq 10.3)}$$

K is a function of temperature, and for the reaction represented in Eq 10.1 is

$$\log K = \frac{-8918}{T} + 9.1148 \qquad \text{(Eq 10.4)}$$

where T is the absolute temperature in degrees Kelvin. The partial pressures of CO and CO_2 required to maintain a given surface austenite carbon content are given by combining Eq 10.2 and Eq 10.3 as follows:

$$\text{wt. \% C} = \frac{1}{Kf_c} P_{CO}^2/P_{CO_2} \qquad \text{(Eq 10.5)}$$

If the CO content of an atmosphere exceeds the partial pressure required to maintain a given carbon content, the reaction represented in Eq 10.1, as written, will go to the left and carburizing will occur until a new equilibrium is reached. This is the case in commercial carburizing where the carbon content of a low-carbon steel is raised to some desirable higher level. On the other hand, if the CO_2 partial pressure is too high relative to the CO content, the reaction in 10.1 will go to the right and decarburization will occur. The latter condition is sometimes purposely introduced in commercial practice if initial carburizing produces too high a carbon content, say 1.2%, and it is desired to reduce surface carbon to a lower level, say 0.9%. This step in a carburizing cycle is referred to as a "diffusion step", since much of initially high carbon in the austenite immediately adjacent to the surface diffuses into interior of the part and produces a deeper case. Equation 10.5 requires a knowledge of the activity coefficient which varies as a function of temperature and the composition of austenite. Harvey (Ref 10.5) tabulates relationships for the activity coefficients in ternary Fe-X-C systems where X may be the elements nickel, silicon, manganese, chromium, molybdenum or vanadium. Harvey also presents a system for evaluating the activity coefficient and carburizing potentials for steels with more than three components.

The preceding discussion described the basic concept of gas equilibrium and carburizing. In conventional gas carburizing practice, carburizing atmospheres are produced by combustion of natural gas or other hydrocarbon gas in exothermic or endothermic gas generators and contain CO, CO_2, CH_4, H_2, H_2O and N_2 (Ref 10.6). Therefore, there are many other reactions that may occur in addition to that represented in 10.1, including the following:

$$CH_4 \rightleftharpoons C + 2H_2 \qquad \text{(Eq 10.6)}$$

$$CO + H_2O \rightleftharpoons CO_2 + H_2 \qquad \text{(Eq 10.7)}$$

Much of the current technology (Ref 10.4) of gas carburizing is based on relationships determined by Harris (Ref 10.7) for Eq 10.1, 10.6 and 10.7 according to the approach described above and with the assumption that the activity of carbon in saturated austenite (austenite with the carbon content given by the A_{cm} at a given temperature) is unity. For carbon concentration in the austenite less than saturation, the activity is assumed to be proportional to the degree of saturation. For example if the carbon content of austenite at saturation is 1.33%, and one wants to know what partial pressures of CO and CO_2 will maintain 1.0% carbon in austenite, the activity is equal to 1.00/1.33 = 0.75. These assumptions appear to be valid for plain carbon and low-nickel steels, and within 10% for more highly alloyed steels (Ref 10.4). A number of curves, based on the Harris approach, relating the CO and CO_2 contents required to maintain various surface carbon contents are given in Ref 10.8 for carburizing temperatures between 825 and 1025 °C (1515 to 1875 °F) in 25 °C (76 °F) increments. Figure 10.5 shows such a curve for 975 °C (1790 °F) conditions. As shown, a much higher CO than CO_2 content is required for carburizing, especially for higher surface carbon contents.

Equation 10.7 gives the equilibrium between CO, CO_2, H_2O and H_2 and is used as the basis of control and determination of the carbon potential of carburizing atmospheres. For the latter purposes, measurements of the contents of H_2O or CO_2 are often used. Carbon dioxide is measured by infrared analyzers. The dew point is defined as the temperature, at a given pressure, at which a gas mixture will precipitate its moisture content and is used to determine the H_2O content of an atmosphere. More recently, carbon potential has been determined by measurement of the partial pressure of oxygen with instruments referred to as oxygen probes. Once the oxygen content is known, the CO and CO_2 contents can be determined by means of the following equilibrium reaction:

$$CO + \tfrac{1}{2}O_2 \rightleftharpoons CO_2 \qquad \text{(Eq 10.8)}$$

The various approaches and instruments for control of carburizing potentials are described in detail in Ref 10.4.

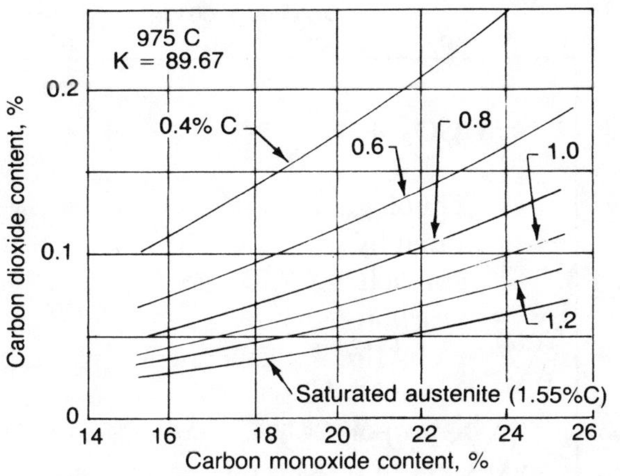

Fig. 10.5. Equilibrium percentages of carbon monoxide and carbon dioxide required to maintain various carbon concentrations at 975 °C (1790 °F) in plain carbon and certain low-alloy steels. (Ref 10.8)

After a given carbon potential is established, the depth of case produced in a given carburizing treatment is determined by the time-dependent diffusion of carbon from the surface to the interior of the steel part. Figure 10.6 shows carbon profiles calculated for an Fe-C alloy carburized at 925 °C (1700 °F) for times between 2 and 16 h, and Fig. 10.7 shows the effect of temperature on carburizing an Fe-C alloy when time is held constant at 8 h. These figures demonstrate the important effect of time and temperature on the depth of case produced by carburizing.

Following Goldstein and Moren (Ref 10.9), the curves of Fig. 10.6 and 10.7 were calculated by means of the Van-Ostrand-Dewey solution to the diffusion equation as follows:

$$\frac{C_c - C_s}{C_o - C_s} = \text{erf}\ (\ X/2\ \sqrt{Dt}\ ) \qquad \text{(Eq 10.9)}$$

where C_s is the surface concentration of carbon as maintained by the carbon potential of the atmosphere; C_o is the initial carbon level in the Fe-C alloy prior to carburizing; D is the diffusion coefficient for carbon in austenite; C_c is the carbon concentration as a function of distance from the surface; and t is the time after the start of carburizing. The diffusion coefficient was assumed to be independent of composition

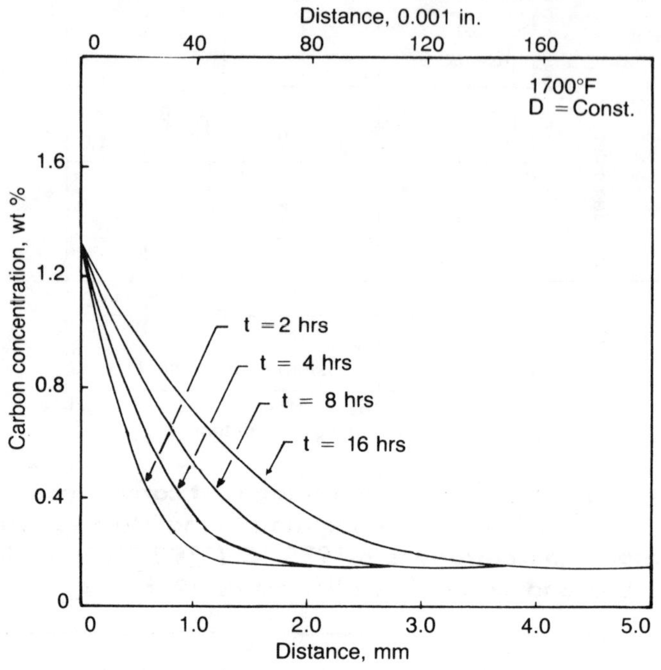

Fig. 10.6. Carbon concentration in an Fe-C alloy as a function of distance calculated for various carburizing times at 925 °C (1700 °F), assuming the diffusion coefficient is independent of composition. (Ref 10.9)

and to have an average value $D_c^\gamma = 0.12 \times \exp(-32\ 000/RT)\ cm^2/s$ (Ref 10.10). The diffusion coefficient in fact varies with carbon concentration of the austenite.

Equation 10.9, while it demonstrates the basic diffusion principles involved in carburizing, is highly idealized with respect to commercial practice. In particular, the diffusion coefficient of carbon in steels varies not only with carbon but also with alloy content, a situation for which Goldstein and Moren (Ref 10.9) have developed mathematical models which incorporate the effect of other alloying elements on the diffusion process. Apart from this recent approach, the equations based on empirical analysis of Eq 10.6 by Harris have proven adequate for plain carbon and alloy steels. At any temperature, Eq 10.9 reduces to:

$$X \text{ (case depth)} = K\sqrt{t} \qquad \text{(Eq 10.10)}$$

where K is a function of temperature and includes the temperature

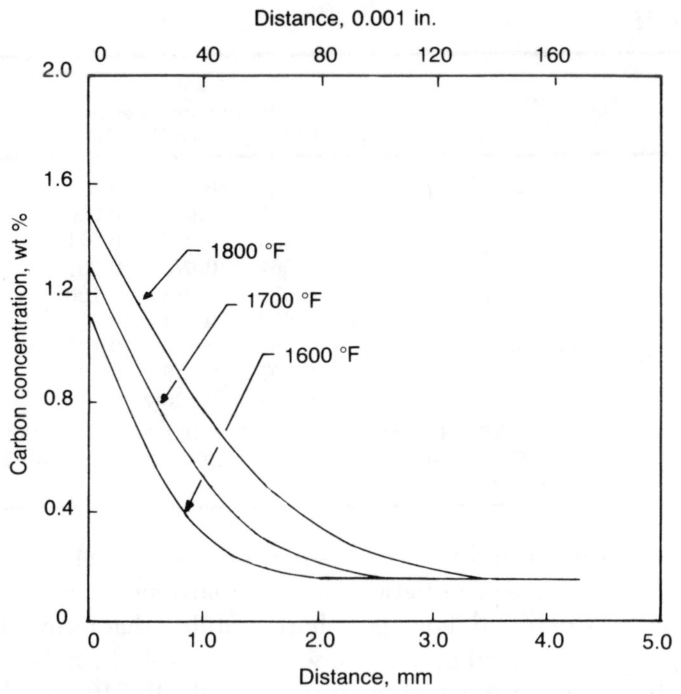

Fig. 10.7. Carbon concentration in an Fe-C alloy as a function of distance calculated for various temperatures for 8-hour carburizing, assuming the diffusion coefficient is independent of composition. (Ref 10.9)

dependence of the diffusion coefficient. Table 10.1 (Ref 10.2) lists values of case depths at various times for three commonly used carburizing temperatures.

Carburizing: Properties and Structure

The objective of carburizing is to obtain a high-carbon martensitic case with good wear and fatigue resistance superimposed on a tough, low-carbon steel core. Carburizing steels usually have base-carbon contents around 0.2%. Therefore, if hardenability is low as is the case for plain carbon steels, the core microstructure will consist of ferrite and pearlite of relatively low strength. Many applications, however, require high core strength to support the case in heavy duty applications. In addition, core strength is required where the stress gradients

Table 10.1. Case Depth Calculated by the Harris Equation (Ref 10.2)

Time, t, h	Case depth (in.) after carburizing at(a): 1600 °F	1650 °F	1700 °F
2	0.025	0.030	0.035
4	0.035	0.042	0.050
8	0.050	0.060	0.071
12	0.061	0.073	0.087
16	0.071	0.084	0.100
20	0.079	0.094	0.112
24	0.086	0.103	0.122
30	0.097	0.116	0.137
36	0.108	0.126	0.150

(a) Case depth = $0.025 \sqrt{t}$ for 927 °C (1700 °F); $0.021 \sqrt{t}$ for 899 °C (1650 °F); $0.018 \sqrt{t}$ for 871 °C (1600 °F). For normal carburizing (saturated austenite at the steel surface while at temperature).

between the surface and interior of a part in service are high enough to cause subsurface crack initiation in an unhardened core. For these reasons, alloy steels with good core hardenability that form martensite throughout a carburized part are in wide use. Table 10.2, taken from a study of fracture resistance of carburized steels (Ref 10.11), shows the compositions and the hardenabilities in D_I of some commonly used alloy carburizing steels. The EX grades of steels are exchange grade steels not yet assigned standard SAE designations (Ref 10.12). In the case of the carburizing grades, these steels have been developed to match different hardenability ranges of standard grades (as shown in Table 10.2) by means of adjusting alloy content.

Once the proper surface carbon concentration and depth of case are established by control of the processing parameters, the carburized steel is quenched to form martensite and tempered. The part may be hardened directly after carburizing or may be cooled and reheated to refine the microstructure. Figure 10.8 shows the microstructure of an EX 24 steel carburized and diffused at 1050 °C (1920 °F) and cooled to 845 °C (1550 °F) prior to quenching in oil (Ref 10.13). This microstructure is typical of that produced in fine-grained steels by direct quenching of carburized cases containing about 1% carbon at the surface. The martensite has a plate morphology, and as much as 20 or 30% retained austenite might be present. Lower carbon contents in a case would produce a finer martensite tending to a lath morphology and would reduce retained austenite content. Figure 10.9 shows the case microstructure produced in an 8620 steel that has been reheated to a temperature below the A_{cm}. The microstructure is highly refined by this treatment, so much so that the matrix martensite and retained

Table 10.2. Compositions, Grain Sizes and Hardenabilities of Some Carburizing Steels (Ref 10.11)

	Group I		Group II		Group III				Group IV
	EX24	SAE 8620	EX29	SAE 4320	20NiMoCr6	SAE 4817	SAE 4820	EX32	EX55
Composition, wt %:									
Carbon	0.20	0.20	0.20	0.21	0.22	0.17	0.19	0.19	0.17
Manganese	0.88	0.89	0.87	0.58	0.58	0.54	0.60	0.82	0.87
Silicon	0.34	0.34	0.34	0.33	0.54	0.33	0.28	0.27	0.28
Phosphorus	0.015(a)	0.015(a)	0.015(a)	0.015(a)	0.021	0.015(a)	0.016	0.017	0.015(a)
Silicon	0.02(a)	0.02(a)	0.02(a)	0.02(a)	0.027	0.02(a)	0.02	0.02	0.02(a)
Chromium	0.51	0.47	0.48	0.52	0.64	NA(b)	NA(b)	0.53	0.49
Molybdenum	0.26	0.21	0.34	0.26	0.31	0.27	0.27	0.52	0.74
Nickel	NA(b)	0.53	0.54	1.76	1.56	3.56	3.48	0.80	1.84
Aluminum	0.08(a)	0.08(a)	0.08(a)	0.08(a)	0.043	0.08(a)	0.075	0.082	0.08(a)
ASTM grain size	9½	9½	9½	9½	7½	9½	9½	9	9½
D_I hardenability:									
Inches	1.6	1.7	2.0	1.9	3.0	2.5	2.7	3.3	4.7
Millimeters	41	43	51	48	76	63	69	84	120

(a) Amount added. (b) NA = none added.

austenite are not resolvable in the light microscope. The fine white particles that are visible are carbides that have been retained because the case was not reheated into the single-phase austenite field.

Figures 10.8 and 10.9 show microstructures that are typical of the results of good commercial practice. Problems due to processing, however, may adversely affect microstructures and properties. One common feature of conventional gas carburizing is the internal oxidation that develops to a depth of about 0.001 in. (0.025 mm) from the carburized surface. Figure 10.10 shows an example of intergranular oxidation in a gas carburized specimen. Although many applications tolerate the presence of such intergranular oxidation, the oxides do adversely affect fatigue resistance (Ref 10.14), and for critical applications the oxide containing layer is sometimes removed by grinding or machining. Since the oxidation is a direct result of the oxygen content of the carburizing gases, partial pressure or vacuum carburizing that significantly reduces oxygen content of a carburizing atmosphere is also used to eliminate the surface oxidation (Ref 10.15). A systematic study (Ref 10.16) of internal oxidation during carburizing showed that certain elements such as chromium, manganese, silicon and titanium enhanced oxidation, and that removal of silicon from carburizing steels tended to eliminate the surface oxidation.

Figures 10.11 and 10.12 show the deleterious effects of too high a surface carbon concentration on case microstructure. The high surface carbon concentration may be a result of too short a diffusion cycle after carburizing has first saturated the austenite, or a result of geometry,

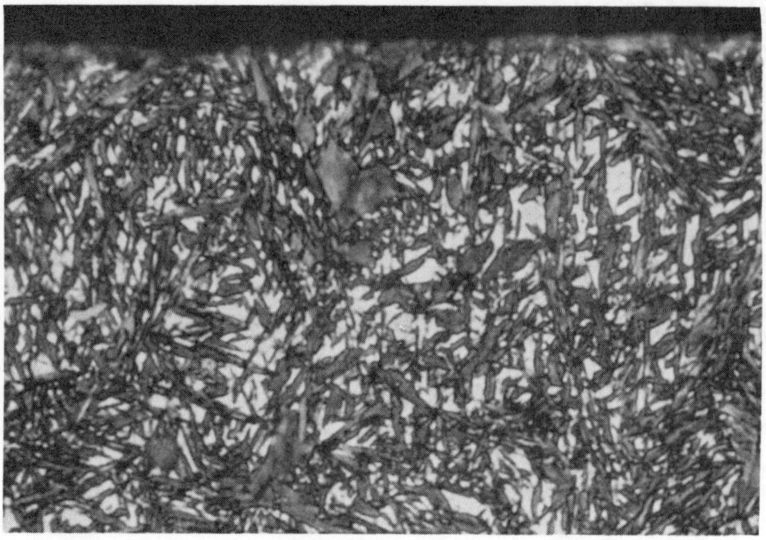

Fig. 10.8. Microstructure adjacent to surface of an EX 24 steel carburized and diffused at 1050 °C (1920 °F), cooled to 854 °C (1550 °F) and oil quenched. Magnification, 1000 ×. (Ref 10.13)

Fig. 10.9. Microstructure adjacent to surface of 8620 steel carburized at 1050 °C (1920 °F), oil quenched from 845 °C (1550 °F) and reheated to 845 °C (1550 °F). Magnification, 1000 ×. (Ref 10.13)

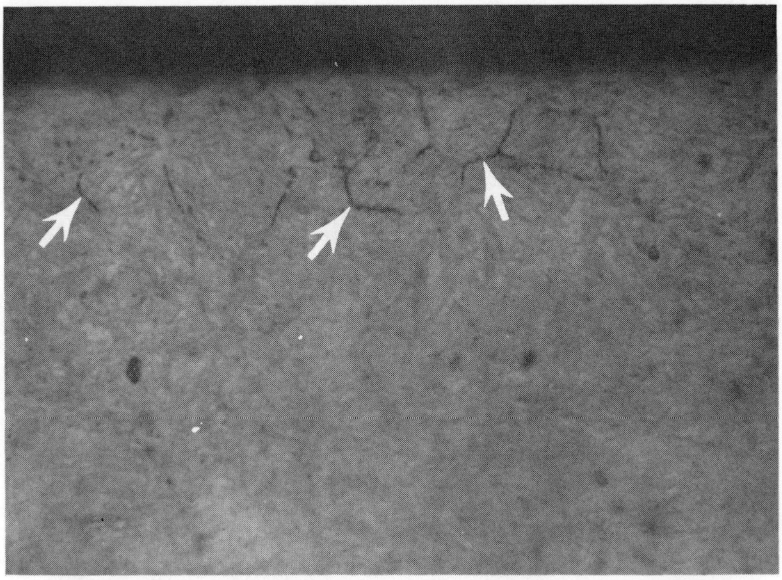

Fig. 10.10. Example of surface intergranular oxidation along austenite grain boundaries in a carburized steel. Magnification, 1000 ×. (G. Krauss, unpublished research, Colorado School of Mines, Golden, Colo.)

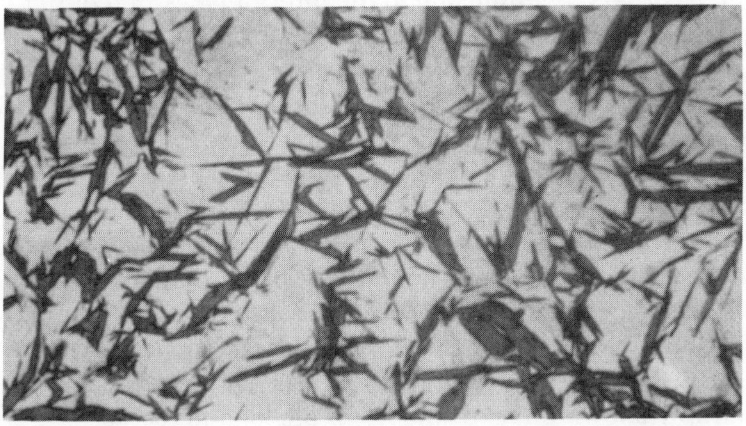

Fig. 10.11. High austenite content in corner of an 8620 steel carburized and diffused at 1050 °C (1920 °F) and cooled to 845 °C (1550 °F) before oil quenching. (Ref 10.17)

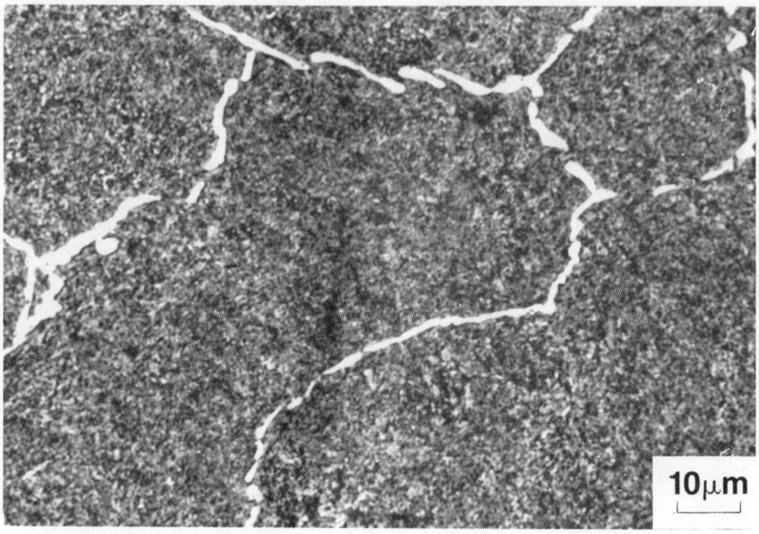

Fig. 10.12. Cementite network in corner of an EX 24 specimen carburized at 1050 °C (1920 °F), oil quenched, and reheated to 845 °C (1550 °F). (Ref 10.13)

where carbon has good surface access to a portion of a steel (especially specimen corners) but diffusion to the interior is restricted (Ref 10.17). Figure 10.11 shows a very high retained austenite content in an 8620 steel and Fig. 10.12 shows grain boundary carbides that have formed in an EX 24 steel that has been reheated for hardening after carburizing. The reheating has caused some agglomeration of the carbides and therefore reduced their detrimental effect on fatigue crack initiation (Ref 10.17).

Another microstructural feature that may adversely affect properties of carburized steels is the microcracking that develops in high-carbon martensitic microstructures because of plate impingement (Ref 10.18). Figure 10.13 shows examples of microcracks that have formed in the case of a coarse-grained 8620 steel (Ref 10.19). The micrograph shows an extreme example of microcracking. Microcracking is much reduced when martensite forms in fine-grain austenite (Ref 10.20); when the martensite morphology approaches that of lath martensite as carbon is reduced below 1% (Ref 10.21); and as a result of tempering (Ref 10.22). Many of the microstructural features described above and their influence on the properties of carburized parts have been extensively reviewed by Parrish (Ref 10.23).

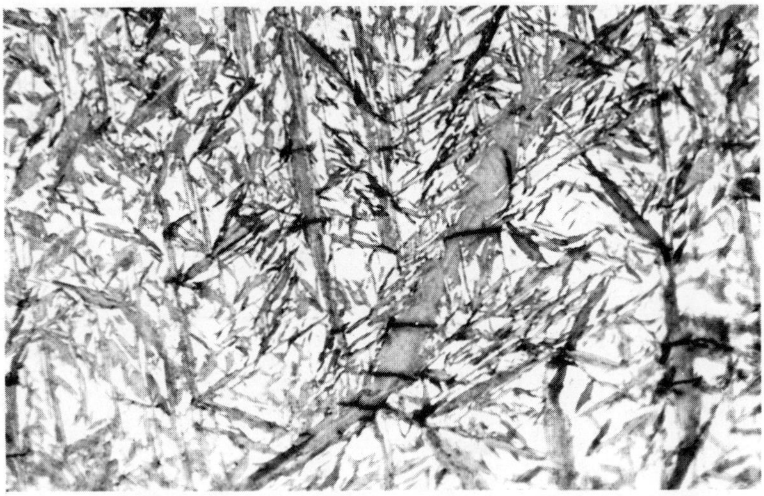

Fig. 10.13. Microcracks in the martensite of a carburized coarse-grained 8620 steel. (Ref 10.19)

Figure 10.14 shows the results of a study of the effect of martensite morphology, including the effects of microcracking, on fatigue resistance of a carburized coarse-grained 8620 steel (Ref 10.19). All specimens were chemically polished and, therefore, the influence of intergranular oxidation on fatigue cracking was removed from the study. The specimens directly quenched from the carburizing temperature had the coarsest structure and the highest density of microcracks, some of which were directly exposed on the specimen surfaces by the chemical polishing. The single reheat specimens had a finer austenite grain structure and therefore finer martensite plates and a lower density of microcracks. Since the retained austenite content and hardness profiles of the direct and single reheat specimens were identical, the improved fatigue resistance of the single reheat specimens is attributed to the smaller size of the microcracks and their lower density in the finer structure. The best fatigue resistance was shown by the double reheat specimens with case microstructure similar to that shown in Fig. 10.9. The virtual elimination of the surface microcracks, the much finer structure of the martensite, and the higher hardness and reduced retained austenite were probably all factors that contributed to the very high fatigue resistance of the double reheat specimens.

The study (Ref 10.19) of chemically polished carburized specimens showed that microcracks, or possibly an associated microstructural

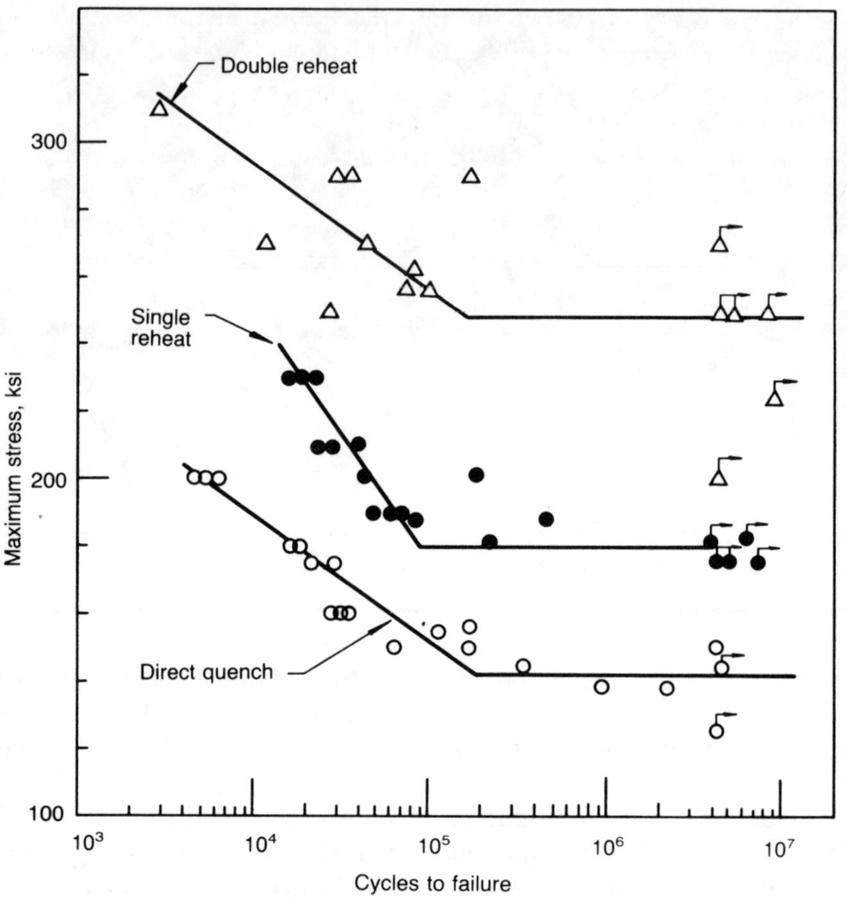

Fig. 10.14. Maximum applied stress versus cycles to failure (S-N curve) for four-point bend fatigue specimens of 8620 steel given three different carburizing treatments. (Ref 10.19)

feature such as an embrittled austenite grain boundary, initiated fatigue fracture in specimens quenched from the austenite field. In specimens reheated to the austenite-cementite field, fatigue cracks were initiated at pits produced by the chemical polishing. A study of fatigue crack origins in a steel carburized to 0.7% carbon showed that cracks in electropolished specimens initiated mainly at prior austenite grain boundaries and sometimes at inclusion particles (Ref 10.24). In commercially carburized specimens not subjected to any surface polishing or removal treatment, fatigue cracks are most probably initiated at intergranular surface oxides or surface roughness in one form or another, such as grooves produced by machining. In fine-grained steels or steels carburized to a carbon level that yields lath martensite, microcracking would not be expected to be a major cause of fatigue crack initiation.

The study of the carburized coarse-grained 8620 steel also revealed several characteristics of the fracture of carburized steels. Figure 10.15 shows fatigue fracture origins in direct quenched and double reheat carburized specimens. The fatigue crack in the direct quenched specimen (Fig. 10.15a) was initiated either at microcracks or an embrittled prior austenite grain boundary. The fatigue crack was quite smooth (area within dashed semicircle) but on overload and rapid propagation became intergranular. In contrast, both the fatigue and overload portions of the fracture surface of the double reheat specimen are transgranular and quite smooth. Figure 10.16 shows higher magnification scanning electron micrographs of overload fracture in direct quench and double reheated specimens. The intergranular fracture (See Fig. 10.16a) is frequently observed in carburized steels and is determined by Auger analysis (Ref 10.23) to be associated with higher phosphorus and carbon concentrations than those found in the bulk of the steel. The shape of the Auger peak indicates that the carbon at the grain boundary is present in the form of cementite, and thus the intergranular fracture appears to be very similar to that associated with TME of medium-carbon steels. In carburized steels, however, the conditions for intergranular TME-type embrittlement appear to be present in as-quenched and tempered conditions that would normally be considered to be immune to TME in medium-carbon steels (Ref 10.25).

In addition to microstructure, the residual stresses developed during quenching of a carburized steel also favorably affect fatigue resistance. Koistinen (Ref 10.27) first explained that the compressive surface stresses formed in carburized steels were due to M_s gradients associated with the carbon gradients in carburized specimens. At the surface, M_s is a minimum, and as carbon decreases with distance into the specimen, M_s increases. On quenching, therefore, temperature is first lowered below the M_s at some point removed from the surface, and martensite forms at that location. Further on in the quenching cycle, the low M_s surface transforms to martensite, and because its expansion is restrained by the already transformed subsurface layer, it is put into compression. Ebert (Ref 10.28) has extensively discussed the development of residual stresses in carburized steels, and Fig. 10.17, taken from his work, compares the surface compressive stresses of a carburized specimen to the surface tensile stresses developed in a through hardened specimen.

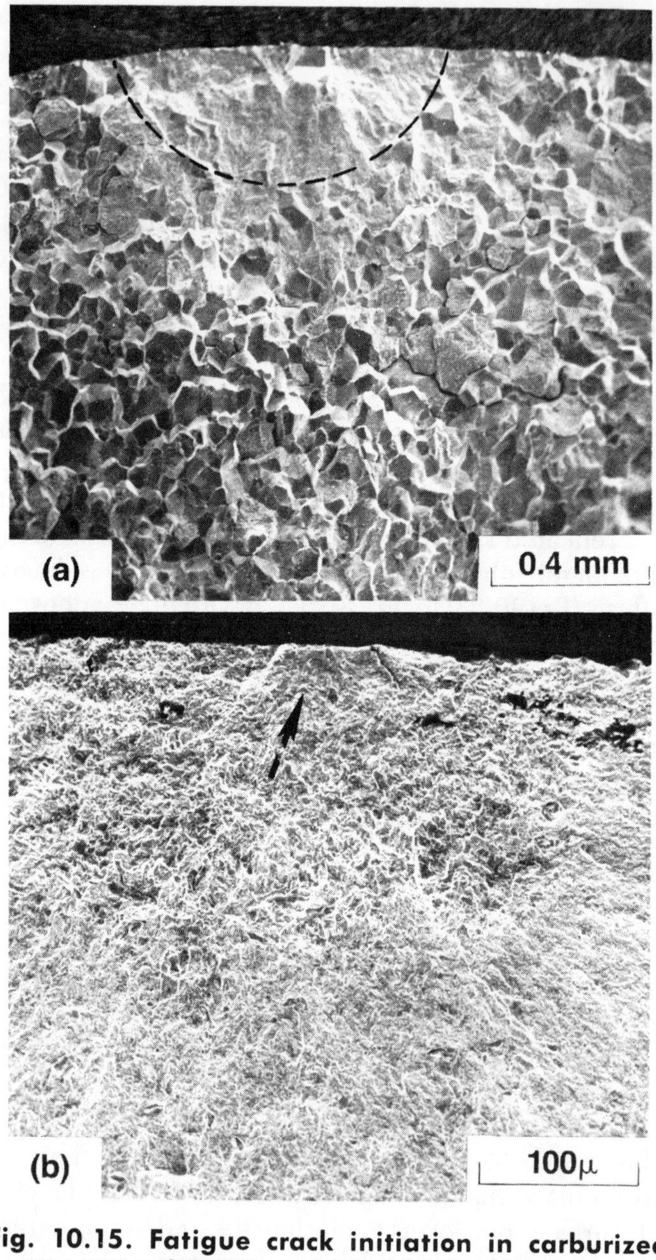

Fig. 10.15. Fatigue crack initiation in carburized coarse-grained 8620 steel (a) quenched directly from carburizing at 927 °C (1700 °F) and (b) reheated after carburizing to 788 °C (1450 °F). Both specimens tempered at 145 °C (300 °F). Scanning electron micrographs. (Ref 10.19)

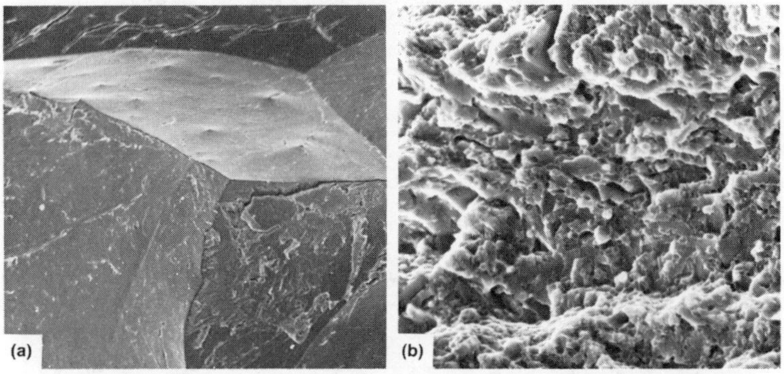

Fig. 10.16. Overload case fracture surfaces in carburized 8620 steel (a) quenched directly after carburizing at 927 °C (1700 °F) and (b) reheated to 708 °C (1450 °F). Both specimens tempered at 145 °C (300 °F). Scanning electron micrographs. (Ref 10.25)

Nitriding

Nitriding is a surface hardening heat treatment that introduces nitrogen into the surface of steel while it is in the ferritic condition. Nitriding, therefore, is similar to carburizing in that surface composition is altered, but different in that the nitrogen is added into ferrite instead of austenite. The fact that nitriding does not involve heating into the austenite phase field and a subsequent quench to form martensite means that nitriding can be accomplished with a minimum of distortion and excellent dimensional control.

Steels that are nitrided are generally medium-carbon steels that contain strong nitride-forming elements such as chromium, aluminum, vanadium and molybdenum. Aluminum especially, as already discussed with respect to austenite grain size control (See Chapter 7), is a very powerful nitride former and is used in amounts between 0.85 and 1.5% in nitriding steels (Ref 10.29). Prior to nitriding, the steels are autenitized, quenched and tempered. Tempering is performed at temperatures between 540 to 750 °C (1000 to 1300 °F), a range above that at which the nitriding is performed. Tempering above the nitriding temperature provides a core structure that will be stable during nitriding.

Gas nitriding is accomplished with ammonia gas which dissociates on the surface of the steel according to the following reaction:

$$NH_3 \rightleftharpoons N + 3H \qquad \text{(Eq 10.11)}$$

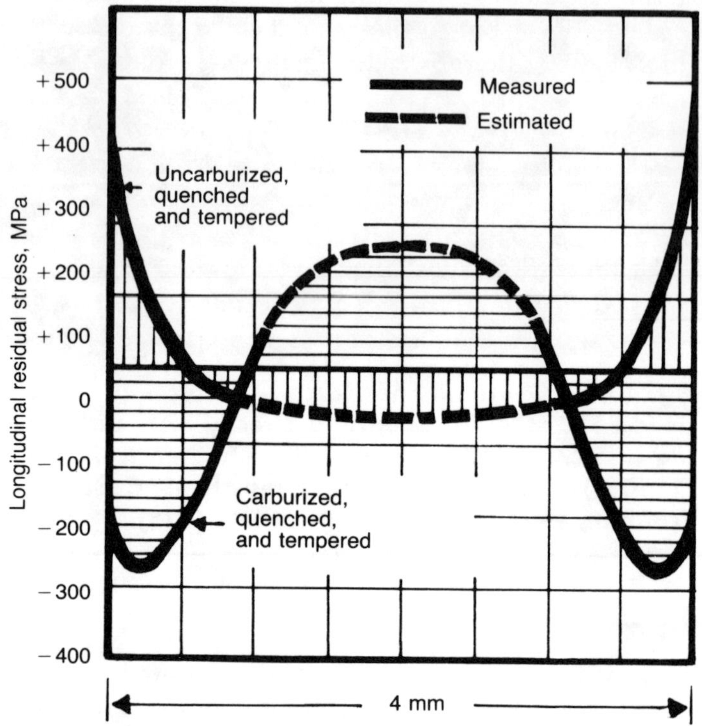

Composition of base steel:
0.26 C, 1.08 Mn, 0.93 Cr, 0.15 Ni, 0.3 Mo, 0.06 Ti, 0.024 S, 0.013P

Heat treatments:
Gas carburize 930-940 °C, quench from 840-850 °C, temper 1 h at 180-200 °C

Specimen dimensions:
110 mm long, 15 mm wide, 4 mm thick

Fig. 10.17. Residual stress as a function of distance through the thickness of a carburized and uncarburized chromium-carbon steel specimens. (Ref 10.28)

The resulting atomic nitrogen is absorbed at the surface of the steel. The nitriding may be accomplished by either a single-stage or double-stage process. The single-stage process is performed at 500 to 525 °C (930 to 975 °F) with 15 to 30% dissociation of the ammonia, i.e., with an atmosphere that contains 70 to 85% NH_3, the source of the nitrogen. This process produces a "white layer" of brittle iron nitride. A patented process developed by Floe (Ref 10.30) uses a two-stage process to minimize the thickness of the white layer. The first stage is similar to that described above, but in the second stage the dissociation is increased to 65 to 85%, thereby reducing the NH_3 content of the atmosphere that supplies nitrogen to the surface according to the reaction represented in Eq 10.11. As a result, the iron nitride does not grow as rapidly, and in fact dissolves as it supplies nitrogen into the interior of the steel. Nitriding times are quite long, anywhere from 10 to 130 h depending on the application (Ref 10.29), and the case depths are relatively shallow, usually less than 0.020 in. (0.5 mm).

The microstructure of a nitrided steel is shown in Fig. 10.18. The white layer is clearly visible. Below the white layer, fine alloy nitrides and/or zone structures have formed but are much smaller than can be resolved in the light microscope. Jack and his colleagues have studied many alloys of iron, nitrogen and other substitutional alloying elements (Ref 10.31). At the temperatures at which commercial nitriding is performed, very fine clusters or precipitate zones of substitutional alloying elements and the interstitial nitrogen form. The substitutional-interstitial zones remain quite fine and do not coarsen readily on heating because of the sluggish diffusion of the substitutional atoms. At lower temperatures only iron nitrides form, and at higher temperatures alloy nitrides form. Nitrided cases are harder than those produced by carburizing and are quite stable in service up to the temperature of the nitriding process. Nitriding, therefore, produces excellent wear, seizing and galling resistance under conditions where heat is generated by friction between moving parts in contact. Improved fatigue life is also an important benefit of nitriding.

Carbonitriding and Ferritic Nitrocarburizing

Carbonitriding is a surface hardening heat treatment that introduces carbon and nitrogen into the austenite of steel. This treatment is therefore similar to carburizing in that the austenite composition is changed and high surface hardness is produced by quenching to form martensite. Carbonitriding surface hardening, however, is dependent to some extent on nitride as well as martensite formation.

The process of carbonitriding utilizes an atmosphere containing ammonia plus a carbon-rich gas or vaporized liquid hydrocarbon that is a source of carbon as in carburizing. The ammonia dissociates on the surface of the steel and introduces atomic nitrogen. The nitrogen inhibits the diffusion of carbon and this factor plus the fact that carbonitriding is performed at lower temperatures (705 to 900 °C, or 1300 to

Fig. 10.18. White layer and diffusion zone in nitrided steel. Steel is Nitralloy 135 Modified containing 0.4% carbon, 1.6% chromium, 0.35% molybdenum and 1.13% aluminum. Base microstructure is tempered martensite of hardness R_c 30. Etched in 1½% nital. Magnification, 500×. (Courtesy of Denis Stratford, Sundstrand Corp., Denver, Colo.)

1650 °F) and shorter times than carburizing, results in relatively shallow case depths, from 0.003 to 0.030 in. (0.075 to 0.75 mm). The lower carbonitriding temperatures are not often used because of the hazard of explosion and the brittle structures formed at the lower temperatures (Ref 10.4). The nitrogen also enhances hardenability, and makes it possible to form martensite in plain carbon and low-alloy steels that initially have low hardenability. The nitrides formed due to the presence of nitrogen also contribute to the high hardness of the case.

Nitrogen, similar to carbon, is an austenite stabilizer. Therefore, considerable austenite may be retained after quenching a carbonitrided part. If the retained austenite content is so high that it reduces hardness and wear resistance, it may be controlled by reducing the ammonia content of the carbonitriding gas either throughout the cycle or during the latter portion of the cycle (Ref 10.4).

Another type of surface hardening that involves the introduction of carbon and nitrogen into a steel is ferritic nitrocarburizing. In contrast to carbonitriding but similar to the nitriding process, the carbon and nitrogen are added to ferrite below the Ac_1 temperature. Bell

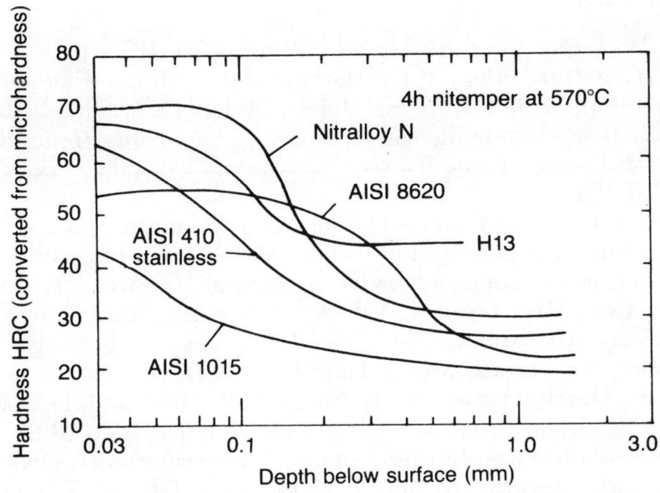

Fig. 10.19. Microhardness profiles in various types of steels after gaseous ferritic nitrocarburizing (Nitemper process). (Ref 10.32)

(Ref 10.32) describes a number of ferritic nitrocarburizing processes, both liquid and gaseous. The common beneficial result in all the processes is a very thin single-phase layer of epsilon carbonitride – a hexagonal ternary compound of iron, nitrogen and carbon – formed between 450 and 590 °C (840 and 1095 °F) (Ref 10.32).

The epsilon carbonitride compound layer has excellent wear and antiscuffing properties and is produced with minimum distortion. The layer can be formed on inexpensive mild steels with ferrite-pearlite microstructures, thereby greatly improving their wear and fatigue resistance. Figure 10.19 shows the hardness profiles of various steels subjected to a gaseous ferritic nitrocarburizing treatment. The very high hardness and shallow case depths produced by the process in alloy steels are apparent, and even the plain steel benefits significantly. Some of the increased hardness of the case is due to a diffusion zone beneath the compound layer, especially in the more highly alloyed steels with strong nitride formers such as Nitralloy N (1.00 to 1.30% chromium, 0.85 to 1.20% aluminum and 0.20 to 0.30% molybdenum). In this diffusion zone, nitrides or precipitate zones similar to those that form as a result of nitriding are developed as nitrogen diffuses into the interior of the steel from the compound layer. As a general rule, the compound layer gives good tribological properties, but a substantial diffusion zone is required for good fatigue resistance.

References

10.1 H. W. Grönegress, *Stahl und Eisen,* Vol 70, 1950, p 192

10.2 *Surface Hardening,* P. D. Harvey, (Ed.), Metals Engineering Institute, American Society for Metals, Metals Park, Oh., 1979

10.3 Induction Hardening and Tempering, in *Metals Handbook,* Vol 2, 8th Ed., American Society for Metals, Metals Park, Oh., 1964, p 167-188

10.4 *Carburizing and Carbonitriding,* ASM Committee on Gas Carburizing, American Society for Metals, Metals Park, Oh., 1977

10.5 F. J. Harvey, Thermodynamic Aspects of Gas-Metal Heat Treating Reactions, *Met Trans A,* Vol 9A, 1978, p 1507-1513

10.6 Furnace Atmospheres, in *Metals Handbook,* Vol 2, 8th Ed., American Society for Metals, Metals Park, Oh., 1964, p 67-84

10.7 F. E. Harris, Reactions Between Hot Steel and Furnace Atmospheres, *Metal Progress,* Vol 47 (No. 1), Jan 1945, p 84-89

10.8 Application of Equilibrium Data, in *Carburizing and Carbonitriding,* American Society for Metals, Metals Park, Oh., 1977, p 14-15.

10.9 J. I. Goldstein and A. E. Moren, Diffusion Modeling of the Carburization Process, *Met Trans A,* Vol 9A, 1978, p 1515-1525

10.10 C. Wells and R. F. Mehl, Rate of Diffusion of Carbon in Austenite in Plain Carbon, in Nickel, and in Manganese Steels, *Trans AIME,* Vol 140, 1940, p 279-306

10.11 D. E. Diesburg and G. T. Eldis, Fracture Resistance of Various Carburized Steels, *Met Trans A,* Vol 9A, 1978, p 1561-1570

10.12 W. T. Groves, How to Select the Right EX Steel, *Metal Progress,* Vol 102, 1972, p 89-99

10.13 K. D. Jones and G. Krauss, Microstructure and Fatigue of Partial Pressure Carburized SAE 8620 and EX 24 Steels, *J Heat Treating,* Vol 1, 1979, p 64-71

10.14 W. J. Ptashnik, Influence of Partial Pressure Carburizing on Bending Fatigue Durability, Report No. MD 77-008, Manufacturing Development, General Motors Corp., Warren, Mich., 1977

10.15 H. C. Child, Vacuum Carburizing, *Heat Treat Met,* Vol 3, 1976, p 60-65

10.16 R. Chatterjee-Fischer, Internal Oxidation During Carburizing and Heat Treating, *Met Trans A,* Vol 9A, 1978, p 1553-1560

10.17 K. D. Jones and G. Krauss, Effects of High-carbon Specimen Corners on Microstructure and Fatigue of Partial Pressure Carburized Steels, *Proceedings of the Metals Society, Heat Treatment '79,* Birmingham, England, 22-24 May 1979

10.18 A. R. Marder and A. O. Benscoter, Microcracking in Fe-C Acicular Martensite, *Trans ASM,* Vol 61, 1968, p 293-299

10.19 C. A. Apple and G. Krauss, Microcracking and Fatigue in a Carburized Steel, *Met Trans,* Vol 4, 1973, p 1195-1200

10.20 R. P. Brobst and G. Krauss, The Effect of Austenite Grain Size on Microcracking in Martensite of an Fe-1.22 C Alloy, *Met Trans,* Vol 5, 1975, p 457-462

10.21 M. G. Mendiratta, J. Sasser and G. Krauss, Effect of Dissolved Carbon on Microcracking in Martensite of an Fe-1.39 C Alloy, *Met Trans,* Vol 3, 1972, p 351-353

10.22 T. A. Balliett and G. Krauss, The Effect of the First and Second Stages of Tempering on Microcracking in Martensite of an Fe-1.22 C Alloy, *Met Trans A,* Vol 7A, 1976, p 81-86

10.23 G. Parrish, The Influence of Microstructure on the Properties of Case-Carburized Components", series of articles in *Heat Treatment of Metals,* Vol 3, 4, 1976-1977; published in book form by American Society for Metals, Metals Park, Oh., 1980

10.24 L. Magnusson and T. Ericsson, Initiation and Propagation of Fatigue Cracks in Carburized Steel, *Proceedings of the Metals Society, Heat Treatment '79,* Birmingham, England, 22-24 May 1979

10.25 G. Krauss, The Microstructure and Fatigue of a Carburized Steel, *Met Trans A,* Vol 9A, 1978, p 1527-1535

10.26 H. K. Obermeyer, The Effects of Heat Treatment and Phosphorus Content on Fracture of a 0.85% Carbon Alloy Steel, MS Thesis, Colorado School of Mines, Golden, 1978

10.27 D. P. Koistinen, The Distribution of Residual Stresses in Carburized Cases and Their Origin, *Trans ASM,* Vol 50, 1958, p 227-241

10.28 L. J. Ebert, The Role of Residual Stresses in the Mechanical Performance of Case Carburized Steel, *Met Trans A,* Vol 9A, 1978, p 1537-1551

10.29 Gas Nitriding, in *Metals Handbook,* Vol 2, 8th Ed., American Society for Metals, Metals Park, Oh., 1964, p 149-163

10.30 C. F. Floe, A Study of the Nitriding Process: Effect of Ammonia on Case Depth and Structure, *Trans ASM,* Vol 32, 1943, p 134

10.31 K. H. Jack, Nitriding, *Proceedings of the Metals Society, Heat Treatment '73,* London, 1973, p 39-50

10.32 T. Bell, Ferritic Nitrocarburizing, *Heat Treatment of Metals,* Vol 2, 1975, p 39-49

Appendix: Temperature Conversions

The general arrangement of this conversion table was devised by Sauveur and Boylston. The middle columns of numbers (in **boldface** type) contain the temperature readings (°F or °C) to be converted. When converting from degrees Fahrenheit to degrees Celsius, read the Celsius equivalent in the column headed "C". When converting from Celsius to Fahrenheit, read the Fahrenheit equivalent in the column headed "F".

F		C	F		C	F		C	F		C
.....	**−458**	−272.22	.....	**−378**	−227.78	.....	**−298**	−183.33	−360.4	**−218**	−138.89
.....	**−456**	−271.11	.....	**−376**	−226.67	.....	**−296**	−182.22	−356.8	**−216**	−137.78
.....	**−454**	−270.00	.....	**−374**	−225.56	.....	**−294**	−181.11	−353.2	**−214**	−136.67
.....	**−452**	−268.89	.....	**−372**	−224.44	.....	**−292**	−180.00	−349.6	**−212**	−135.56
.....	**−450**	−267.78	.....	**−370**	−223.33	.....	**−290**	−178.89	−346.0	**−210**	−134.44
.....	**−448**	−266.67	.....	**−368**	−222.22	.....	**−288**	−177.78	−342.4	**−208**	−133.33
.....	**−446**	−265.56	.....	**−366**	−221.11	.....	**−286**	−176.67	−338.8	**−206**	−132.22
.....	**−444**	−264.44	.....	**−364**	−220.00	.....	**−284**	−175.56	−335.2	**−204**	−131.11
.....	**−442**	−263.33	.....	**−362**	−218.89	.....	**−282**	−174.44	−331.6	**−202**	−130.00
.....	**−440**	−262.22	.....	**−360**	−217.78	.....	**−280**	−173.33	−328.0	**−200**	−128.89
.....	**−438**	−261.11	.....	**−358**	−216.67	.....	**−278**	−172.22	−324.4	**−198**	−127.78
.....	**−436**	−260.00	.....	**−356**	−215.56	.....	**−276**	−171.11	−320.8	**−196**	−126.67
.....	**−434**	−258.89	.....	**−354**	−214.44	.....	**−274**	−170.00	−317.2	**−194**	−125.56
.....	**−432**	−257.78	.....	**−352**	−213.33	−457.6	**−272**	−168.89	−313.6	**−192**	−124.44
.....	**−430**	−256.67	.....	**−350**	−212.22	−454.0	**−270**	−167.78	−310.0	**−190**	−123.33
.....	**−428**	−255.56	.....	**−348**	−211.11	−450.4	**−268**	−166.67	−306.4	**−188**	−122.22
.....	**−426**	−254.44	.....	**−346**	−210.00	−446.8	**−266**	−165.56	−302.8	**−186**	−121.11
.....	**−424**	−253.33	.....	**−344**	−208.89	−443.2	**−264**	−164.44	−299.2	**−184**	−120.00
.....	**−422**	−252.22	.....	**−342**	−207.78	−439.6	**−262**	−163.33	−295.6	**−182**	−118.89
.....	**−420**	−251.11	.....	**−340**	−206.67	−436.0	**−260**	−162.22	−292.0	**−180**	−117.78
.....	**−418**	−250.00	.....	**−338**	−205.56	−432.4	**−258**	−161.11	−288.4	**−178**	−116.67
.....	**−416**	−248.89	.....	**−336**	−204.44	−428.8	**−256**	−160.00	−284.8	**−176**	−115.56
.....	**−414**	−247.78	.....	**−334**	−203.33	−425.2	**−254**	−158.89	−281.2	**−174**	−114.44
.....	**−412**	−246.67	.....	**−332**	−202.22	−421.6	**−252**	−157.78	−277.6	**−172**	−113.33
.....	**−410**	−245.56	.....	**−330**	−201.11	−418.0	**−250**	−156.67	−274.0	**−170**	−112.22
.....	**−408**	−244.44	.....	**−328**	−200.00	−414.4	**−248**	−155.56	−270.4	**−168**	−111.11
.....	**−406**	−243.33	.....	**−326**	−198.89	−410.8	**−246**	−154.44	−266.8	**−166**	−110.00
.....	**−404**	−242.22	.....	**−324**	−197.78	−407.2	**−244**	−153.33	−263.2	**−164**	−108.89
.....	**−402**	−241.11	.....	**−322**	−196.67	−403.6	**−242**	−152.22	−259.6	**−162**	−107.78
.....	**−400**	−240.00	.....	**−320**	−195.56	−400.0	**−240**	−151.11	−256.0	**−160**	−106.67
.....	**−398**	−238.89	.....	**−318**	−194.44	−396.4	**−238**	−150.00	−252.4	**−158**	−105.56
.....	**−396**	−237.78	.....	**−316**	−193.33	−392.8	**−236**	−148.89	−248.8	**−156**	−104.44
.....	**−394**	−236.67	.....	**−314**	−192.22	−389.2	**−234**	−147.78	−245.2	**−154**	−103.33
.....	**−392**	−235.56	.....	**−312**	−191.11	−385.6	**−232**	−146.67	−241.6	**−152**	−102.22
.....	**−390**	−234.44	.....	**−310**	−190.00	−382.0	**−230**	−145.56	−238.0	**−150**	−101.11
.....	**−388**	−233.33	.....	**−308**	−188.89	−378.4	**−228**	−144.44	−234.4	**−148**	−100.00
.....	**−386**	−232.22	.....	**−306**	−187.78	−374.8	**−226**	−143.33	−230.8	**−146**	−98.89
.....	**−384**	−231.11	.....	**−304**	−186.67	−371.2	**−224**	−142.22	−227.2	**−144**	−97.78
.....	**−382**	−230.00	.....	**−302**	−185.56	−367.6	**−222**	−141.11	−223.6	**−142**	−96.67
.....	**−380**	−228.89	.....	**−300**	−184.44	−364.0	**−220**	−140.00	−220.0	**−140**	−95.56

(continued on the next page)

Temperature Conversions (contd.)

F		C	F		C	F		C	F		C
-216.4	**-138**	-94.44	+35.6	**+2**	-16.67	287.6	**142**	61.11	539.6	**282**	138.89
-212.8	**-136**	-93.33	+39.2	**+4**	-15.56	291.2	**144**	62.22	543.2	**284**	140.00
-209.2	**-134**	-92.22	+42.8	**+6**	-14.44	294.8	**146**	63.33	546.8	**286**	141.11
-205.6	**-132**	-91.11	+46.4	**+8**	-13.33	298.4	**148**	64.44	550.4	**288**	142.22
-202.0	**-130**	-90.00	+50.0	**+10**	-12.22	302.0	**150**	65.56	554.0	**290**	143.33
-198.4	**-128**	-88.89	+53.6	**+12**	-11.11	305.6	**152**	66.67	557.6	**292**	144.44
-194.8	**-126**	-87.78	+57.2	**+14**	-10.00	309.2	**154**	67.78	561.2	**294**	145.56
-191.2	**-124**	-86.67	+60.8	**+16**	-8.89	312.8	**156**	68.89	564.8	**296**	146.67
-187.6	**-122**	-85.56	+64.4	**+18**	-7.78	316.4	**158**	70.00	568.4	**298**	147.78
-184.0	**-120**	-84.44	+68.0	**+20**	-6.67	320.0	**160**	71.11	572.0	**300**	148.89
-180.4	**-118**	-83.33	+71.6	**+22**	-5.56	323.6	**162**	72.22	575.6	**302**	150.00
-176.8	**-116**	-82.22	+75.2	**+24**	-4.44	327.2	**164**	73.33	579.2	**304**	151.11
-173.2	**-114**	-81.11	+78.8	**+26**	-3.33	330.8	**166**	74.44	582.8	**306**	152.22
-169.6	**-112**	-80.00	+82.4	**+28**	-2.22	334.4	**168**	75.56	586.4	**308**	153.33
-166.0	**-110**	-78.89	+86.0	**+30**	-1.11	338.0	**170**	76.67	590.0	**310**	154.44
-162.4	**-108**	-77.78	+89.6	**+32**	±0.00	341.6	**172**	77.78	593.6	**312**	155.56
-158.8	**-106**	-76.67	+93.2	**+34**	+1.11	345.2	**174**	78.89	597.2	**314**	156.67
-155.2	**-104**	-75.56	+96.8	**+36**	+2.22	348.8	**176**	80.00	600.8	**316**	157.78
-151.6	**-102**	-74.44	+100.4	**+38**	+3.33	352.4	**178**	81.11	604.4	**318**	158.89
-148.0	**-100**	-73.33	+104.0	**+40**	+4.44	356.0	**180**	82.22	608.0	**320**	160.00
-144.4	**-98**	-72.22	107.6	**42**	5.56	359.6	**182**	83.33	611.6	**322**	161.11
-140.8	**-96**	-71.11	111.2	**44**	6.67	363.2	**184**	84.44	615.2	**324**	162.22
-137.2	**-94**	-70.00	114.8	**46**	7.78	366.8	**186**	85.56	618.8	**326**	163.33
-133.6	**-92**	-68.89	118.4	**48**	8.89	370.4	**188**	86.67	622.4	**328**	164.44
-130.0	**-90**	-67.78	122.0	**50**	10.00	374.0	**190**	87.78	626.0	**330**	165.56
-126.4	**-88**	-66.67	125.6	**52**	11.11	377.6	**192**	88.89	629.6	**332**	166.67
-122.8	**-86**	-65.56	129.2	**54**	12.22	381.2	**194**	90.00	633.2	**334**	167.78
-119.2	**-84**	-64.44	132.8	**56**	13.33	384.8	**196**	91.11	636.8	**336**	168.89
-115.6	**-82**	-63.33	136.4	**58**	14.44	388.4	**198**	92.22	640.4	**338**	170.00
-112.0	**-80**	-62.22	140.0	**60**	15.56	392.0	**200**	93.33	644.0	**340**	171.11
-108.4	**-78**	-61.11	143.6	**62**	16.67	395.6	**202**	94.44	647.6	**342**	172.22
-104.8	**-76**	-60.00	147.2	**64**	17.78	399.2	**204**	95.56	651.2	**344**	173.33
-101.2	**-74**	-58.89	150.8	**66**	18.89	402.8	**206**	96.67	654.8	**346**	174.44
-97.6	**-72**	-57.78	154.4	**68**	20.00	406.4	**208**	97.78	658.4	**348**	175.56
-94.0	**-70**	-56.67	158.0	**70**	21.11	410.0	**210**	98.89	662.0	**350**	176.67
-90.4	**-68**	-55.56	161.6	**72**	22.22	413.6	**212**	100.00	665.6	**352**	177.78
-86.8	**-66**	-54.44	165.2	**74**	23.33	417.2	**214**	101.11	669.2	**354**	178.89
-83.2	**-64**	-53.33	168.8	**76**	24.44	420.8	**216**	102.22	672.8	**356**	180.00
-79.6	**-62**	-52.22	172.4	**78**	25.56	424.4	**218**	103.33	676.4	**358**	181.11
-76.0	**-60**	-51.11	176.0	**80**	26.67	428.0	**220**	104.44	680.0	**360**	182.22
-72.4	**-58**	-50.00	179.6	**82**	27.78	431.6	**222**	105.56	683.6	**362**	183.33
-68.8	**-56**	-48.89	183.2	**84**	28.89	435.2	**224**	106.67	687.2	**364**	184.44
-65.2	**-54**	-47.78	186.8	**86**	30.00	438.8	**226**	107.78	690.8	**366**	185.56
-61.6	**-52**	-46.67	190.4	**88**	31.11	442.4	**228**	108.89	694.4	**368**	186.67
-58.0	**-50**	-45.56	194.0	**90**	32.22	446.0	**230**	110.00	698.0	**370**	187.78
-54.4	**-48**	-44.44	197.6	**92**	33.33	449.6	**232**	111.11	701.6	**372**	188.89
-50.8	**-46**	-43.33	201.2	**94**	34.44	453.2	**234**	112.22	705.2	**374**	190.00
-47.2	**-44**	-42.22	204.8	**96**	35.56	456.8	**236**	113.33	708.8	**376**	191.11
-43.6	**-42**	-41.11	208.4	**98**	36.67	460.4	**238**	114.44	712.4	**378**	192.22
-40.0	**-40**	-40.00	212.0	**100**	37.78	464.0	**240**	115.56	716.0	**380**	193.33
-36.4	**-38**	-38.89	215.6	**102**	38.89	467.6	**242**	116.67	719.6	**382**	194.44
-32.8	**-36**	-37.78	219.2	**104**	40.00	471.2	**244**	117.78	723.2	**384**	195.56
-29.2	**-34**	-36.67	222.8	**106**	41.11	474.8	**246**	118.89	726.8	**386**	196.67
-25.6	**-32**	-35.56	226.4	**108**	42.22	478.4	**248**	120.00	730.4	**388**	197.78
-22.0	**-30**	-34.44	230.0	**110**	43.33	482.0	**250**	121.11	734.0	**390**	198.89
-18.4	**-28**	-33.33	233.6	**112**	44.44	485.6	**252**	122.22	737.6	**392**	200.00
-14.8	**-26**	-32.22	237.2	**114**	45.56	489.2	**254**	123.33	741.2	**394**	201.11
-11.2	**-24**	-31.11	240.8	**116**	46.67	492.8	**256**	124.44	744.8	**396**	202.22
-7.6	**-22**	-30.00	244.4	**118**	47.78	496.4	**258**	125.56	748.4	**398**	203.33
-4.0	**-20**	-28.89	248.0	**120**	48.89	500.0	**260**	126.67	752.0	**400**	204.44
-0.4	**-18**	-27.78	251.6	**122**	50.00	503.6	**262**	127.78	755.6	**402**	205.56
+3.2	**-16**	-26.67	255.2	**124**	51.11	507.2	**264**	128.89	759.2	**404**	206.67
+6.8	**-14**	-25.56	258.8	**126**	52.22	510.8	**266**	130.00	762.8	**406**	207.78
+10.4	**-12**	-24.44	262.4	**128**	53.33	514.4	**268**	131.11	766.4	**408**	208.89
+14.0	**-10**	-23.33	266.0	**130**	54.44	518.0	**270**	132.22	770.0	**410**	210.00
+17.6	**-8**	-22.22	269.6	**132**	55.56	521.6	**272**	133.33	773.6	**412**	211.11
+21.2	**-6**	-21.11	273.2	**134**	56.67	525.2	**274**	134.44	777.2	**414**	212.22
+24.8	**-4**	-20.00	276.8	**136**	57.78	528.8	**276**	135.56	780.8	**416**	213.33
+28.4	**-2**	-18.89	280.4	**138**	58.89	532.4	**278**	136.67	784.4	**418**	214.44
+32.0	**±0**	-17.78	284.0	**140**	60.00	536.0	**280**	137.78	788.0	**420**	215.56

F		C	F		C	F		C	F		C
791.6	**422**	216.67	1130.0	**610**	321.11	2390.0	**1310**	710.00	3650.0	**2010**	1098.9
795.2	**424**	217.78	1148.0	**620**	326.67	2408.0	**1320**	715.56	3668.0	**2020**	1104.4
798.8	**426**	218.89	1166.0	**630**	332.22	2426.0	**1330**	721.11	3686.0	**2030**	1110.0
802.4	**428**	220.00	1184.0	**640**	337.78	2444.0	**1340**	726.67	3704.0	**2040**	1115.6
806.0	**430**	221.11	1202.0	**650**	343.33	2462.0	**1350**	732.22	3722.0	**2050**	1121.1
809.6	**432**	222.22	1220.0	**660**	348.89	2480.0	**1360**	737.78	3740.0	**2060**	1126.7
813.2	**434**	223.33	1238.0	**670**	354.44	2498.0	**1370**	743.33	3758.0	**2070**	1132.2
816.8	**436**	224.44	1256.0	**680**	360.00	2516.0	**1380**	748.89	3776.0	**2080**	1137.8
820.4	**438**	225.56	1274.0	**690**	365.56	2534.0	**1390**	754.44	3794.0	**2090**	1143.3
824.0	**440**	226.67	1292.0	**700**	371.11	2552.0	**1400**	760.00	3812.0	**2100**	1148.9
827.6	**442**	227.78	1310.0	**710**	376.67	2570.0	**1410**	765.56	3830.0	**2110**	1154.4
831.2	**444**	228.89	1328.0	**720**	382.22	2588.0	**1420**	771.11	3848.0	**2120**	1160.0
834.8	**446**	230.00	1346.0	**730**	387.78	2606.0	**1430**	776.67	3866.0	**2130**	1165.6
838.4	**448**	231.11	1364.0	**740**	393.33	2624.0	**1440**	782.22	3884.0	**2140**	1171.1
842.0	**450**	232.22	1382.0	**750**	398.89	2642.0	**1450**	787.78	3902.0	**2150**	1176.7
845.6	**452**	233.33	1400.0	**760**	404.44	2660.0	**1460**	793.33	3920.0	**2160**	1182.2
849.2	**454**	234.44	1418.0	**770**	410.00	2678.0	**1470**	798.89	3938.0	**2170**	1187.8
852.8	**456**	235.56	1436.0	**780**	415.56	2696.0	**1480**	804.44	3956.0	**2180**	1193.3
856.4	**458**	236.67	1454.0	**790**	421.11	2714.0	**1490**	810.00	3974.0	**2190**	1198.9
860.0	**460**	237.78	1472.0	**800**	426.67	2732.0	**1500**	815.56	3992.0	**2200**	1204.4
863.6	**462**	238.89	1490.0	**810**	432.22	2750.0	**1510**	821.11	4010.0	**2210**	1210.0
867.2	**464**	240.00	1508.0	**820**	437.78	2768.0	**1520**	826.67	4028.0	**2220**	1215.6
870.8	**466**	241.11	1526.0	**830**	443.33	2786.0	**1530**	832.22	4046.0	**2230**	1221.1
874.4	**468**	242.22	1544.0	**840**	448.89	2804.0	**1540**	837.78	4064.0	**2240**	1226.7
878.0	**470**	243.33	1562.0	**850**	454.44	2822.0	**1550**	843.33	4082.0	**2250**	1232.2
881.6	**472**	244.44	1580.0	**860**	460.00	2840.0	**1560**	848.89	4100.0	**2260**	1237.8
885.2	**474**	245.56	1598.0	**870**	465.56	2858.0	**1570**	854.44	4118.0	**2270**	1243.3
888.8	**476**	246.67	1616.0	**880**	471.11	2876.0	**1580**	860.00	4136.0	**2280**	1248.9
892.4	**478**	247.78	1634.0	**890**	476.67	2894.0	**1590**	865.56	4154.0	**2290**	1254.4
896.0	**480**	248.89	1652.0	**900**	482.22	2912.0	**1600**	871.11	4172.0	**2300**	1260.0
899.6	**482**	250.00	1670.0	**910**	487.78	2930.0	**1610**	876.67	4190.0	**2310**	1265.6
903.2	**484**	251.11	1688.0	**920**	493.33	2948.0	**1620**	882.22	4208.0	**2320**	1271.1
906.8	**486**	252.22	1706.0	**930**	498.89	2966.0	**1630**	887.78	4226.0	**2330**	1276.7
910.4	**488**	253.33	1724.0	**940**	504.44	2984.0	**1640**	893.33	4244.0	**2340**	1282.2
914.0	**490**	254.44	1742.0	**950**	510.00	3002.0	**1650**	898.89	4262.0	**2350**	1287.8
917.6	**492**	255.56	1760.0	**960**	515.56	3020.0	**1660**	904.44	4280.0	**2360**	1293.3
921.2	**494**	256.67	1778.0	**970**	521.11	3038.0	**1670**	910.00	4298.0	**2370**	1298.9
924.8	**496**	257.78	1796.0	**980**	526.67	3056.0	**1680**	915.56	4316.0	**2380**	1304.4
928.4	**498**	258.89	1814.0	**990**	532.22	3074.0	**1690**	921.11	4334.0	**2390**	1310.0
932.0	**500**	260.00	1832.0	**1000**	337.78	3092.0	**1700**	926.67	4352.0	**2400**	1315.6
935.6	**502**	261.11	1850.0	**1010**	543.33	3110.0	**1710**	932.22	4370.0	**2410**	1321.1
939.2	**504**	262.22	1868.0	**1020**	548.89	3128.0	**1720**	937.78	4388.0	**2420**	1326.7
942.8	**506**	263.33	1886.0	**1030**	554.44	3146.0	**1730**	943.33	4406.0	**2430**	1332.2
946.4	**508**	264.44	1904.0	**1040**	560.00	3164.0	**1740**	948.89	4424.0	**2440**	1337.8
950.0	**510**	265.56	1922.0	**1050**	565.56	3182.0	**1750**	954.44	4442.0	**2450**	1343.3
953.6	**512**	266.67	1940.0	**1060**	571.11	3200.0	**1760**	960.00	4460.0	**2460**	1348.9
957.2	**514**	267.78	1958.0	**1070**	576.67	3218.0	**1770**	965.56	4478.0	**2470**	1354.4
960.8	**516**	268.89	1976.0	**1080**	582.22	3236.0	**1780**	971.11	4496.0	**2480**	1360.0
964.4	**518**	270.00	1994.0	**1090**	587.78	3254.0	**1790**	976.67	4514.0	**2490**	1365.6
968.0	**520**	271.11	2012.0	**1100**	593.33	3272.0	**1800**	982.22	4532.0	**2500**	1371.1
971.6	**522**	272.22	2030.0	**1110**	598.89	3290.0	**1810**	987.78	4550.0	**2510**	1376.7
975.2	**524**	273.33	2048.0	**1120**	604.44	3308.0	**1820**	993.33	4568.0	**2520**	1382.2
978.8	**526**	274.44	2066.0	**1130**	610.00	3326.0	**1830**	998.89	4586.0	**2530**	1387.8
982.4	**528**	275.56	2084.0	**1140**	615.56	3344.0	**1840**	1004.4	4604.0	**2540**	1393.3
986.0	**530**	276.67	2102.0	**1150**	621.11	3362.0	**1850**	1010.0	4622.0	**2550**	1398.9
989.6	**532**	277.78	2120.0	**1160**	626.67	3380.0	**1860**	1015.6	4640.0	**2560**	1404.4
993.2	**534**	278.89	2138.0	**1170**	632.22	3398.0	**1870**	1021.1	4658.0	**2570**	1410.0
996.8	**536**	280.00	2156.0	**1180**	637.78	3416.0	**1880**	1026.7	4676.0	**2580**	1415.6
1000.4	**538**	281.11	2174.0	**1190**	643.33	3434.0	**1890**	1032.2	4694.0	**2590**	1421.1
1004.0	**540**	282.22	2192.0	**1200**	648.89	3452.0	**1900**	1037.8	4712.0	**2600**	1426.7
1007.6	**542**	283.33	2210.0	**1210**	654.44	3470.0	**1910**	1043.3	4730.0	**2610**	1432.2
1011.2	**544**	284.44	2228.0	**1220**	660.00	3488.0	**1920**	1048.9	4748.0	**2620**	1437.8
1014.8	**546**	285.56	2246.0	**1230**	665.56	3506.0	**1930**	1054.4	4766.0	**2630**	1443.3
1018.4	**548**	286.67	2264.0	**1240**	671.11	3524.0	**1940**	1060.0	4784.0	**2640**	1448.9
1022.0	**550**	287.78	2282.0	**1250**	676.67	3542.0	**1950**	1065.6	4802.0	**2650**	1454.4
1040.0	**560**	293.33	2300.0	**1260**	682.22	3560.0	**1960**	1071.1	4820.0	**2660**	1460.0
1058.0	**570**	298.89	2318.0	**1270**	687.78	3578.0	**1970**	1076.7	4838.0	**2670**	1465.6
1076.0	**580**	304.44	2336.0	**1280**	693.33	3596.0	**1980**	1082.2	4856.0	**2680**	1471.1
1094.0	**590**	310.00	2354.0	**1290**	698.89	3614.0	**1990**	1087.8	4874.0	**2690**	1476.7
1112.0	**600**	315.56	2372.0	**1300**	704.44	3632.0	**2000**	1093.3	4892.0	**2700**	1482.2

(continued on the next page)

Temperature Conversions (contd.)

F		C	F		C	F		C	F		C
4910.0	**2710**	1487.8	5270.0	**2910**	1598.9	6152.0	**3400**	1871.1	8402.0	**4650**	2565.5
4928.0	**2720**	1493.3	5288.0	**2920**	1604.4	6242.0	**3450**	1898.8	8492.0	**4700**	2593.3
4946.0	**2730**	1498.9	5306.0	**2930**	1610.0	6332.0	**3500**	1926.6	8582.0	**4750**	2621.1
4964.0	**2740**	1504.4	5324.0	**2940**	1615.6	6422.0	**3550**	1954.4	8672.0	**4800**	2648.8
4982.0	**2750**	1510.0	5342.0	**2950**	1621.1	6512.0	**3600**	1982.2	8762.0	**4850**	2676.6
5000.0	**2760**	1515.6	5360.0	**2960**	1626.7	6602.0	**3650**	2010.0	8852.0	**4900**	2704.4
5018.0	**2770**	1521.1	5378.0	**2970**	1632.2	6692.0	**3700**	2037.7	8942.0	**4950**	2732.2
5036.0	**2780**	1526.7	5396.0	**2980**	1637.8	6782.0	**3750**	2065.5	9032.0	**5000**	2760.0
5054.0	**2790**	1532.2	5414.0	**2990**	1643.3	6872.0	**3800**	2093.3	9122.0	**5050**	2787.7
5072.0	**2800**	1537.8	5432.0	**3000**	1648.9	6962.0	**3850**	2121.1	9212.0	**5100**	2815.5
5090.0	**2810**	1543.3	5450.0	**3010**	1654.4	7052.0	**3900**	2148.8	9302.0	**5150**	2843.3
5108.0	**2820**	1548.9	5468.0	**3020**	1660.0	7142.0	**3950**	2176.6	9392.0	**5200**	2871.1
5126.0	**2830**	1554.4	5486.0	**3030**	1665.6	7232.0	**4000**	2204.4	9482.0	**5250**	2898.8
5144.0	**2840**	1560.0	5504.0	**3040**	1671.1	7322.0	**4050**	2232.2	9572.0	**5300**	2926.6
5162.0	**2850**	1565.6	5522.0	**3050**	1676.7	7412.0	**4100**	2260.0	9662.0	**5350**	2954.4
5180.0	**2860**	1571.1	5540.0	**3060**	1682.2	7502.0	**4150**	2287.7	9752.0	**5400**	2982.2
5198.0	**2870**	1576.7	5558.0	**3070**	1687.8	7592.0	**4200**	2315.5	9842.0	**5450**	3010.0
5216.0	**2880**	1582.2	5576.0	**3080**	1693.3	7682.0	**4250**	2343.3	9932.0	**5500**	3037.7
5234.0	**2890**	1587.8	5594.0	**3090**	1698.9	7772.0	**4300**	2371.1	10022.0	**5550**	3065.5
5252.0	**2900**	1593.3	5612.0	**3100**	1704.4	7862.0	**4350**	2398.8	10112.0	**5600**	3093.3
			5702.0	**3150**	1732.2	7952.0	**4400**	2426.6			
			5792.0	**3200**	1760.0	8042.0	**4450**	2454.4			
			5882.0	**3250**	1787.7	8132.0	**4500**	2482.2			
			5972.0	**3300**	1815.5	8222.0	**4550**	2510.0			
			6062.0	**3350**	1843.3	8312.0	**4600**	2537.7			

Index

Emphasis in the index, as in the book, is placed on the crystallographic, microstructural and phase transformation phenomena underlying heat treating processes, rather than on specific methods or specific materials. The letter (D), (M) or (T) following an entry signifies that information on the subject is presented in a diagram, micrograph or table.